From Chemical Philosophy to Theoretical Chemistry

"I'M ON THE VERGE OF A MAJOR BREAKTHROUGH BUT I'M ALSO AT THAT POINT WHERE CHEMISTRY LEAVES OFF AND PHYSICS BEGINS, SO I'LL HAVE TO DROP THE WHOLE THING."

From Chemical Philosophy to Theoretical Chemistry

Dynamics of Matter and Dynamics of Disciplines, 1800–1950

Mary Jo Nye

University of California Press
Berkeley / Los Angeles / London

University of California Press
Berkeley and Los Angeles, California

University of California Press
London, England

Library of Congress Cataloging-in-Publication Data

Nye, Mary Jo.
From chemical philosophy to theoretical chemistry : dynamics of matter and dynamics of disciplines, 1800–1950 / Mary Jo Nye.
p. cm.
Includes bibliographical references (p.) and index.
ISBN 0-520-08210-9
1. Chemistry, Physical and theoretical—History—19th century. 2. Chemistry, Physical and theoretical—History—20th century.
I. Title.
QD452.N94 1993
541′.094′09034—dc20 92-43114
CIP

Printed in the United States of America

1 2 3 4 5 6 7 8 9

The paper used in this publication meets the minimum requirements of American National Standard for Information Sciences—Permanence of Paper for Printed Library Materials, ANSI Z39.48-1984 ∞

To Bob

Contents

Plates

Abbreviations

Journals

Annalen der Physik	*Ann. Phys.*
Annales de Chimie et de Physique	*Ann. Chim. Phys.*
Annuaire des Anciens Elèves de l'Ecole Normale Supérieure	*Ann. AEENS*
Annual Review of Physical Chemistry	*Ann. Rev. P. Chem.*
Archive for History of Exact Sciences	*Arch. Ex. Sci.*
Archives Internationales d'Histoire des Sciences	*AIHS*
Berichte. Deutsche Chemische Gesellschaft	*Berichte*
British Association for the Advancement of Science Reports	*BAAS Rep.*
British Journal for the History of Science	*BJHS*
British Journal for the Philosophy of Science	*BJPS*
Bulletin de la Société Chimique de France	*Bull. SCF*
Bulletin de la Société des Amis de l'École Normale Supérieure	*Bull. SAENS*
Chemical and Engineering News	*CENews*
Comptes Rendus Hebdomadaires de l'Académie des Sciences	*CR*
Historical Studies in the Physical and Biological Sciences	*HSPS*
Journal de Chimie Physique	*JCP*

Journal de Physique	*JP*
Journal of Chemical Education	*JChem. Ed.*
Journal of Chemical Physics	*J. Chem. Physics*
Journal of the American Chemical Society	*JACS*
Journal of the Chemical Society (London)	*JCS*
Journal of the Royal Institute of Chemistry	*J. Royal Inst. Chem.*
Liebig's Annalen der Chemie	*Liebig's Ann.*
Memoirs and Proceedings of the Chemical Society (London)	*Mem. London CS*
Memoirs and Proceedings of the Manchester Literary and Philosophical Society	*Mem. Manchester LPS*
Philosophical Magazine	*Phil. Mag.*
Physikalische Zeitschrift	*PZ*
Proceedings of the National Academy of Sciences	*Proc. NAS*
Proceedings of the Royal Society of London	*Proc. RSL*
Studies in the History and Philosophy of Science	*SHPS*
Transactions of the Faraday Society	*Trans. Far. Soc.*
Zeitschrift für Physik	*ZP*
Zeitschrift für Physikalische Chemie	*ZPC*

Archives

Archives Nationales de France	AN
Bancroft Library, University of California at Berkeley	BL.UCB
Bodleian Library, Duke Humphreys Manuscript Collection, Oxford	Bod. Oxford
California Institute of Technology Archives	Caltech
Ecole Normale Supérieure Archives	ENS
Imperial College, London, Archives	ICL
Laboratoire de Chimie de l'Ecole Normale Supérieure Bibliothèque	Lab. ENS
Lincoln College, Oxford, Library	L. Coll. Oxford
Royal Society of London Library	RSL
Sources for the History of Quantum Physics, Office for the History of Science and Technology, University of California at Berkeley	SHQP
University College, London, Archives	UCL

Acknowledgments

It is a pleasure to acknowledge and thank those who have helped and supported me in my work on this project since 1985, when I first began research for this study in Paris with the aid of a grant from the University of Oklahoma Research Council. I have enjoyed generous support from the National Endowment for the Humanities (RH-20758-86), the National Science Foundation (DIR-8911578), the University of Oklahoma Senior Faculty Summer Research Award, and the Southwestern Bell Summer Fellowship Program.

Absolutely crucial was the Rutgers Center for Historical Analysis at Rutgers University in New Brunswick which provided a stimulating and congenial environment for thinking about "The Historical Construction of Identities" during the 1989–90 academic year. For this, I am especially grateful to John Gillis.

Colleagues and friends have been generous and kind in their criticisms, commentary, and advice on various parts of the manuscript. I am grateful to Alexi Assmus, Ted Benfey, Micheline Charpentier-Morize, Mi Gyung Kim, Roald Hoffmann, Karl Hufbauer, Peter Loewenberg, John Servos, and especially Sam Schweber. Terry Shinn offered provocative comments on my early thinking about models and metaphors, as did John Heilbron on my initial reflections about the historical demarcation of physics and chemistry.

I also want to thank students and colleagues for their good conversation and university libraries at the California Institute of Technology, the University of California, Berkeley, Harvard University, Princeton University, Rutgers University, and, of course, the University of Oklahoma. Among my students at Oklahoma, I am particularly grateful to Samantha McClintock, Kuang-Tai Hsu, Michael Keas, Shawn Smith, and JoAnn Palmeri for help in various ways.

I also acknowledge with gratitude the hospitality of the Bibliothèque Nationale, Archives Nationales de France, Faculté de Pharmacie de l'Université de Paris, and the British Library.

Over the past years, some symposia and colloquia groups have provided especially valuable insights at critical times, and I would like to thank colleagues at the New Trends in the History of Science conference in Utrecht in August 1986, the Beckman Center symposium, Chemical Sciences in the Modern World, in May 1990, the International Summer School in History of Science in Uppsala in June 1990, and the conference on research schools in New Haven in December 1990.

Some of the materials in this book have appeared in somewhat different form in the following articles: ''Explanation and Convention in Nineteenth-Century Chemistry,'' *New Trends in the History of Science*, ed. R. P. W. Visser et al. (Amsterdam: Editions Rodopi, 1989): 171–186; ''Chemical Explanation and Physical Dynamics: Two Research Schools at the First Solvay Chemistry Conferences, 1922–1928,'' *Annals of Science* 46 (1989): 461–480; ''Physics and Chemistry: Commensurate or Incommensurate Sciences?'' *The Invention of Physical Science: Intersections of Mathematics, Theology and Natural Philosophy Since the Seventeenth Century. Essays in Honor of Erwin N. Hiebert*, ed. Mary Jo Nye et al. (Dordrecht: Kluwer, 1992): 105–224; ''National Styles? Research Schools in French and English Chemistry, 1880–1930,'' *Osiris* [2]8 (1993): 30–49; and ''Philosophies of Chemistry since the Eighteenth Century,'' in Seymour Mauskopf, ed., *Chemical Sciences in the Modern World* (Philadelphia: University of Pennsylvania Press, in press). I am grateful for permissions to publish materials previously published.

For their gracious aid in my research on the Paris School of theoretical chemistry, I express appreciation to Pierre Petitmengin and Mme. Dauphragne at the library and archives of the Ecole Normale Supérieure; Mlle. Françoise Ridard at the library of the Laboratoire de Chimie of the Ecole Normale Supérieure; to Yves Noel, former student of Albert Kirrmann and professor of chemistry at the University of Caen; Pierre Piganiol, former director of research at Gobelins and French science policy adviser; Constantin Georgoulis, former student of Charles Prévost and director of research in the CNRS at the University of Paris; and Mme. Noemie Guern-Prévost, daughter of Charles Prévost. I am grateful to Mme. Paule René-Bazin for help some years ago with obtaining photocopies of materials then housed in the Archives de l'Université de Paris I (Sorbonne).

For their equally gracious aid in my research on the London-Manchester School, I express appreciation to Jennifer Seddon Curtis, Joe Marsh, and Rajkumari Williamson at the University of Manchester Institute of Science and Technology and to Stella Butler of the Manchester Museum of Science and Industry. I am grateful to Mrs. C. J. Anderson at the Library of University College, London; to Jeanne Pingree and Mrs. M. Felton at the Archives of

Imperial College, London; to Anne Delahaye at the Library of Lincoln College, Oxford; Colin Harris at the Duke Humphreys Manuscript Collection of the Bodleian Library, Oxford; and to N. H. Robinson and Sandra Cumming at the Library of the Royal Society of London. I very much enjoyed the hospitality of Ruth Harris and Iain Pears in Oxford and London and of Rhoda Lee in Manchester.

For permission to quote from the G. N. Lewis papers, I thank the Bancroft Library, University of California, Berkeley; Henry Armstrong papers, Imperial College Archives; Robert Robinson Papers, Royal Society of London; Nevil Sidgwick Papers, Lincoln College, Oxford; Charles Coulson and Frederick Soddy Papers, Bodleian Library, Oxford, and Eileen F. Coulson, Keith U. Ingold, and H. Christopher Longuet-Higgins; F. G. Donnan Papers, University College, London; and various interviews, the Sources for the History of Quantum Physics at Berkeley.

For help with chemical figures, I am grateful to my Oklahoma colleague, Glenn Dryhurst, and for their encouragement and assistance at the University of California Press, I thank Michelle Nordon, Sheila Berg, and, especially, Elizabeth Knoll.

As always, a special debt of gratitude is due Bob Nye, who helped me conceptually in thinking about disciplinary identity, who read and criticized the manuscript despite its sometimes forbidding technical aspects, and who agreed to my plan to work in English and American archives along with our customary French haunts. Thanks to him, and to Lesley, I at least have worked out my own identity.

Introduction

Chemistry: That Most Excellent Child of Intellect and Art
Cyril Hinshelwood

When the *Journal of Chemical Physics* was founded in 1933, its first editor, Harold Urey, then associate professor of chemistry at Columbia University, wrote that "the boundary between the sciences of physics and chemistry has been completely bridged."[1] The appearance of the journal reflected the outcome of scientific developments during the last decades in which there had been a reshaping and reforging of links between two disciplines that had been largely autonomous since the early nineteenth century.

By common agreement among many historians of science, "chemistry" and "physics" became fairly well demarcated communities or disciplines around 1830, some hundred years before the founding of the *Journal of Chemical Physics*.[2] This was about the time that Auguste Comte was embarking on his *Cours de philosophie positive*, in which he laid out a hierarchy of the positive sciences as he observed them in contemporary Paris. In this hierarchy, the mastery of mathematics and physics was historically and foundationally prior to chemistry.

Not coincidentally, the 1830s was the period when organic chemistry became a subfield or subdiscipline within chemical science—but a clearly dominant one, dealing with the material substances of biology and physiology,

1. See Harold Urey, editorial, *J.Chem.Physics* 1 (1933); and J. W. Stout, "The *Journal of Chemical Physics*: The First 50 Years," *Ann.Rev.P.Chem.* 37 (1986): 1–23.

2. In the postscript to *The Structure of Scientific Revolutions*, Thomas Kuhn claimed there was no physics community before the mid-nineteenth century and that it "was then formed by the merger of parts of two previously separate communities, mathematics and natural philosophy (*physique expérimentale*)." See Thomas S. Kuhn, *The Structure of Scientific Revolutions*, 2d ed. (Chicago: University of Chicago Press, 1970): 182–187.

agriculture and industry, mineralogy and crystallography.[3] At the end of the nineteenth century, physical chemistry established itself as the practice of an articulate, identifiable scientific community, as did quantum chemistry around 1940, about the same time that physical organic chemistry became a coherent field of specialization.[4] Together, physical chemistry, physical organic chemistry, and quantum chemistry have comprised a modern theoretical chemistry.

For chemistry as a whole, and for each of these chemical disciplines, there developed a historical (indeed, genealogical) legacy and a core literature, as well as a set of shared problems, practices, principles, and values. Thomas Kuhn has treated such disciplinary components as categories of the ''paradigm'' or the ''disciplinary matrix,'' which are useful in understanding normal science before its transformation during a period of revolution.[5] My concern is not revolution but the evolution of eighteenth-century chemical philosophy, whose practitioners aspired to understand the dynamics of matter, into twentieth-century theoretical chemistry, whose practitioners claimed to do so.

I originally conceived this book as a study of epistemological questions about differences between chemistry and physics, namely, how chemists' aims and methods in scientific explanation have been different from physicists' and how these aims and methods have overlapped. This problem remains a principal focus of the book: is there a way of seeing and describing the natural world that has been consistently ''chemical''? If so, how has it compared to a ''physical'' way of understanding the natural world? Have chemistry and physics been commensurable or incommensurable sciences? Answering these questions historically leads to sets of answers that are specific to time and place; to distinct texts, individuals, schools, and traditions; to disciplines and disciplinary histories.

Sociologists, drawing on scientists' publications, have used the categories of disciplines, subfields, specialties, and subspecialties to formulate a model of the hierarchy of scientific organization.[6] They, and we, speak of networks, clusters, and ''work groups.''[7] In this study, I use the word ''discipline'' both

3. See Frederic L. Holmes, ''The Complementarity of Teaching and Research in Liebig's Laboratory,'' 121–164, and Jeffrey A. Johnson, ''Hierarchy and Creativity in Chemistry, 1871–1914,'' 214–240, in Kathryn M. Olesko, ed., *Science in Germany: The Intersection of Institutional and Intellectual Issues* 5 (2d ser.), *Osiris* (1989); and the March 1992 issue of the *British Journal for the History of Science* 25, no. 84, entitled *Organic Chemistry and High Technology, 1850–1950.*

4. On physical chemistry, with emphasis on its transplantation from Europe to America, see John W. Servos, *Physical Chemistry from Ostwald to Pauling: The Making of a Science in America* (Princeton: Princeton University Press, 1990).

5. Kuhn, *Structure.*

6. Daryl Chubin illustrates the relationships of these categories by the example: (1) discipline = physics; (2) subfield = high energy or elementary particle physics; (3) specialty = weak interactions; (4) subspecialty = experimental, rather than theoretical, studies. Daryl E. Chubin, ''State of the Field: The Conceptualization of Scientific Specialties,'' *The Sociological Quarterly* 17 (1976): 448–476, esp. 450, 456–457.

7. Nicholas C. Mullins, *Theory and Theory Groups in Contemporary American Sociology*

to denote chemistry as a whole and to characterize some of its leading "sub-fields," namely, organic chemistry, physical chemistry, physical organic chemistry, quantum chemistry, and theoretical chemistry. All these fields or subjects of study have regularly been called "disciplines" by their practitioners. I focus on some research schools within these disciplines that addressed particular problems and on individuals who have been part of schools and disciplines, participating in their traditions.

In what follows, I do not present a comprehensive survey of the history of chemistry, nor do I construct a systematic model of discipline formation. This is a historical narrative and analysis of aspects of the *idea* of chemistry, the *idea* of "theoretical" chemistry, and the institutional and intellectual processes that constructed *disciplines* in chemistry in their various historical forms. While considerable attention is paid to the ways in which chemists demarcated their domain *against* other domains, as well as to their statements of a chemical epistemology and their development of systems of chemical language and imagery, the overall study is guided by two premises.

The first premise is that we risk overlooking the crux of a scientific discipline if we do not understand the discipline's values and problems. These provide a unity that no merely institutional history can explain. The second premise is that a programmatic conceptual core of chemical thought from the eighteenth century through the twentieth century was what I call the problem of the dynamics of matter (What holds a substance together? What makes it change?). An early expression for this problem of dynamics was the concept of "chemical affinity."

Dynamics, namely, the mechanism of chemical reactivity, was not the only conceptual core to chemistry. We might focus as well on the concepts of chemical "species" and chemical "constitution," and indeed these concepts figure in the history that follows. However, the dynamics of matter was a kernel at the heart of chemistry, with varying paces of growth. It constituted both disputed and common territory for practitioners of chemical philosophy and natural philosophy. More recently, it provided a point of controversy and an area of compromise for practitioners of the disciplines of physics and chemistry. Thus, the dynamics of matter is a theme providing especially important insights into the relations between chemistry and physics as intellectual systems, at the same time that the social dynamics of individuals and groups also helps to explain disciplinary development.[8]

The plan of the study is the following. Part One analyzes aspects of the construction of chemistry as a well-defined discipline in the nineteenth century,

(New York: Harper and Row, 1973). Also, Eugene Garfield, *Citation Indexing—Its Theory and Application in Science, Technology and Humanities* (Philadelphia: ISI, 1979).

8. On the centrality of "affinity" and "matter" to chemistry, see the book by that name: Trevor H. Levere, *Affinity and Matter: Elements of Chemical Philosophy* (Oxford: Oxford University Press, 1971).

showing how its history can be explained in terms of elements of disciplinary identity. In chapter 1, I lay out a schema for the construction of identity consisting of six elements or components, which briefly may be summarized as (1) genealogy and family descent, including historical mythology of heroic origins and heroic episodes; (2) core literature defining the group's archetypical language and imagery; (3) practices and rituals; (4) physical homeland, including institutions based on citizenship rights and responsibilities; (5) external recognition; and (6) shared values and unsolved problems. Among its several virtues, the identity schema can be used to distinguish scientific disciplines from ideological movements and religious cults.

I discuss in chapter 2 how the physics/chemistry demarcation, previously discussed principally from the point of view of historians of nineteenth-century physics, profitably can be seen from a more peculiarly chemical point of view. I emphasize that Antoine Lavoisier's generation of *chimistes* and *physiciens* pursued problems of common interest, often in collaboration. It was Lavoisier, himself a member of the well-established *chimie* section of the Académie Royale des Sciences, who helped found a separate section of *physique générale* at the Académie in the late eighteenth century. By 1830, chemistry and physics had become distinct disciplines not only in Paris but in academies and universities throughout Europe, Great Britain, and the United States, beginning an era of specialization often characterized by mutual incomprehension of adjacent scientific domains. Thus, in writing about chemistry and physics in 1864, Lothar Meyer could be found pleading for the need to reunite "the now severed sciences."[9]

What often goes unheeded in traditional and "positivist" histories of science is the well-developed character of the disciplinary identity of chemistry *before* physics, and this is a point I especially emphasize. An important reason for the earlier disciplinary identity of chemistry lay in its general recognition as an art of laboratory practice useful in medicine, industry, manufactures, and agriculture. The necessities that drove many chemists to arrive at serviceable results, before they could boast of having a theoretically based science, led some academicians to scorn the "philosophical" status of chemistry. Indeed, late in the nineteenth century, self-consciously "physical chemists" like Wilhelm Ostwald and Svante Arrhenius claimed to be founding a theoretical chemistry for the first time, contrasting their methods and aims with their colleagues in organic chemistry.

But it is not true, of course, that organic chemists lacked theories, even though they often described their methodology in the nineteenth century as one of descriptive analogy and classification. In chapters 3 and 4, I turn to the aims and methods of nineteenth-century chemistry and to the conventions of

9. Lothar Meyer, introduction to the 1864 edition of *Modern Theories of Chemistry*, 5th ed., trans. Phillips Bedson and W. C. Williams (London: Longmans, Green & Co., 1888): xxv.

language, symbol, and imagery that formed its classical laws and literature, especially in the dominant discipline of organic chemistry.

What we find is a considerably complex set of interlocking theories. Their iconic and visual imagery notwithstanding, nineteenth-century chemists for the most part avoided mechanical models and naive realism, although they often were characterized as simple-minded in their use of pictures. They certainly mostly eschewed algebraic or rigorous mathematical representations. What is striking in this chemical literature is the use, in keeping with a pragmatic epistemology, of complementary metaphors, images, classifications, and conventions. The aim was theoretical explanation of the constitution and multiple functions, of the complex *behavior*, of chemical substances. This behavior could not be reduced to the linear relationships of mathematical physics, nor could it be envisioned to be the result of static structures in three-dimensional space.

Nineteenth-century scientists and philosophers grew accustomed to claiming that chemistry is a science of taxonomy and classification resting on empirical and descriptive foundations, in contrast to mechanical physics resting on axiomatic and mathematical foundations. The methods and theories of chemical explanation owed as much to natural history as to natural philosophy. As a consequence, in his post-Newtonian *Metaphysical Foundations of Natural Science*, Immanuel Kant expressed the generalization about chemistry that it could never be a genuine science, that is, true knowledge, because its subject matter is intractable to the method of deductive systemization. The ubiquity of this judgment is exemplified by its place in William Paley's *Natural Theology*, the mandatory textbook read by every Cambridge gentleman throughout the nineteenth century.[10]

During the course of the nineteenth century, organic chemists developed a complex and integrated theory of chemical identity and constitution that included theories of chemical atoms and equivalents, radicals, substitution, types, and valence, eventually unified in the so-called structure theory of constitutional formulas. But by midcentury, there was some discontent among chemists about the limitations of the overall theoretical framework within which they were working.

James Muspratt and Wilhelm von Hofmann, for example, expressed dissatisfaction in the journal of the London Chemical Society with the failure of organic chemists to understand the *mechanisms* of chemical change. They wrote,

10. Paley wrote that while chemistry might afford an argument of God's design "in a high degree satisfactory," it could not "afford the same species of argument as that which mechanism affords." In William Paley, *Natural Theology*, rev. Amer. ed. (New York: Sheldon, 1854): 48. I am indebted to Kuang-Tai Hsu for this reference. On Kant, in the *Gesammelte Schriften* (Berlin, 1911), IV: 470–471, quoted in Frederick Gregory, "Romantic Kantianism and the End of the Newtonian Dream in Chemistry," *Archives Internationales d'Histoire des Sciences*, 34 (1984): 108–123, on 109, 113.

> Notwithstanding the laborious endeavors that have been made of late years to enable us to foretell the metamorphoses which a given compound should sustain under the influence of different chemical agents, still it must be admitted that we have by no means succeeded in establishing antecedently, with absolute certainty, the *modus operandi* of a decomposition.[11]

By the end of the nineteenth century, the elusive dream of a mechanical or dynamical chemistry, supplementing the static chemistry of structural formulas, increasingly began to haunt the chemist's waking hours. This revival of concern with dynamics, or what often was called ''affinity,'' is discussed in Part Two, beginning with chapter 5. Perhaps more than anything else, the reformulation in mid-nineteenth-century physics of Newton's laws of motion, substituting calculations of energy for presuppositions about force, revitalized attempts to solve the problems of chemical activation and chemical mechanics. Analyzing the problem of chemical ''affinity'' proved more congenial to the use of ''energy'' than of ''force.''

The pursuit of chemical thermodynamics, especially its applications in solution chemistry and electrochemistry, resulted in the establishment of a distinct scientific discipline of physical chemistry by the so-called Ionists in the late nineteenth century. These practitioners defined their aim as ''theoretical chemistry,'' bridging the domains of physics and chemistry. Ostwald contrasted the aims and methods of the new discipline with the practices of organic chemists, whom he characterized as powerful, hidebound, fact-mongering opponents of the young band of Ionists.

In analyzing the emergence of this new chemical discipline in chapter 5, I focus on three research programs among the many that became characteristic of the new field. First, chemical thermodynamics, from the 1880s through the 1920s, provided a mathematical mechanics of matter, determining exact relations between energy values and properties of macroscopic systems. However, this thermodynamics was unsatisfactory to many chemists, especially to organic chemists, because it did not picture molecules, mechanisms, or the intermediate steps that determine the time a reaction will take.

Second, the radiation hypothesis, pursued from the early 1900s through the 1920s by physicists and physical chemists, proposed a radiation- and energy-based mechanism for activation of intermediate steps in the chemical reaction mechanism, molecule by molecule. But, in the short run, this theory proved wrong. Nonetheless, the radiation hypothesis is interesting for our study because it demonstrates the development of an abstract mathematical and phys-

11. Quoted in Michael Keas, ''The Structure and Philosophy of Group Research: August Wilhelm von Hofmann's Laboratory Research Program in London (1845–1865)'' (Ph.D. dissertation, University of Oklahoma, 1992): 50. See James S. Muspratt and A. W. Hofmann, ''On Certain Processes in which Aniline Is Formed,'' *Mem. London CS* 2 (1845): 249–254, on 249.

ical approach for dealing with reaction mechanisms, flawed by lack of detailed familiarity with chemical laboratory experience, especially organic chemistry.

Third, the electron theory of valence, cultivated mainly by Anglo-American physicists and physical chemists in the first two decades of the twentieth century, offered mechanical models for chemical affinity on the molecular level. These models combined data from structural chemistry with insights about physical mechanisms involving ions and electrons from the rapidly developing work of radiation physicists and spectroscopists. The further application of this third approach is the subject of chapters 6, 7, and 8, as it was developed by specific research schools in different national traditions and became part of the fundamental framework of the new subdiscipline of physical organic chemistry.

If some organic chemists resisted the program of the late-nineteenth-century Ionists, especially German organic chemists in reaction to Ostwald, others proved open-minded and even enthusiastic after 1900 in using ion and electron theories for a new approach to activation and orientation of molecular sites in chemical reactions. The first really successful theories of chemical reaction mechanisms were made independently of quantum theory, using ideas about molecular ions and electron dynamics, although these theories subsequently received a deeper rationale from quantum theory.[12] In the development of nonquantum mechanical theories about the internal mechanics of chemical reactions and activation, English and American chemists led the way.

It is curious that while German-speaking physicists contributed strongly, even disproportionately, to the new quantum physics and wrote some of the first mathematical papers that attempted to apply the electron theory to the problem of chemical bonding (Richard Abegg and Guido Bödlander), German chemists, on the whole, with the exception of Fritz Arndt, were not initially interested in these developments.[13] It may be that this disinterest is the very result of an attitude toward Ostwald and physical chemistry expressed by Richard Willstätter, a distinguished organic chemist who made his reputation by establishing the structure of cocaine and preparing relatively pure chlorophyll. Of Ostwald, he said,

> [He] had no knowledge or understanding of the development, content, or effect of our organic chemical views which had proved themselves to be uncommonly fruitful. . . . Ostwald also lacked sufficient depth in the theoretical physical foun-

12. Chemistry used to be less "deep," less "fundamental" than physics, said Charles Coulson in his inaugural lecture of an Oxford chair in theoretical chemistry in 1973, but "this is no longer the case." C. A. Coulson, "Theoretical Chemistry: Past and Future," lecture delivered before the University of Oxford on February 23, 1973, ed. S. L. Altman (Oxford: Clarendon, 1974): 6–9.

13. See J. C. Smith, *The Development of Organic Chemistry at Oxford*, 2 pts., typescript in the Robinson Papers, RSL.

dations of organic chemistry, and, more important, lacked the modesty of the natural scientist.[14]

In chapters 6 through 8, I concentrate in considerable detail on two research schools that sought to unify organic chemistry and physical chemistry with theoretical foundations built on the ion and electron theory. These schools are loosely designated the ''Paris'' and the ''London-Manchester'' schools, where ''school'' connotes a network of personal and professional associations over several generations at the Ecole Normale Supérieure, in the first case, and at London University and the University of Manchester, in the second case.

The Paris school included Robert Lespieau (1864–1947), Georges Dupont (1884–1958), Charles Prévost (1899–1983), and Albert Kirrmann (1900–1974). Principal figures in the London-Manchester school were Arthur Lapworth (1872–1941), Thomas Martin Lowry (1874–1936), Robert Robinson (1886–1975), Jocelyn Thorpe (1872–1940), and Christopher Ingold (1893–1970). A broadly defined German research school pursuing ionic and electronic theories of reaction mechanisms in organic chemistry does not enter into this history, because it did not exist.

For the Paris and the London-Manchester schools, the local culture, educational tradition, laboratory research programs, theoretical systems, and personal networks that helped forge a school identity and a disciplinary identity are worked out in some detail. In addition to constructing an account of a new branch of theoretical chemistry, focused on organic reaction mechanisms, these chapters suggest important differences in national traditions within the disciplinary field of physical organic chemistry.

In France, the application of electron theories to chemical problems was slower than in England and in the United States, and it was indebted to these developments. While some French chemists claimed to have contributed to the theory of reaction mechanisms, they were perceived by others to have lagged behind in developing theoretical chemistry. Lespieau's school of physical or ''theoretical'' organic chemistry at the Ecole Normale was rooted in a tradition more oriented toward classical thermodynamics, radiation theory, and spectroscopy than was typical of organic chemists in England. Its practitioners were suspicious of a theoretical chemistry based in a speculative and model-oriented approach rather than in a descriptive or logically rigorous one. These differences in the Paris and London-Manchester schools suggest the importance of national traditions in the formation of disciplines and disciplinary specialties.

Part Two concludes with an analysis in chapter 8 of Ingold's theory of organic reaction mechanisms, his construction of a new chemical language,

14. Richard Willstätter, *From My Life*, trans. Lili S. Hornig (New York and Amsterdam: W. A. Benjamin, 1965): 94–95. Diana Barkan notes that Nernst's ''contact with traditional chemical problems was minimal'' (''Walther Nernst,'' 1990, 132).

and the integration of his approach with quantum wave mechanics during the 1920s and 1930s. Ingold succeeded in founding a new subdiscipline through his research laboratories at London, then at Leeds, and then at London again. He became the *Vater* figure for more than one generation of physical organic chemists, for many of whom his 1953 text, *Structure and Mechanism in Organic Chemistry*, served as a classic text of chemical literature.

Part Three concludes the book with a look in chapter 9 at some of the founding ideas and founding figures of quantum chemistry, a discipline dominated by the Americans Linus Pauling, John Slater, and Robert Mulliken, with considerable debts to German-speaking theoretical physicists. Similarities and differences between the aims of chemists and physicists are demonstrated in their applications of quantum wave mechanics to simple molecules in the 1920s and 1930s. International meetings, following the Second World War, at Paris and at Shelter Island provided occasions for quantum chemists and chemical physicists to cement the foundations of a discipline of theoretical chemistry, including its unique language and notation.

In the conclusion, the relationship of disciplines to schools and traditions is discussed, both in the particular instances of the Paris and London-Manchester schools and in the more general history of Continental and Anglo-American science, using elements of the disciplinary identity schema developed at the beginning of the book. I end with an inquiry into the views of mid-twentieth-century theoretical chemists and chemical physicists on the commensurability of their disciplines, the reducibility of chemical explanation to physical principles and laws, and the distinctiveness of chemistry from physics.

Part One

Discipline-Building in Chemistry

1

Scientific Disciplines

The Construction of Identity

The Study of Scientific Disciplines, Schools, and Traditions

The historical problem of the development of physics and chemistry as separate disciplines has been insufficiently analyzed. While there have been some studies of specialized disciplinary formation, for example, in physical chemistry, radioastronomy, and molecular biology, few historians or sociologists have tackled the historical bifurcation of physics and chemistry or the nature of their interactions over long periods of time.[1]

The category of "discipline" is useful because it carries conceptual, social, and psychological meaning. The French historian and philosopher Michel Foucault especially focused attention on the idea of the relation of knowledge to "disciplines," much as the American historian and philosopher of science Thomas Kuhn earlier developed multiple meanings of the category "paradigm." Indeed, the two categories, "discipline" and "paradigm," share the common strengths and weaknesses of carrying a gridwork of conceptual, so-

1. In general, on scientific disciplines, see Gerard Lemaine et al., eds., *Perspectives on the Emergence of Scientific Disciplines* (The Hague: Mouton, 1976), including R. G. A. Dolby, "The Case of Physical Chemistry," 63–73; David O. Edge and Michael J. Mulkay, *Astronomy Transformed: The Emergence of Radio Astronomy in Britain* (New York: Wiley, 1976); Paul Hoch, "Institutional versus Intellectual Migrations in the Nucleation of New Scientific Specialties," *SHPS* 18 (1987): 481–500; John Law, "The Formation of Specialties in Science: The Case of X-ray Protein Crystallography," *Science Studies* 3 (1973): 275–303; Nicholas Mullins, "The Development of Scientific Specialties: The Phage Group and the Origins of Molecular Biology," *Minerva* 10 (1972): 51–82; and for a more historical approach, Robert Kohler, *From Medical Chemistry to Biochemistry: The Making of a Biomedical Discipline* (Cambridge: Cambridge University Press, 1982).

cial, and psychological meanings. Kuhn, after all, eventually substituted the term "disciplinary matrix" for "paradigm" after reflecting on his critics.[2]

Foucault, like his French predecessor and mentor, Gaston Bachelard, paid particular attention to the primacy in history of discursive breaks and ruptures in knowledge or belief systems.[3] In this and in Foucault's emphasis on the relative coercion that disciplines exercised on their practitioners, he made arguments already familiar to Anglo-American scholars acquainted with Kuhn's characterizations of "normal science" and the reasons for a scientific community's coherent outlook. However, unlike Kuhn, Foucault declined to dissect the so-called hard sciences as objects of inquiry, restricting himself to discourses and power relationships in the medical, biological, and social sciences.[4] However, Foucault did see the potential in the application of his method for the destruction of the demarcation between scientific and nonscientific spheres of action and belief.

Constructivist sociologists of science, more than historians, have entered where Foucault declined to tread, dealing not only with the "big science" of modern times but with the classical period of Robert Boyle and the early Royal Society. Bruno Latour, Steven Shapin, and Simon Schaffer have emphasized social and discursive strategies of coercion and persuasion, marshaled by scientific protagonists who mimic codes of behavior ordinarily associated with the battlefield or the salon.[5] This is a sociology of confrontation and conquest that emphasizes the contingent personal and cultural forms that scientific discourse can assume.

Traditional historical sociology of science, by contrast, has tended to emphasize growth and evolution, rather than confrontation and conquest. Sociological accounts like those of Michael Mulkay identify scientific innovation not with radical conflict but with the gradual emergence of new disciplines or specializations. This approach is in tune with traditional sociology, dating back

2. Thomas S. Kuhn, *Structure.*

3. See Simon Schaffer on Bachelard's denial that natural philosophy was scientific. Simon Schaffer, "Natural Philosophy," 55–92, in G. S. Rousseau and Roy Porter, eds., *The Ferment of Knowledge: Studies in the Historiography of Eighteenth-Century Science* (Cambridge: Cambridge University Press, 1980), on 73. See Gaston Bachelard, *La psychoanalyse du feu* (Paris: Gallimard, 1938) and *La formation de l'esprit scientifique* (Paris: Vrin, 1938).

4. See Michel Foucault, *Les mots et les choses: Une archéologie des sciences humaines* (Paris: Gallimard, 1966). Also see, *Discipline and Punish: The Birth of the Prison*, trans. Alan Sheridan (New York: Vintage, 1977): 184–185, 225–227. In an interview, Foucault spoke of science as the "disciplinary police of knowledge"; in Michel Foucault, *Politics, Philosophy, Culture: Interviews and Other Writings, 1977–1984*, ed. L. Kritzman and trans. Alan Sheridan (New York: Routledge, 1988).

5. See Bruno Latour, *The Pasteurization of France*, trans. Alan Sheridan and John Law (Cambridge: Harvard University Press, 1988); Steven Shapin and Simon Schaffer, *Leviathan and the Air-Pump: Hobbes, Boyle and the Experimental Life* (Princeton: Princeton University Press, 1985). Also see, Bruno Latour and Steve Woolgar, *Laboratory Life: The Construction of Scientific Facts*, 2d ed. (Princeton: Princeton University Press, 1986).

to Auguste Comte, and also with the semi-Durkheimian sociology of Robert Merton and Joseph Ben-David.[6]

In the Durkheimian disciplinary analysis, evolutionary and biological models explain scientific growth on the basis of conceptual and especially social migrations, hybridizations, or speciations. Philosophers have used this model, too. Kuhn suggested it toward the conclusion of *The Structure of Scientific Revolutions*, Stephen Toulmin proposed it some twenty years ago for the competition of ideas and theories, and David Hull and Robert Richards have recently updated the model. Hull, for example, has taken into account how it is that scientific concepts pass through research groups (or cliques) and play a role in scientists' social struggles for students, funds, and journal space.[7]

Accounts written by historians, in contrast to sociologists, often have continued to emphasize the raw power of ideas and laboratory methods in shaping schools and disciplines. William Coleman argued that the

> cognitive elements of experimental physiology were to [Claude] Bernard a decisive instrument in translating bold explanatory ideals and proposed methods of a science in the making into the worldly needs of laboratory space, financial support, and provision for the training of students, each a function of social organization and together constituting the disciplinary domain par excellence.[8]

Similarly, Charles Rosenberg has written that knowledge is "a central element in shaping the structure of disciplinary cultures and subcultures."[9] In this interpretation, ideas are the engine, not the accoutrements, of discipline building.

If there is a middle way between the newer sociological construction of discipline bounding and the more traditional narrative of biography and ideas, it may lie in the analysis of research or teaching schools.[10] In 1981, Gerald Geison published an essay in *History of Science* that argued the centrality of

6. Michael Mulkay, "Three Models of Scientific Development," *Sociological Review* 23 (1975): 509–526, 535–537; also see discussions of this tradition by S. T. Keith and Paul K. Hoch, "Formation of a Research School: Theoretical Solid State Physics at Bristol, 1930–1954," *BJHS* 19 (1986): 19–44; and Gerald L. Geison, "Scientific Change, Emerging Specialties, and Research Schools, " *History of Science* 19 (1981): 20–37.

7. Thomas Kuhn, *Structure*, 172–173. Stephen E. Toulmin, "The Evolutionary Development of Natural Science," *American Scientist* 55 (1967): 456–471. David L. Hull, *Science as a Process: An Evolutionary Account of the Social and Conceptual Development of Science* (Chicago: University of Chicago Press, 1988); and Robert J. Richards, *Darwin and the Emergence of Evolutionary Theories of Mind and Behavior* (Chicago: University of Chicago Press, 1987).

8. William Coleman, "The Cognitive Basis of the Discipline: Claude Bernard on Physiology," *Isis* 76 (1985): 49–70.

9. Charles Rosenberg, "Toward an Ecology of Knowledge: On Discipline, Context, and History," in Alexandra Oleson and John Voss, eds., *The Organization of Knowledge in Modern America* (Baltimore: Johns Hopkins University Press, 1979): 447.

10. On sociological constructivism, see Steven Shapin, "Discipline and Bounding: The History and Sociology of Science as Seen through the Externalism-Internalism Debates," 203–208, in *Conference on Critical Problems and Research Frontiers in History of Science and History of Technology, 30 October–3 November* (Madison: History of Science Society, 1991).

"research schools" to charting the course of scientific development. The study of research schools or research groups is a fruitful focus of analysis in the work of Joseph Fruton on the contrasting chemical schools of Emil Fischer and Franz Hofmeister as well as in studies on Justus Liebig by both Fruton and F. L. Holmes, following up an earlier path-breaking study by J. B. Morrell.[11]

The research "tradition" is understood to manifest itself across many chronological eras and geographic boundaries, in association with ongoing "themata" of both a conceptual and methodological nature.[12] In contrast, the research school is usually associated with a specific locale or a person, for example, Liebig's school, the Giessen school. But the more important the school, the more likely it transcends the local and idiosyncratic. Thus, Jean-Baptiste Dumas's school of chemistry came to be thought of as the Parisian school of chemistry, which produced a typical confusion of Parisian institutional identities, since Dumas held six positions in Paris during 1824–1868, several of them simultaneously.

Dumas's Parisian school also became known as the *French* school of chemistry. Théophile Jules Pelouze complained about Dumas that he was using the theory of substitution to elevate himself to the position of "Chef de l'Ecole" of the new organic chemistry. Adolphe Wurtz praised Dumas in saying that "the basic lines [of the substitution theory] are indelibly drawn, and it was the French School that drew them."[13]

Similarly, Ernest Rutherford's laboratory research group at Manchester, and later at Cambridge, came to be identified with the "English school" of physics, despite the fact that Rutherford was a New Zealander who came to Manchester from McGill University in Montréal. Nevil Mott, in remembering the work of

11. Geison, "Scientific Change." J. S. Fruton, "Contrasts in Scientific Style: Emil Fischer and Franz Hofmeister, Their Research Schools and Their Theories of Protein Structure," *Proceedings of the American Philosophical Society* 129 (1985): 313–370; also see, Fruton, "The Liebig Research Group, a Reappraisal," ibid., 132 (1988): 1–66; Holmes, "Complementarity," 121–164; and J. B. Morrell, "The Chemist Breeders: The Research Schools of Liebig and Thomas Thomson," *Ambix* 19 (1972): 1–46.

Examples of other work on research schools: M. Eckert, "Sommerfeld's School and the Electron Theory of Metals," *HSPS* 17 (1987): 191–234; Gerald Geison, *Michael Foster and the Cambridge School of Physiology: The Scientific Enterprise in Late Victorian Society* (Princeton: Princeton University Press, 1978); L. J. Klosterman, "A Research School of Chemistry in the Nineteenth Century: Jean Baptiste Dumas and His Research Students," *Annals of Science* 43 (1985): 1–80; H. A. M. Snelders, "J. H. van't Hoff's Research School in Amsterdam (1877–1895)," *Janus* 71 (1984): 1–30; F. L. Holmes, "The Formation of the Munich School of Metabolism," in William Coleman and F. L. Holmes, eds., *The Investigative Enterprise: Studies on Nineteenth-Century Physiology and Medicine* (Berkeley, Los Angeles, and London: University of California Press, 1988).

12. See Gerald Holton, *Thematic Origins of Scientific Thought: Kepler to Einstein*, rev. ed. (Cambridge: Harvard University Press, 1988).

13. On Pelouze, quoted in F. L. Holmes, *Claude Bernard and Animal Chemistry: The Emergence of a Scientist* (Cambridge: Harvard University Press, 1974): 77; and on Wurtz, in éloge for Dumas, *CR*, 1884, quoted in Klosterman, "A Research School of Chemistry," 4–5.

solid-state physicist Ronald Gurney at Bristol, recalled the influence of Rutherford's school:

> It was typical of [Gurney] that, although he could use mathematics as well as another he never trusted it; no theory was a theory to him unless he could grasp it intuitively, and, above all, draw diagrams of it. In this he was among theorists perhaps the truest representative of Rutherford's school, and of the British tradition.[14]

In later chapters, I analyze two broadly defined research schools, one in France and another in England, and their roles in the development of the discipline of a theoretical chemistry distinct from physical chemistry and theoretical physics. One group, which I call the Paris school, established the field of theoretical chemistry at the Ecole Normale Supérieure. It was allied with organic chemistry, on the one hand, and physical chemistry, on the other. The second school, which I call the London-Manchester school, similarly combined problems and approaches from organic and physical chemistry but more daringly dabbled in the physics of electron theory and quantum mechanics. Thus, the discipline of theoretical chemistry took different forms in the two national traditions.

In his definition of research schools, Geison made them largely identical to laboratory-based research groups,[15] like Fruton, emphasizing the group's local nature and strictly enumerable membership.[16] My use of the term "research school" is considerably broader because it allows for a "school" of individuals to be linked through a network of institutions and generations. I recognize the broadening of the strictly defined research school as a process in which, as Kathryn Olesko has noted, the identity of a school becomes diluted by the generalization of its style over separate locales and generations.[17]

As Steven Turner has argued, the use of the word "school" in the nineteenth century often carried a pejorative meaning, in which schools were associated with controversy and polemics.[18] This is not surprising, since the struggle against scholasticism and against the skeptical schools of the seventeenth century was by no means forgotten in the early nineteenth century, and subservience to disciplinary "masters" could be viewed as a vestige of the old tyrannies of scholasticism and clericalism. However, a shift from pejorative

14. Nevil F. Mott, "Obituary, Dr. R. W. Gurney," *Nature* 171 (1953): 910, quoted in Keith and Hoch, "Formation of a Research School," 37.

15. Geison, "Scientific Change," 23.

16. See discussion of "groups" and "schools" in Keas, "The Structure and Philosophy of Group Research."

17. See Kathryn Olesko, "Tacit Knowledge and School Formation: Exact Experimental Physics at Göttingen," in *Osiris*, 2d ser. Vol. VIII, *Research Schools* (to appear in 1993), ed. Gerald Geison and F. L. Holmes.

18. See Steven Turner, "Scientific Schools and Scientific Controversy: The Case of Vision Studies in Germany," ibid.

to positive connotation seems to have occurred when the individual or local school became conflated with a national school, in a century characterized by burgeoning nationalism.

In thinking about "traditions," "disciplines," and "schools," it seems that there should be a logical relationship or order among the three categories. One might posit a temporal order, for example, of the "school" creating the "discipline," which begets the "tradition." But it is easy to sense that a linear order will not work. One counterexample is the argument by S. T. Keith and Paul Hoch that their study of solid-state physics reveals the emergence of a new discipline from the synthesis of older traditions and new skills, in which there is no simple correspondence between a research school and a single discipline or specialty.[19] Similarly, Holmes has emphasized the problem-oriented nature of scientific investigation, which often defies straightforward disciplinary classification.[20]

As a consequence, I would like to develop a working explanation or, perhaps more accurately, a schematic taxonomy for the particular kind of "knowledge-discipline" we find in science. My scheme is explicitly not a sociological model. In contrast, I draw on the notion of the construction of historical identities. In practice, national or ethnic identity may be conflated with professional or disciplinary identity, in a way that actually integrates the notions of "tradition" and "school" with that of our historical understanding of "discipline." In this explanation, the school is best understood by analogy to an extended family rather than to a nuclear group at one site.

To respect these historical developments, I have accordingly merged the category of "discipline" with the categories of "subfields," "specialties," and "subspecialties," so that I use the word "discipline" both to denote chemistry as a whole and to characterize some of its leading subfields, namely, organic chemistry, physical chemistry, physical organic chemistry, quantum chemistry, and theoretical chemistry. As I will demonstrate, this telescoping procedure is not unwarranted, since virtually all fields have regularly been called "disciplines" by their practitioners.[21]

Nor is this conflation unwarranted from a sociological point of view. We need only keep in mind Pierre Bourdieu's influential notion of an "intellectual field" composed of individuals, small groups, schools, and disciplines distributed as a network of agents and powers. Fieldlike relationships exist within subfields, which in turn occupy particular regions within the broader intellectual field.[22]

19. Keith and Hoch, "Formation of a Research School," 20.

20. See Frederic Lawrence Holmes, *Eighteenth-Century Chemistry as an Investigative Enterprise* (Berkeley: University of California Office for History of Science and Technology, 1989).

21. See n. 6 to Introduction, above.

22. See Pierre Bourdieu, "Intellectual Field and Creative Project," *Social Science Information* 8 (1969): 89–119, and more recently, "The Social Space and the Genesis of Groups," *Social*

The elements I shall employ have been used by ethnologists, anthropologists, and historians to define collectivities associated with ethnic or national identity, such as in Basque, Breton, Russian, or Hungarian identity, but it seems clear that such an approach may be used for many highly integrated groups.[23] Among historians and sociologists of science, Pnina Abir-Am has most nearly taken the approach I am employing, in her work on the Bloomsbury "biotheoretical gathering" and the development of molecular biology.[24] The great advantage of the identity approach is its capacity to join the separate analyses of schools, disciplines, and traditions into an integrated analytical matrix, which respects the conceptual and discursive practices of contemporary participants.

Elements of Identity in the History of Scientific Disciplines

My "identity" conception of a scientific discipline has six elements, which I will briefly enumerate and then illustrate. These are (1) a genealogy and family descent, including historical mythology of heroic origins and heroic episodes; (2) a core literature defining archetypical language and imagery; (3) practices and rituals that are codified and performed; (4) a physical homeland, including institutions based on citizenship rights and responsibilities; (5) external recognition; and (6) shared values and unsolved problems.

It should be noted at the outset that this identity schema for scientific disciplines runs the risk of superficially obscuring the differences among scientific communities and Seventh-Day Adventists, Daughters of the American Revolution, or Freudian analysts. In this, the construction of identity may look like an example of the constructivist sociology of science. However, the identity schema is capable of providing important demarcation criteria for science, especially in focusing on the problems scientists choose to study and the values by which scientists organize their solutions.

Thus, while an identity conception of scientific disciplines in many respects accentuates the similarities of organized scientific disciplines to other organ-

23. Jacqueline Urla, "Inventing the Future: Cultural Difference and Modernity"; Martine Segalen, "Breton Identity: Local Types and Regional Stereotype"; Seymour Becker, "National Self-Definition in a Multiethnic Society: The Nineteenth-Century Russian Intelligentsia"; Tamas Hofer, "Peasant Culture and National Symbolism in Eastern Europe, 1880–1960." Papers on file at the Rutgers Center for Historical Analysis, Rutgers University.

24. Pnina Abir-Am, "The Biotheoretical Gathering, Transdisciplinary Authority and the Incipient Legitimation of Molecular Biology in the 1930s," *History of Science* 25 (1987): 1–70. Also see Karin D. Knorr-Cetina, "The Ethnographic Study of Scientific Work: Towards a Constructivist Interpretation of Science," 115–140, in Karin D. Knorr-Cetina and Michael Mulkay, eds., *Science Observed* (Hollywood: Sage, 1983).

ized communities, the core of the scientific discipline is missed if the discipline's particular values and characteristic problems are not noted and understood. In discipline building, as in discipline bounding, practitioners have referred to a common set of problems that they seek to resolve through an appropriate, if flexible and pragmatic, set of methods that they believe will produce adequately good or right solutions. While these problems and scientific "values" are, like my more obviously sociological categories, relative to time and place, they are more directly related to the natural world. I do not agree with Harry Collins that "the natural world has a small or nonexistent role in the construction of scientific knowledge."[25]

The "identity" definition of a scientific discipline draws an analogy from national or ethnic identity to professional or scientific identity by taking into account recent concerns in the history and sociology of science. It is useful to compare elements of the identity concept with some arguments made in the 1930s by the bacteriologist, Ludwik Fleck. His monograph, *Entstehung und Entwicklung einer wissenschaftlichen Tatsache* (1935), which marshals examples primarily from the history of medicine and especially theories about syphilis, influenced Kuhn's thinking in the late 1950s. Fleck's book now often is cited by sociologists of science interested in his denial of the objective existence of a scientific fact and his argument that the true creator of a new scientific idea is not an individual but the thought-collective.[26]

Fleck discussed characteristics of what he called the "thought-collective," "thought-community," or "thought-commune" (*Denkkollektiv, Denkgemeinschaft, Denkgemeinde*), drawing comparisons among the ancient guild, the church community, the trade union, and modern science. For Fleck, a "thought-style" (*Denkstil*) characterizes a thought-commune, which is "isolated formally, but also absolutely bonded together, through statutory and customary arrangements, sometimes a separate language, or at least special terminology."[27] Fleck noted the period of apprenticeship required "for every trade, every religious community, every field of knowledge" and the exclusiveness and intolerance that results in a "stylized limitation upon the problems admitted" for study and action."[28]

Fleck's characterization of the thought-style of "modern scientific thinking, especially in the natural sciences," suggests that he considered scientific practice to differ in some important respects from other social or ideological practices:

25. Harry Collins, "Stages in the Empirical Program of Relativism," *Social Studies of Science* 11 (1981): 3–10, on 3.

26. I am grateful to Peter Loewenberg for making this connection. For Fleck and a brief discussion by Thomas Kuhn, see Ludwik Fleck, *Genesis and Development of a Scientific Fact*, trans. Fred Bradley and Thaddeus J. Trenn, foreword by Thomas S. Kuhn (Chicago: University of Chicago Press, 1979).

27. Ibid., 103.

28. Ibid., 104.

> The specific intellectual mood of modern scientific thinking . . . is expressed as a common *reverence* for an ideal—the ideal of objective truth, clarity, and accuracy.[29]

This definition, however, leaves nature out of the picture, despite Fleck's references to "natural science," perhaps because of his own cultural conditioning within the German tradition in which Wissenschaft included all forms of knowing. Fleck's characterization of science clearly contains the radical potential used in recent constructivist sociology of science to argue that scientists differ little in their behavior from acolytes of strong religious or political persuasion, and Fleck sometimes might be thought to be describing a nationalist or Volk movement rather than scientific practice:

> [The thought-style of science] consists in the *belief* that what is being revered can be achieved only in the distant, perhaps infinitely distant, future; in the *glorification* of dedicating oneself to its service; in a definite *hero worship* and a distinct *tradition*.[30]

Some of Fleck's elements of scientific Denkstil can be included in my categories of disciplinary identity: hero worship (corresponding to 1, as defined above), specialized language (2), apprenticeship (3), statutory and customary arrangements (4), exclusiveness (5), and shared ideals (6). Like Fleck's, my approach is one that argues the existence of specific scientific "styles" that have been culturally and historically conditioned. However, the identity approach to scientific disciplines is not meant to be "Fleckian," any more than it is "Kuhnian" or "Foucaultian," although it shares characteristics with these analyses and is informed by them.

We turn now to particular historical examples illustrating elements of disciplinary identity, with special attention to chemistry and physics.

1. The matter of genealogy is the first crucial element in the definition of a discipline. Whether one is constructing an Irish-American identity, a Lithuanian identity, or a physical chemistry identity, the authority of continuity is compelling. Thus, those chemists who prided themselves on founding physical chemistry at the end of the nineteenth century marshaled links to figures in the early nineteenth century or late eighteenth century.

Ostwald, the late-nineteenth-century co-founder of modern physical chemistry, claimed that physical chemistry is the descendant of Humphry Davy's and Michael Faraday's work in electrochemistry at the beginning of the cen-

29. Ibid., 142.
30. Ibid.

tury.[31] At the beginning of this century, the mathematical physicist and master of thermodynamics, Pierre Duhem, identified the spirit of physical chemistry with Pierre Joseph Macquer's mid-eighteenth-century notion of *chimie-physique moderne* and with Lavoisier's late-eighteenth-century concern

> with clarity and logical rigor; . . . with the help of the physicist's instruments—the balance, the thermometer, the calorimeter—foreign until then to the chemist, he introduced into chemical experiments an unthought of precision.[32]

Chemists still take pride in tracing their heritage back to Liebig and to Giessen, or to Lavoisier and to Paris, constructing genealogies to the *Doktor-Vater* in the manner of lineages of European royalty or the Daughters of the American Revolution. There is the well-known genealogical table at the Deutsches Museum in Munich which depicts Liebig as the originator of a scientific family tree, bearing fruit in Nobel Prize winners.[33] A typical family tree was constructed by the late Jerald Zuckermann, a specialist in inorganic chemistry, whose chemistry education at the University of Pennyslvania and Harvard University connected him to the eighteenth-century fathers Antoine Lavoisier and Geoffroy Rouelle in Paris and Torbern Bergman in Uppsala.[34]

These genealogies, of course, take their point of origin not in ordinary people but in heroic figures who fought enemies, even villains, and who won their battles by wit and dexterity. Lavoisier's triumph over the evil forces of Georg Stahl and phlogiston theory is well known, and unlike most heroic scientists, Lavoisier met a martyr's end in the epic events of the French Revolution. This is historical mythology, a crucial constitutive element of the discipline.[35]

Borrowing from Eric Hobsbawm, we may say that scientists, like others, have "invented traditions, legitimating and promoting their new scientific field."[36] But chemists, more than natural philosophers and physicists, inte-

31. Wilhelm Ostwald, *Elektrochemie. Ihre Geschichte und Lehre* (Leipzig: Verlag von Veit, 1896): p. 1148, quoted in Diana Barkan, "Walther Nernst and the Transition to Modern Physical Chemistry," Ph.D. dissertation, Harvard University, 1990, 44.

32. Pierre Duhem, "Une science nouvelle: La chimie physique," *Revue Philomatique de Bordeaux et du Sud-Ouest* (1899): 205–219, 260–280, on 215; also quoted in Barkan, "Walther Nernst," 21.

33. See R. Sachtleben, "Nobel Prize Winners Descended from Liebig," *JChem.Ed.* 35 (1958): 73–75.

34. J. J. Zuckerman, "The Chemist as Teacher of History," *JChem.Ed.* 64 (1987): 828–835, on 828.

35. On Lavoisier, see Bernadette Bensaude-Vincent, "A Founder Myth in the History of Science? The Lavoisier Case," 53–78, in Loren Graham et al., eds., *Functions and Uses of Disciplinary Histories. Sociology of the Sciences Yearbook* VII (Dordrecht: Reidel, 1983).

36. Eric Hobsbawm and Terence Ranger, eds., *Invented Traditions* (Cambridge: Cambridge University Press, 1983). For examples of invented traditions, see John Gillis, "The Cultural Production of Family Identities in Nineteenth-Century Britain," RCHA Paper, 1989. Also see, Barkan, "Walther Nernst," 28.

grated the history of their discipline into the formal training of their students. Nineteenth-century courses in "philosophical chemistry" often dealt largely with the history of chemistry.[37]

Not surprisingly, these histories differed markedly from one another according to their country of origin. Thomas Thomson's two-volume history of chemistry and Dumas's lecture course in philosophical chemistry, both published in the 1830s, present somewhat different accounts by chemists of the formation of chemistry as a discipline. Thomson gives a more favorable reading to the work of German chemists Johann J. Becher and Stahl and considerably more detail than Dumas to British chemists.[38]

Perhaps the most outstanding history of chemistry in the nineteenth century was the four-volume work by German chemist Hermann Kopp at midcentury. The most controversial history was Wurtz's 1869 book with its infamous prolegomenon:

> Chemistry is a French science. It was founded by Lavoisier, of immortal memory; . . . . he was at once the author of a new theory and the creator of the true method in chemistry; and the superiority of the method gave wings to the theory.[39]

The history of chemistry continued to be a component of chemical studies in the late nineteenth and early twentieth century. At Manchester, the Final Honors Chemistry Examination for 1905 included sections of inorganic chemistry, organic chemistry, physical chemistry, and the history of chemistry and chemical philosophy, each section three hours long.[40]

For physicists, too, historical mythology was important, one of the first exemplars being Johann Karl Fischer's eight-volume *Geschichte der Physik* of 1801–1808, which was a source for the better-known histories completed

37. On chemistry courses in nineteenth-century France, see Bensaude-Vincent, "A Founder Myth in the History of Science?" n. 3, 76–77.

38. J. B. Dumas, *Leçons sur la philosophie chimique*, ed. M. Bineau (Paris: Ebrard, 1937), and Thomas Thomson, *The History of Chemistry*, 2 vols. (London: H. Colburn and R. Bentley, 1830–31).

39. Hermann Kopp, *Geschichte der Chemie*, 4 vols. (Braunschweig: Vieweg, 1843–1847). Adolphe Wurtz, *A History of Chemical Theory*, trans. Henry Watts (London: Macmillan, 1869), on 1. As so often happens in historical mythologies, Wurtz's account had meaning for a contemporary quarrel in his own immediate scientific community. See Alan J. Rocke, "The 'Quiet Revolution' of the 1850s: Scientific Theory as Social Production and Empirical Practice," in Seymour Mauskopf, ed., *Chemical Sciences in the Modern World* (Philadelphia: University of Pennsylvania Press, in press).

40. See Addendum I in G. N. Burkhardt, *Arthur Lapworth and Others*, typescript in Robert Robinson papers, Library of the Royal Society of London. Students preparing the *diplôme d'études supérieures* in chemistry at the École Normale Supérieure in the 1920s were asked to discuss questions in the oral part of the juried examination. Topics included catalysis, stereochemistry of salt complexes, and the origins of atomic notation. See bound copy of examination memoirs, presented to Albert Kirrmann, in archives of École Normale Supérieure Laboratoire de Chimie.

around 1880 by Johann Christian Poggendorff and Ferdinand Rosenberger.[41] Special histories appeared earlier than general histories of physics, for example, Joseph Priestley's *History and Present State of Electricity* in 1767.[42] A brief historical introduction to a physics doctoral thesis was ordinary in the late nineteenth century, both for setting up the writer's claim to originality and for setting up the claim to belonging within a tradition of legitimate disciplinary scientific study. Of course, the tradition of scientists' citing earlier work is in itself a tradition of giving a history of the problem and its practitioners.

2. An important role of disciplinary history was to introduce scientists not just to family history but to the core and classic literature of the field. For physicists like Rosenberger, the classic literature literally meant the classics of Greek philosophy, including Aristotle, Plato, Euclid, and Archimedes, since "Die ersten Physiker sind greichische Naturphilosophen."[43]

As with so many literary classics, the scientific classics, even if unread, nonetheless form the basis for a common language and pattern of images shared by practitioners of the discipline. Isaac Newton's *Principia* and *Optics* and Jean-Baptiste Fourier's *Théorie analytique de chaleur* are among the classics for early-nineteenth-century physics, establishing a basic vocabulary and methods of argument for common discourse and debate. Lavoisier's *Traité élémentaire de chimie* and Jöns Jakob Berzelius's *Essai sur la théorie des proportions chimiques* perform a similar role for general chemistry. The classic texts from which evolves a contemporary core are few in number and convey instant psychological and semiotic meaning for disciplinary identity, just as Shakespeare's tragedies, Dante's *Divine Comedy*, and Goethe's *Faust* carry English, Italian, and German national identity, respectively. It is hard to imagine that a scientific discipline can exist without a textual core. Such an exemplary core text, to be discussed later in this study, is Ingold's *Structure and Mechanism in Organic Chemistry* (1953), which creates a technical language and representation of the world shared and promulgated by a new community in physical organic chemistry.

3. Another element of identity is constructed by the practice of ritual and tacit knowledge. Historians and sociologists of science who have emphasized the roots of science in craft practices have long noted the tradition of apprenticeship at the heart of the scientific enterprise. This apprenticeship is practiced in the laboratory, where the construction and use of instruments must be learned;

41. J. C. Poggendorff, *Geschichte der Physik* (Leipzig: Barth, 1879), and Ferdinand Rosenberger, *Die Geschichte der Physik*, 2 pts. (Braunschweig: Vieweg, 1882–1884).

42. Joseph Priestley, *History and Present State of Electricity, with Original Experiments* (London: Dodsley, 1767), and Emil Wilde, *Geschichte der Optik von Ursprunge dieser Wissenschaft bis auf die gegenwärtige Zeit*, 2 vols (Berlin: Rücker und Püchler, 1838–1843).

43. F. Rosenberger, *Geschichte*, I: 4.

it also is practiced in the university lecture hall, the seminar, and the tutorial, where forms of argument and representation are learned. Laboratory and lecture practices are rituals, with rites of passage codified in the oral and written examinations, in the construction of a research project, and in the defense before the masters.

Similarly, the Festschrift, the conference and congress, and the afternoon colloquium with sherry or tea are rituals that reenact the past, reaffirm the legitimacy of the discipline and its practitioners, and reenact traditions. Rituals are double edged because they reopen wounds from the past and demonstrate unsolved problems in the present. Like family dinners at Thanksgiving or reunions of old schoolmates, rituals are fraught with tension as well as with the cozy sense of belonging to a group, for better or worse.

Attendance at key meetings is part of the rite of passage for admission to a discipline, and in many cases, these are meetings of scientific societies or scientific sections organized around the discipline-at-large or the disciplinary specialty. The founding of physics, chemistry, and physical chemistry societies often turned informal meetings of like-minded friends into formal ritual occasions.

Other kinds of regular meetings have also become constitutive of disciplinary identity: for example, the Solvay Physics Congress, which first met in 1911, and the Solvay Chemistry Congress, which first met in 1922. For both, there are volumes of published proceedings, including accounts of the extemporaneous discussions following each paper, and ritual formal photographic portraits of the participants.

Some of the most famous ritual occasions in physical science are the celebrations of the retirement of the leader of a school or the founder of a discipline, a subject discussed in some detail by Abir-Am for molecular biology. On these occasions, the elders of the disciplinary tribe reminisce about the founders of the tribe, associating themselves with the heroes and confirming their special magic.[44] Two such occasions, one a celebration of Auguste Kekulé and the other of Marcellin Berthelot at the turn of this century, are remarkable instances of the degree to which the ritual of the scientific community may also become part of a larger ritual of the nation-state, much like the state funeral of Newton in 1727. The details of these two occasions deserve attention.

In 1890, Kekulé was honored by his friends, his former students, and the German Chemical Society at a "Benzol Fest" held at the Berlin City Hall. The date chosen, March 11, 1890, was the anniversary of the publication date of the first benzene structure paper, *and*, coincidentally, it was the kaiser's birthday. The celebration was well attended by the press, industrialists, and government representatives as well as by colleagues from Germany and

44. See Abir-Am, "The Biotheoretical Gathering," and Paul Hoch, in review of Rajkumari Williamson's *The Making of Physicists* (1987), cited above.

abroad. John Wotiz and Susanna Rudofsky, who have written at length about the occasion, found accounts in twenty-eight newspapers.[45]

After students and emissaries presented papers and remarks, Kekulé indulged in personal reminiscences, giving advice to students and reciting some poetry. He also gave an account of the origins of his hypothesis of the benzene ring, a hypothesis associated with the conceptual origins of structural aromatic chemistry: it came to him in a dream first on a bus in London, then while dozing in front of a fireplace in Ghent.

Kekulé did not mention Auguste Laurent's 1854 book in which cyclic hexagon structures are used to represent benzoyl chloride and ammonia, a book he apparently knew before his magical dreams. Wotiz and Rudofsky suggest that Kekulé himself assisted in creating "the notion that Kekulé's contributions were an act of genius . . . in an effort to demonstrate that aromatic chemistry had unique German foundations on which the monopolistic German dye and pharmaceutical industry of the late nineteenth century securely rested."[46]

The second occasion, the fiftieth anniversary of the publication of Berthelot's first memoir, was celebrated, a year late, on November 24, 1901, in the Grand Amphithéâtre of the Sorbonne, which, according to newspaper accounts, had never known such a numerous and distinguished audience. The president of the republic, Emile Loubet, was one of the many public figures present when a commemorative medal was presented to Berthelot. On one of its faces, the medal portrays Berthelot in profile with the legend "Chemical Synthesis, Science Guides Humanity." On the medal's other engraved face, Berthelot sits at a table looking at his laboratory apparatus, with two female figures behind him—Verité unveiling herself and Patrie coiffed in the Phrygian bonnet and presenting to Berthelot a laurel wreath. The inscription reads, "1851–1901, For Fatherland and Truth."[47]

As a recent unsympathetic biographer has revealed, Berthelot's Parisian colleague, Henri Moissan, complained to a friend about the "unmitigated hubris" of the heroes present at the celebration, although Moissan nonetheless took the podium to praise Berthelot on this ritual occasion. Homages were read by foreign colleagues Adolf von Baeyer, Stanislao Cannizzaro, and Jacobus Henricus van't Hoff, all of whose chemical work had combated Berthelot's in the last three decades. On this occasion, Berthelot had become, as his biographer and recalcitrant chemical heir Jean Jacques writes, "the instrument and support of a promotional operation of the ideology for power of . . .

45. John H. Wotiz and Susanna Rudofsky, "The Unknown Kekulé," 21–34, in James G. Traynham, ed., *Essays on the History of Organic Chemistry* (Baton Rouge and London: Louisiana State University Press, 1987).

46. Ibid., 30–31.

47. See Jean Jacques, *Berthelot: Autopsie d'un mythe* (Paris: Belin, 1987): 253–260.

a laic and center-left regime, believing in Progress and Science,'' for whom Berthelot literally became an icon.[48]

The account of these rituals indicates how it is that disciplinary identity is constituted within a network of institutions that are educational, economic, and political in nature. These factors play a role in the next two elements of discipline identity: namely, homeland and external recognition.

4. The institutions of the discipline in many—but not in all—cases structure the ''homeland'' for the discipline. It is commonplace now to describe science in substantial part as the activities of scientific academies, societies, and editorial boards that codify membership criteria, relationships, and responsibilities. As in the body politic at large, collective identity has been established not through the enjoyment of equal rights and responsibilities among members but through a structure of well-differentiated rights and responsibilities. In scientific disciplines, typical structures have been those of the associate, full, and honorary memberships; or director, panel, and referee; or division into ''applied'' and ''fundamental'' research.

The buildings of scientific groups, for example, the national academies of science, confer a physical sense of homeland that is all the more enhanced if, as in the case of the Paris Academy of Sciences, this was the original site occupied by heroic paternal or maternal figures. Visiting heroic sites like the gardens of Trinity College or Berzelius's home in Stockholm confers a symbolic, even psychological, power of identification with the heroic origins of the discipline and a recommitment to its pursuit.

At the same time, homeland is established by the physical objects that give a physical space a special identity: the balances, spectrometers, chromatographs, test tubes, electrometers, computers, even the office and library spaces, lecture halls, and classrooms that the member of a discipline instantly recognizes as his or her peculiar place. There is a sense of recognition, of comfort, of familiarity of sounds and smells in entering a chemistry building, a physics building, and an electrical engineering building, each with its own feel, as if, following an absence, one were a ''native'' returning to view the blue line of the Vosges or the Golden Gate Bridge spanning the bay between San Francisco and Marin County.

The homeland is a mixture of the natural and the constructed, of seemingly changeless and changing realities. For scientific disciplines, there are different *natural* landscapes, just as there are different *built* landscapes. For the physicist in the field of optics, the natural is the beam of light that always is to be found in his or her homeland; for the astrophysicist, it is the particular light that

48. Ibid. See the account, too, by Harry Paul, ''The Debate over the Bankruptcy of Science,'' *French Historical Studies* 5 (1968): 299–327.

emanates from the stars and nebulas; for the electrical physicist, it is the electrical current and the electrical charge; for the organic chemist, it is the array of hydrocarbon substances; for the physical chemist, it is homogeneous and heterogeneous solutions. The constructed landscape of laboratories and instruments might be the one that first strikes the eye of the visitor to the research site; the natural landscape is the one that rivets the gaze of the scientists who have worked there.

5. The homeland, like the discipline as a whole, is a terrain that is recognized for what it is both by its native members and by outsiders. External recognition is an important element of disciplinary identity. Chemistry can be said to have existed as a clearly defined discipline earlier than physics partly because it was a domain clearly recognized by others as a well-defined body of knowledge and of laboratory practice, useful in medicine, agriculture, and industry. Chemistry was of more immediate interest than natural philosophy or physics to civic and entrepreneurial interests.

Among the chemical disciplines, it is frequently said that the formation of organic chemistry as an independent discipline took place in the first half of the nineteenth century. It may be more accurate to say that during the course of the nineteenth century, chemists doing organic chemistry took over the discipline of chemistry.

In the late eighteenth century, Lavoisier not untypically treated organic substances, like tartaric acid, malic acid, and acetic acid, at the conclusion of his treatise on chemistry, without actually organizing the properties of hydrocarbons into a separate section.[49] In contrast, in the last volume of his four-volume history of chemistry, Kopp included a separate section on ''die Ausbildung der organische Chemie'' (1847). Carl Schorlemmer's *Rise and Development of Organic Chemistry*, published in 1879, was the first history of organic chemistry per se.[50]

By the 1880s and 1890s, some scientists began to define their work in chemistry as ''physical chemistry'' *against* organic chemistry and *against* physics. At stake were not only methodological and theoretical issues but also educational and commercial interests, for example, the allotment of university chairs, advertisement of university courses, and funding for institutes and laboratories. Also at stake was an increasing number of consulting positions with industries, for example, in metallurgy, textiles, electricity, gas, and fossil fuels.[51]

When Ernest Solvay first established the Solvay International Institute for

49. Antoine Lavoisier, *Traité élémentaire de chimie*, I (Paris: Cuchet, 1789): 278 ff.

50. Hermann Kopp, *Geschichte der Chemie*; Carl Schorlemmer, *The Rise and Development of Organic Chemistry* (London: Macmillan, 1879; rev. ed., ed. Arthur Smithells, 1894).

51. See Jeffrey A. Johnson, ''Academic Chemistry in Imperial Germany,'' *Isis* 76 (1985): 500–524.

Physics in Brussels, his aim was to further the "progress of physics and physical chemistry." Jagdish Mehra has speculated that the emphasis on physical chemistry was cosmetic, rather than real, in order to honor Walther Nernst,[52] but this seems unlikely. Like many industrialists, Solvay viewed the problems studied by physical chemists to have value for industrial production, especially with respect to the metallurgy of alloys and to industrial electrochemistry.[53] The Solvay Company contributed half a million francs for the building of the Institute of Electrotechology at Nancy, which opened in 1900 facing the Chemical Institute.[54]

If external recognition like Solvay's was important to establishing the discipline of physical chemistry, another source of recognition, as argued by sociologist of science Elisabeth Crawford, was the legitimation within the larger scientific community conferred by the award of Nobel Prizes. Crawford demonstrates that there was a self-conscious effort on the part of Arrhenius, director of the Nobel Institute in Physical Chemistry, to influence the selection of the first Nobel recipients so as to bring recognition to this new field of physical chemistry. The very first prize, in 1901, went to van't Hoff; Arrhenius himself received the 1903 award; and three other physical chemists were Nobel Laureates in the first decade of the prizes.[55]

6. The last category of definition of disciplinary identity is that of shared values and unsolved problems. It is misleadingly easy to conclude that shared values and problems are simply the end result of the other elements enumerated thus far. I would argue, inversely, that the sharing of values and problems is the starting point for genealogies and family descents, historical myths of heroic origins, classic literatures and languages, common practices and rituals, homelands associated with citizenship and institutions, and external recognition.

Shared values include commitment to the worthiness of peer review and judgment by rational consensus. Shared values also include concepts like the definition of proof, the principle of economy, and accepted schema of analogies, models, and abstract representation that are means of scientific explanation. These shared values are further imparted and reinforced through the ongoing traditions and specific research schools of the discipline. Shared prob-

52. Jagdish Mehra, *The Solvay Conference in Physics: Aspects of the Development of Physics since 1911* (Dordrecht: Reidel, 1975): xv.

53. See John Servos on the relation of physical chemistry to industry, in *Physical Chemistry from Ostwald to Pauling.* Also see G. Dubpernell and J. H. Westbrook, eds., *Selected Topics in the History of Electrochemistry, Proceedings of the Electrochemical Society* 78 (Princeton: Electrochemical Society, 1978).

54. See M. J. Nye, *Science in the Provinces: Scientific Communities and Provincial Leadership in France, 1870–1930* (Berkeley, Los Angeles, and London: University of California Press, 1986): 44.

55. Elisabeth Crawford, "Arrhenius, the Atomic Hypothesis, and the 1908 Nobel Prizes in Physics and Chemistry," *Isis* 75 (1984): 503–522.

lems and the collective will to solve them are the cement that keeps the members of the disciplinary group together.

As analyzed by philosophers, sociologists, and historians, perhaps *the* most striking characteristic of scientific work is its penchant for turning up new problems and its elaboration of evaluative criteria for choosing one theory over another. Scientific disciplines identify new problems and solve them. This happens within well-established disciplines and within areas of investigation that become new specialties or disciplines. It is the shared problem solving activity, then, that initially establishes and ultimately prolongs the disciplinary identity.

In the eighteenth century and early nineteenth century, there was a considerable focus in natural philosophy on the goal of a mechanical philosophy that would solve the problem of the dynamics of matter. The idea of a dynamics of matter was a unifying theme for physics and chemistry in the work of scientists like Pierre-Sadi Laplace, Claude-Louis Berthollet, John Dalton, and J. J. Berzelius. However, while physicists and natural philosophers continued to concern themselves in the course of the nineteenth century with universal principles and laws common to corpuscles in motion, chemists in the nineteenth century increasingly focused on establishing patterns of classification for the constitution and behavior of individually unique and idiosyncratic substances. The attitude of nineteenth-century chemists toward collections of amines or esters or fatty acids reminds us not of the physicist James Clerk Maxwell's reflections on the electromagnetic ether but of the biologist Charles Darwin's reaction to barnacles. For example, writing to Liebig in 1850, von Hofmann reported on some recent work:

> I have now begun a more careful investigation of diethylamine and triethylamine—an investigation which brings me much pleasure. I have learned many new things, but I must gradually part with this investigation. . . . Even so, only superhuman strength could enable one to leave these magnificent salts unanalyzed.[56]

In distinguishing physicists and chemists on the basis of the identity criterion of values and problems, it is appropriate to emphasize the molecule as the heart of the chemists' problem-solving concerns. Historically, the chemist's molecules have differed from the physicist's objects of study not just in linear scale but in the multifunctionality of the chemical molecule's character and behavior and in its capacity for generating wholly new objects, in addition to moving through space. The natural history tradition of chemistry is just as important as its natural philosophy tradition.

In what follows, it will be useful to recall the disciplinary identity schema

56. A. W. Hofmann to Justus von Liebig, #52, 29 January 1850, Brock ed. of Liebig, quoted in Keas, "The Structure and Philosophy of Group Research," 145.

from time to time, although it should not be taken to be a single, overriding framework of analysis for this study of a historical evolution from eighteenth-century chemical philosophy to twentieth-century theoretical chemistry. However, thinking about the problem of disciplinary identity as a variant of the constitution of national identity seems particularly appropriate in the historical era of the nineteenth century when both nation building and discipline building were characteristic activities of industrial societies. As we shall see, contemporaries found it natural to read significance into disciplinary identities that had evolved in particular local and national settings.

Thus, French chemists were not engaging in empty rhetoric in claiming that chemistry was a French science. English physicists had a larger nation-building agenda in touting the special significance of Newton for mechanical physics and modern industrial society. Just like nation building, discipline building at some times has been exclusionary and colonizing, at other times, open and cooperative. Just as Chancellor Bismarck was the hero-Vater of German unity, so have scientific heroes, together with their antagonisms and idiosyncrasies, contributed to the formation and fortunes of disciplines.

Like national and ethnic collective identities, disciplinary identities may be permeated by outsiders and their ideas, which become the sources for innovation and creativity. And the elements of conflict that often help establish disciplinary identity in the first place may lead to immigrations and emigrations capable of transforming the substructure and outer perimeter of the discipline. We can see many aspects of these changes as they unfold in the historical record, for example, revisions of the hagiography of the discipline, creation of new institutions and homelands, or, perhaps more immediately, transformations in everyday language.

We turn now in the next several chapters to specific aspects of the working out of the fundamental organizational and conceptual principles defining chemistry in the nineteeth century. We then turn to the development of new specializations as a result of the work of distinctive research schools, in particular, the formation of the disciplines, or subfields, of physical chemistry, physical organic chemistry, and quantum chemistry. Together these new disciplines achieved the status and the goals of those eighteenth-century chemical philosophers who had dreamed of a true "theoretical chemistry."

2

The Historical Demarcation of Chemistry and Physics

Founder Myths and Social Realities

The Historical Problem of the Relation of Chemistry to Physics

On the whole, late-nineteenth-century physicists tended to think of chemistry as a descriptive science in need of physical foundations. Maxwell characterized chemistry as a science lacking clarity and rigor, that is, insufficiently abstract. [1] Hermann von Helmholtz expressed the opinion that chemistry as a science progresses "not quite rationally."[2] These views were in keeping with late Enlightenment traditions in the classification of knowledge. As noted in the introduction, Kant wrote that theoretical reason includes mathematics and physics (Physik), the principles of which are a priori synthetic judgments. One encounters "genuine science only to the extent that one encounters mathematics in it."[3] This Kantian definition excluded late-eighteenth-century chemistry from genuine science.

1. See Keith Nier's characterization of Maxwell's attitude: "Maxwell admitted, almost grudgingly, that chemistry is a physical science. But he could find nothing agreeable to say about it. He passed over it with a perfunctory acknowledgment of high rank and an implied slur regarding lack of clarity, organization, and so forth." In Keith Alfred Nier, "The Emergence of Physics in Nineteenth-Century Britain as a Socially Organized Category of Knowledge: Preliminary Studies" (Ph.D. dissertation, Harvard University, 1975): 102–105. James Clerk Maxwell, "Physical Science," *Encyclopedia Britannica*, 9th ed., 1875–1889, vol. 19, 1–3.

2. Hermann von Helmholtz, quoted in Henry Edward Armstrong, "Presidential Address of the Chemical Section," *BAAS Rep.*, Winnipeg, 1909, 420–454, on 423.

3. See Immanuel Kant, *Critique of Pure Reason* (Norton ed.): 54; and preface to *Metaphysical Foundations of Natural Science*, quoted in Frederick Gregory, "Romantic Kantianism and the End of the Newtonian Dream in Chemistry," *Archives Internationales d'Histoire des Sciences* 34 (1984): 108–123, on 109.

Such attitudes highlight a long-standing tension between the disciplines of physics and chemistry. How did boundaries come to be drawn, and redrawn, between the two fields or disciplines? Who drew them? What changing aims and strategies were used to distinguish chemistry and physics? What should we make of the claim that physics has pride of place over chemistry in the logic or in the history of the sciences?

The interpretation of the priority of physics over chemistry in the logic and history of the scientific disciplines has been strongly rooted in Comte's positivist history, the general outline of which still dominates much of popular scientific teaching as well as the classification of scientific knowledge.

In the early decades of the nineteenth century, Comte taught that the sciences are six in number, each more complicated in turn and each less certain or "positive" in proportion to its scope. Mathematics, astronomy, physics, chemistry, biology, and sociology (or social physics) were said to represent the whole fabric of human knowledge and the history of human progress, built up over time like the cells, tissues, and organs of organic life itself. The aim of each of the sciences was to emulate, insofar as possible, the *mécanique analytique* accomplished by 1788 by the much admired mathematician and physician Joseph-Louis Lagrange.

Comte's account of disciplinary formation, with its conflation of logical structure and historical order of the sciences, both reflected and influenced the science of its time. In reflecting the science that he personally learned in Paris around 1814–1818, Comte's positivist history drew on the claims of Lavoisier's colleagues and acolytes that Lavoisier had *created* the science of chemistry, where none existed before. Comte also relied on this group's demarcation between "physique" and "chimie," a distinction developed more clearly in the French scientific community of the late eighteenth century than elsewhere.

In influencing the history and philosophy of science of later decades, Comte's positivist classification created the conviction that the constitution of mathematics and physics was historically prior to chemistry and conceptually *more fundamental* than chemistry. [4] But the positivist history we often have accepted from Comte is flawed. "Chemistry" as a discipline preceded "physics." In the next two chapters, I will deal with the claim that physics is conceptually a more fundamental science than chemistry and will analyze the characteristic aims and methods of nineteenth-century chemistry, particularly as reinforced through the hegemony of organic chemistry.

Here I first examine historical interpretations of the development of chemistry and physics as separate scientific disciplines and offer a new view of this history by reexamining Lavoisier's role in setting up distinctions among nat-

4. Comte's account may work for a history of transcendental ideas but not for a history of disciplines. Comte's *Cours de philosophie positive* was published in six volumes in Paris during 1830–1842. See the abridgement by Harriet Martineau, in her edition, *The Positive Philosophy of Auguste Comte*, 2 vols. (London: J. Chapman, 1853).

ural philosophy, physics, and chemistry from the standpoint of an established chemical discipline. I then examine the increasingly clear-cut disciplinary demarcation between physics and chemistry as a series of boundaries that were gradually socially constructed. I discuss aspects of the organization of scientific academies and societies; the founding of physical and chemical journals; and the proliferation of teaching specialties that resulted in disciplinary textbooks and specialized laboratories. In doing this, I endeavor to avoid the traps of presentism, on the one hand, or anachronism, on the other.

The concerns of this chapter fit squarely within the framework of disciplinary identity discussed previously. Here we can see some of the mechanisms by which individuals and groups have created or reinforced distinctions and identities among themselves. I largely leave the conceptual aims, language, and strategies characteristic of nineteenth-century chemistry for discussion in chapters 3 and 4. This is not to say that the epistemological aims and technical content of chemistry are completely ignored in this chapter, but they are not emphasized.

Historical Accounts of the Origins of Chemistry and Physics as Disciplines

On the whole, historians have tended to identify the beginnings of the ''modern'' discipline of chemistry with a ''Chemical Revolution,'' with Lavoisier and his circle, and more particularly with pneumatics, the oxygen theory, the balance, and the new chemical nomenclature of the late eighteenth century. This traditional historical interpretation is rooted in nineteenth-century histories of chemistry, particularly those written by the French chemists Dumas, Berthelot, and Wurtz.[5]

In contrast, in the early twentieth century, Hélène Metzger called into question whether a ''revolution'' could occur which did not measure itself against a preexisting community of opinion and practice. She emphasized the coherence, rather than incoherence, of the pre-Lavoisian state of eighteenth-century chemistry.[6] More recently, Evan Melhado has described chemistry as a well-established eighteenth-century discipline ordered around two clusters of phenomena (wet chemistry and combustion chemistry) and two theoretical motifs

5. See Jean-Baptiste Dumas, who argued that Lavoisier's experiment of 1772 led to the overthrow of the phlogiston theory and founded modern chemistry, in *Leçons sur la philosophie chimique* (Paris: Ebrard, 1837); Marcellin Berthelot, *La révolution chimique: Lavoisier* (Paris: Alcan, 1890); and Adolphe Wurtz, ''Histoire des doctrines chimiques depuis Lavoisier,'' introductory essay to *Dictionnaire de chimie pure et appliquée*, 3 vols. (Paris: Hachette, 1868–1878), trans. Henry Watts, *History of Chemical Theory* (London: Macmillan, 1869).

6. On Metzger, see essays in Gad Freudental, ed., *Etudes sur Hélène Metzger* (Leiden: E. J. Brill, 1990), along with H. Metzger, ''Introduction à l'étude du rôle de Lavoisier dans l'histoire de la chimie,'' *Archeion* 14 (1932): 21–50, and *La philosophie de la matière chez Lavoisier* (Paris: Hermann, 1935).

(affinities and phlogiston) [7] Karl Hufbauer and Christoph Meinel have each argued the late-eighteenth-century development of a cohesive German chemistry discipline.[8] And, more broadly, Holmes has described the character of early to mid-eighteenth-century chemistry as an investigative enterprise with research programs that included the chemistry of salts and the chemistry of plants, pursued within university and academic settings.[9]

Thus, there seems to be recent consensus that chemistry was a well-established discipline at least a generation before Lavoisier and that some important continuities, for example, in the chemistry of neutral salts, persisted from the early eighteenth century into the nineteenth century.[10] However, the place of Lavoisier remains crucial in the history and genealogy of chemistry, and for our purposes, it is important to look carefully at one particular aspect of the historical mythology about Lavoisier and the "origins" of modern chemistry.[11] Lavoisier wrote in the *Opuscules physiques et chimiques* (1774) that he "applied to chemistry not only the apparatus and methods of experimental physics but also the spirit of precision and calculation which characterizes that science."[12] In a letter to Benjamin Franklin in 1790, Lavoisier wrote that he had brought chemistry "much closer than heretofore to experimental physics."[13] Here is our particular interest: the relation of chemistry and physics.

Henry Guerlac, Evan Melhado, Anders Lundgren, and others have argued the thesis that Lavoisier's importance for chemistry lay in bringing into it the aims and methods of physics. Historians of science frequently have referred to the "physicalist" tradition in chemistry,[14] and Maurice Crosland has

7. Evan Melhado, "Metzger, Kuhn and Eighteenth-Century Disciplinary History," 111–134, in Freudenthal, ed. *Metzger.*

8. Karl Hufbauer, *The Formation of the German Chemical Community, 1720–1795* (Berkeley, Los Angeles, and London: University of California Press, 1982). And Christoph Meinel, "Zur Socialgeschichte des chemischen Hochschulfaches im 18 Jahrhundert," *Berichte zur Wissenschaftsgeschichte* 10 (1987): 147–168; and "Theory of Practice? The Eighteenth-Century Debate on the Scientific Status of Chemistry," *Ambix* 30 (1983): 121–132.

9. F. L. Holmes, *Eighteenth-Century Chemistry.*

10. See ibid., 107–109. In his book on science and the Enlightenment, Thomas Hankins hedged on this interpretation, writing, "Before 1750, chemistry could not be regarded as an independent discipline." In *Science and the Enlightenment* (Cambridge: Cambridge University Press, 1985): 305.

11. For the classical "origins" statement in the history of science, see Herbert Butterfield's chapter, "The Postponed Scientific Revolution in Chemistry," in his *The Origins of Modern Science, 1300–1800*, rev. ed. (New York: Free Press, 1957).

12. Quoted in Maurice Crosland, "The Development of Chemistry in the Eighteenth Century," *Studies on Voltaire and the Eighteenth Century* 24 (1963): 369–441, on 409.

13. Quoted in Holmes, *Eighteenth-Century Chemistry*, 104.

14. The physicalist tradition in English chemistry is usually dated from Boyle. Robert Schofield argues that Priestley was more a physicist or natural philosopher than a chemist (R. E. Schofield, "Joseph Priestley, Natural Philosopher," *Ambix* 14 (1967): 1–15. On Lavoisier, Henry Guerlac, "Chemistry as a Branch of Physics: Laplace's Collaboration with Lavoisier," *HSPS* 7 (1976): 193–276; Evan M. Melhado, "Chemistry, Physics, and the Chemical Revolution," *Isis* 76 (1985): 195–211; and Anders Lundgren, "The Changing Role of Numbers in Eighteenth-Century Chemistry," 245–266, in Tore Frängsmyr, J. L. Heilbron, and Robin E. Rider, eds. *The*

claimed that both pneumatic chemistry and chemical affinity had their origins in physics, or at least in natural philosophy.[15] Melhado has argued more radically that Lavoisier perceived himself as a ''physicien'' in the tradition of experimental physics and that the *Traité élémentaire de chimie* is a work ''subordinating chemistry to experimental physics.''[16]

This interpretation that Lavoisier was a ''physicist'' who brought a new physical program into a previously incoherent chemical tradition is in keeping with the Comtian positivist legacy. But what could this mean in an eighteenth-century context?

The French word *physique*, like its counterparts, is rooted in the Greek word for [N]ature, in the sense of the nature of things around us. Aristotle defined physics as theoretical knowledge, different from mathematics and theology, in that physics concerns itself with things that are always changing. Aristotle's ''physics'' included the study of those changes that we now identify with ''chemistry,'' a word rooted in Egyptian and Arabic words that came into Western parlance only in the Middle Ages. In its seventeenth- and eighteenth-century usage, ''physics'' was theoretical by definition, whereas ''chemistry'' had both theoretical and practical parts, the practical part having to do with instruments of heat and solution, the theoretical part being ''philosophical,'' sometimes historical, and always systematic.[17] In the mid-eighteenth century, Macquer wrote the *Elémens de chymie théorique* (1749) and the *Elémens de chymie pratique* (1751). There was no analogue in physics to practical chemistry; physics was by definition theoretical.[18]

Quantifying Spirit in the Eighteenth Century (Berkeley, Los Angeles, and Oxford: University of California Press, 1990). Also see articles by C. E. Perrin, Arthur Donovan, and Evan Melhado in the ''Critiques and Contentions'' section, *Isis* 81 (1990): 259–276.

15. Crosland, ''The Development of Chemistry,'' 371. As Crosland notes, the concept of chemical ''affinity'' also had a history preceding Newtonian ideas of ''force.''

16. Melhado, ''Chemistry, Physics, and the Chemical Revolution,'' 209–210. But it should not be forgotten that originators of the Lavoisier as physicist argument were Germans incensed by Wurtz's claim that chemistry is a French science, See Jakob Volhard, ''Die Begrundung der Chemie durch Lavoisier,'' *Journal für praktische Chemie* 110 (1870): 1–47; and Hermann Kolbe, ''Ueber den Zustand der Chemie in Frankreich,'' ibid.: 173–183. I am grateful to Alan Rocke for this insight.

17. See Antoine de Fourcroy, *Système des connaissances chimiques, et de leurs applications aux phénomènes de la nature et de l'art*, 11 vols. (Paris: Baudoin, 1800), I: xxx–xxxi.

18. There is a counterpart in astronomy to the theoretical/practical division in chemistry, e.g., Jean Delambre, *Astronomie théorique et pratique* (Paris: V. Courcier, 1814). Examples in chemistry include Georg Wolfgang Wedel, *Compendium Chimiae Theoreticae et Practicae, Methodo Analytica. Propositae* (Jena: J. F. Bielckii, 1715); Hermann Boerhaave, *A New Method of Chemistry, including the Theory and Practice* (London: Printed for J. Osborn and T. Longman, 1727); and P. J. Macquer, *Elemens de chymie pratique* (Paris: Hérissant, 1749), and *Elemens de chymie théorique* (Paris: Hérissant, 1751). In contrast to chemistry, texts in natural philosophy were ''mathematical and experimental'' or ''mechanical and experimental.'' For example, J. T. Desaguliers, *Physico-Mechanical Lectures, or An Account of What Is Explained and Demonstrated in the Course of Mechanical and Experimental Philosophy, given by J. T. Desaguliers* (London: Printed for the Author, 1717); J. B. Biot, *Traité de physique expérimentale et mathématique* (Paris: Deterville, 1816).

But what were the contents of "physique" in the mid-eighteenth century? "Natural Philosophy" was a well-known subject, popularly understood to include mechanics, astronomy, and optics ("mathematical physics" or "physique générale") and almost everything else that has to do with the nonliving world ("physique particulière").[19] So broadly defined was this tradition that presentist and positivist-oriented contemporary scholars sometimes have designated it as "unscientific." Few have been as harsh as the epistemologist Gaston Bachelard, who judged natural philosophy to have been the nonrational, incoherent, murky underground of modern science. For Bachelard, eighteenth-century "natural philosophy" was overly speculative and insufficiently reined in by analogical reasoning and mathematical rigor.[20]

More recently, historians of physics have argued that physics became a rigorous and coherent discipline only when the empirically based principle of energy replaced the metaphysically suspect principle of force as the basis for a unified field of experimental study.[21] Donald Cardwell, Peter Harman, and Crosbie Smith all have dated the coming into being of "modern physics" by the rejection of the principles of matter and force as fundamental principles of natural philosophy or physique générale. They suggest that the emerging concept of "energy" defined modern physics sometime after 1830, freeing scientists from mechanical philosophy, action-at-a-distance, and subtle fluids. David Knight, partly on the same grounds, dates the modern, narrow meaning of physics even later, around the mid-nineteenth century.[22]

What should we take the modern, narrow meaning of physics to be? Its late-nineteenth-century meaning is given by the categories of the subject index

19. Exemplars are Adam Walker, *Analysis of a Course of Lectures on Natural and Experimental Philosophy*, 1st ed. (London: Printed for the Author, 1766), 20th ed., 1827; Joseph Priestley, *Experiments and Observations relating to Various Branches of Natural Philosophy* (London: J. Johnson, 1779); William Nicolson, *Introduction to Natural Philosophy*, 2 vols, 2d ed. (London: J. Johnson, 1787); Antoine Libes, *Traité élémentaire de physique* (Paris: Crapelet, 1801); and F. A. C. Gren, *Grundriss der Naturlehre in seinem mathematischen und chemischen Theile* (Halle, 1793; first published 1788).

20. See Gaston Bachelard, *La formation de l'esprit scientifique* (Paris: Vrin, 1938), and *L'activité rationaliste de la physique contemporaine* (Paris: Vrin, 1951). Also, Simon Schaffer, "Natural Philosophy," 55–92, in Rousseau and Porter, eds., *The Ferment of Knowledge*; and "Scientific Discoveries and the End of Natural Philosophy," *Social Studies of Science* 16 (1986): 387–420.

21. See the classic argument by Pierre Duhem, in *The Aim and Structure of Physical Theory*, trans. from 2d French ed., Philip Wiener (Princeton: Princeton University Press, 1954).

22. Peter M. Harman, *Energy, Force and Matter: The Conceptual Development of Nineteenth-Century Physics* (Cambridge: Cambridge University Press, 1982); D. S. L. Cardwell, *From Watt to Clausius* (Ithaca: Cornell University Press, 1971): 293; Crosbie Smith, "A New Chart for British Natural Philosophy: The Development of Energy Physics in the Nineteenth Century," *History of Science* 16 (1978): 231–279; and David Knight, *The Age of Science: The Scientific World-View in the Nineteenth Century* (Oxford: Basil Blackwell, 1986): 161. Susan F. Cannon dates physics as an "invention" by the French around 1810–1830. She attributes the invention to Ampère, Carnot, Fourier, and Fresnel. In Susan Faye Cannon, *Science and Culture: The Early Victorian Period* (New York: Dawson and Science History Publications, 1978): 114. I am indebted to Richard Beyler for originally calling my attention to some of these arguments.

of the Royal Society catalog which lists scientific papers published during the nineteenth century. Only the first three volumes of the index were completed: they are mathematics, mechanics, and physics; the next volumes were to be chemistry and astronomy; and there were to be seventeen volumes in all. Mechanics includes some 21,295 entries, covering measurement of dynamical quantities, geometry and kinematics of particles and solid bodies, principles of rational mechanics, statics of particles and rigid bodies, and so on. Physics is divided into two parts: Part I covers "generalities" (including "molecular physics"), heat, light (including "invisible radiation"), and sound (33,344 entries); and Part II covers electricity and magnetism (23,300 entries).[23]

In contrast to this classification at the turn of the twentieth century, the eighteenth-century content of natural philosophy was broader still. Lavoisier's predecessor, Macquer, thought of chemistry as a specialized part of physics, writing in 1749:

> Physics [physique] has made more progress in one hundred fifty years than in thousands. But if this is true in regard to the other part of Physics, it is even more strongly the case in relation to Chemistry.[24]

In one of the most successful and typical popular texts of natural philosophy, which went through twenty editions from 1766 to 1827, Adam Walker arranged the contents in sections of mechanics, astronomy, light (including optics), fluids (including air), chemistry (including heat), magnetism, and electricity.[25]

Indeed, with respect to the meaning of natural philosophy, we must look at Lavoisier's role in the history of physics and chemistry in a new light. Whereas Melhado and others have suggested that Lavoisier was *really* a physicist, I see it differently. In reconstituting the foundations, language, and boundaries of the chemical discipline toward the end of the eighteenth century, Lavoisier broke decisively with the view that chemistry was *a part* of physics. Lavoisier and his colleagues helped establish a clearer distinction between "chimie" and "physique" (e.g., in the *Opuscules physiques et chimiques*). For Lavoisier,

23. See *Royal Society of London: Catalogue of Scientific Papers, 1800–1900, Subject Index.* Vol. I (1908); Vol. II 1909); Vol. III, Pt. I (1912) and Pt. II (1914) (Cambridge: Cambridge University Press).

24. Macquer, *Elemens de chymie théorique*, viii.

25. Adam Walker, *Analysis of a Course of Lectures on Natural and Experimental Philosophy.* For an early-nineteenth-century critique of this tradition, see E.-G. Fischer, *Lehrbuch der mechanischen Naturlehre*, 3d ed., 2 vols. (Berlin: G. C. Nauck's Buchhandlung, 1826–1827), I: vii–viii, 4–5, who argued that what belonged to Physik proper were investigations of natural phenomena on the basis of the laws of mechanics, including heat, light, and electricity but not chemistry.

this distinction entailed the view that physics ''precedes'' chemistry but without determining the course chemistry takes.[26]

The statement that physique precedes chimie made no sense to some members of the old school of physique générale. ''How,'' asked Jean André DeLuc, in response to Antoine de Fourcroy, can physique, ''which is the *assembly* [my emphasis] of natural knowledge, *precede* chemistry?''[27] Voicing a sense of competition and rivalry with the chemists, Jean Delambre complained that chemistry was diverting physicists from their usual investigations, and Antoine Libes wrote that the sphere of physics now was divided in France between chemists and mathematicians.[28]

Paradoxically, Lavoisier's and his colleagues' reformulation of the relationship between chemistry and physics helped effect the emergence of a narrower, specialized physics in France, just as it more concretely and tightly defined the conceptual and institutional parameters of chemistry. The stronger demarcation between physics and chemistry raised the prestige of the already constituted chemical discipline, which was directed toward discovering the composition, changes in composition, and properties of bodies.[29] According to chemists, physics was concerned with the laws of mechanics underlying matter in motion, and the Lavoisier school of chemists now took physics to be prefatory, but not central, to their own disciplinary domain.

Lavoisier's program and the new chemistry were seen by many to constitute a statement of the centrality of chemistry, rather than ''physics'' or ''natural philosophy,'' to the sciences as a whole. The notion of chemistry as a ''central science'' is found in the hierarchy of the sciences outlined by Comte in the early nineteenth century. It can be observed later in the mapping of the structure of twentieth-century science, in which the sociologist Eugene Garfield, for example, identifies the three disciplinary poles of physics, chemistry, and biomedicine, in which ''chemistry functions as a critical point of integration for much of the natural sciences—a hypothesis that is supported further by the very weak connection between *physics* and *biomedicine.*''[30]

26. See Bensaude-Vincent's analysis of Lavoisier's views on the need to reorganize the teaching of chemistry, including previously unpublished texts, in ''A View of the Chemical Revolution through Contemporary Textbooks: Lavoisier, Fourcroy and Chaptal,'' *BJHS* 23 (1990): 435–460.

27. For these comments, see J. A. Deluc, *Introduction à la physique terrestre . . . précedé de deux mémoires sur la nouvelles théories chimiques* (Paris: Nyon, 1803): 151–152, 163; quoting Fourcroy, on p. 164, that ''la physique . . . doit précéder ses recherches, sans pouvoir diriger sa marche,'' in Antoine F. de Fourcroy, *Système des connoissances chimiques, et de leurs application aux phénomènes de la nature et de l'art*, I (Paris: Baudouin, Brumaire An IX). And for last quotation, ibid., 166.

28. Quoted in Robert H. Silliman, ''Fresnel and the Emergence of Physics as a Discipline,'' *HSPS* 4 (1972): 137–162, on 142–143.

29. From a sampling of nineteenth-century statements about the aims of chemistry as a scientific discipline, Robert Friedel found the key words to be ''composition,'' ''properties,'' and ''change.'' In Robert Friedel, ''Defining Chemistry: Origins of the Heroic Chemist,'' in Seymour Mauskopf, ed., *Chemical Sciences in the Modern World.*

30. Eugene Garfield, *Citation Indexing—Its Theory and Application in Science, Technology and Humanities* (Philadelphia: ISI, 1979): 113.

But we must keep in mind that Lavoisier helped create the positivist legacy that ''physics'' is prior to ''chemistry.'' The demarcation, as well as the ordering, was part of his self-constituted founding mythology of chemistry, in which Lavoisier was the founder of chemistry, just as Galileo and Newton were the founders of physics. The establishment of clearly separate disciplines of ''physics'' and ''chemistry'' then was accomplished not only by conceptual and methodological distinctions, which will be discussed in the next chapters, but by social means. These included the establishment of scientific academies and societies, the founding of journals, and the standardization of teaching traditions, including the construction of disciplinary texts and specialized laboratories. The examples below are taken mostly from developments in France, Great Britain, and northern Europe, and they demonstrate that discipline bounding occurred earlier on the European continent than in Great Britain and the United States.

Academies and Societies

The Académie Royale des Sciences, first established in 1666, set up six sections in 1699: geometry, mechanics, astronomy, chemistry, botany, and anatomy. Lavoisier, as a member of the chemistry section, helped create a seventh section, ''physique générale,'' in 1785. Previously, ''physique'' might be used as an umbrella term for the sciences; now it was a separate science.[31]

While Lavoisier's role in the establishment of a physics section is significant, it is made even more intriguing by Arthur Donovan's discovery that Lavoisier actually drew up a proposal for a physics section and a list of nominees in 1766, two years before he was elected to the Academy. Donovan notes that Lavoisier's list of eight nominees included six men whom he likely perceived to be his rivals in upcoming elections to the chemistry section. Was Lavoisier simply devising a method to ensure his own imminent membership in the Academy?[32]

Be that as it may, by 1803, specialization within the disciplines had proceeded far enough that the Academy set up two classes for its sections, the ''physical sciences'' and ''mathematical sciences,'' each class with its own secretary. In this classification, physics became a mathematical science and chemistry a physical science.[33]

31. The Académie Royale des Sciences was established in 1666 with fifteen members but no classes or sections. See Maurice Crosland, ''The French Academy of Sciences in the Nineteenth Century,'' *Minerva* 16 (Spring 1978): 73–102, on 75; and *Oeuvres de Lavoisier*, IV (Paris: Imprimerie nationale, 1868): 559. A useful source for the Academy is the *Index biographique de l'Académie des Sciences, 1666–1978* (Paris: Gauthier-Villars, 1979), and for the Academy in the nineteenth century, Maurice Crosland, *Science Under Control: The French Academy of Sciences 1795–1914* (Cambridge: Cambridge University Press, 1992).

32. Arthur Donovan, ''Antoine Lavoisier, Academician,'' paper read at the annual meeting of the History of Science Society, October 25–28, 1990, Seattle.

33. See Crosland, ''The French Academy of Sciences.''

The Societas Regia Scientiarum, established in Berlin in 1700, was devoted to "Physik, Chemie, Astronomie, Geographie, Mechanik, Optik, Algebra, Geometrie, und dergleichen nützlichen Wissenschaften." By the early 1800s, the Berlin Akademie included four classes: the Physikalische, Mathematische, Historisch-philologische und Philosophisch-historische, and Philosophische Klasse. Similarly, the Scientific Society at Göttingen had three divisions in the early nineteenth century: Historisch-philosophische, Mathematische, and Physikalische.[34]

In contrast, the Royal Society of London had no disciplinary sections. Its 1662 letters of patent established the Royal Society to support philosophical studies, "especially those which endeavor by solid Experiments either to reform or improve Philosophy."[35] It was only in 1838 that the Royal Society established committees for special branches of science. The "permanent committees" included astronomy, chemistry, geology and mineralogy, mathematics, physics, and "Physiology, including the Natural History of organized beings."[36] According to Keith Nier, who looked at the records of these committees, electricity fell under the purview of the chemistry committee, which was initially chaired by Faraday. Mechanics was addressed by the mathematics committee,[37] and the physics committee concerned itself largely with meteorology and the "physics of the earth."

The nonspecialized scientific tradition that was pursued until the 1840s at the Royal Society was characteristic of many British and American scientific academies and societies in the eighteenth and early nineteenth century. The Edinburgh Philosophical Society (founded in 1737) set up two broad divisions, for example. There was a section for chemistry, anatomy, botany, medicine, metals and minerals, natural history, and "what is of a more particular nature"; and a section for "the general parts" of science, namely, geometry, astronomy, mechanics, optics, and geography.[38]

Natural philosophy or physics does not enter into this Edinburgh classification as *one* of the sciences, presumably because it still carried the connotation

34. See Adolf Harnack, *Geschichte der Königlich Preussischen Akademie der Wissenschaften zu Berlin*, 3 vols. (Berlin, 1900), I: 74, and *Inhaltsverzeichniss der Abhandlungen der Königlich Akademie der Wissenschaften zu Berlin aus den Jahren 1822 bis 1872* (Berlin: Harrwitz und Gossman, 1873). Also see, John Merz, *A History of European Thought in the Nineteenth Century*, 4 vols. (1904–1912; New York: Dover reprint ed., 1965), I: n. 2, 170–171.

35. Thomas Sprat, *History of the Royal Society*, ed. Jackson I. Cope and Harold W. Jones. (St. Louis: Washington University and London: Routledge, Kegan and Paul, 1959): 134–138.

36. See Marie Boas Hall, *All Scientists Now: The Royal Society in the Nineteenth Century* (Cambridge: Cambridge University Press, 1984): 68.

37. See Keith Nier, "Emergence of Physics," 202.

38. Roger L. Emerson, "The Philosophical Society of Edinburgh, 1737–1747," *BJHS* 12 (1979): 154–191, and "The Philosophical Society of Edinburgh, 1748–1768," *BJHS* 14 (1981): 133–176. On the Royal Society of Edinburgh, see Steven Shapin, "Property, Patronage and the Politics of Science: The Founding of the Royal Society of Edinburgh," *BJHS* 7 (1974): 1–41. I am grateful to Kerry Magruder for the Emerson references.

of an umbrella science that included the general and the particular. Similarly, the subject or field of physics fails to appear in William Whewell's late 1830s classification of the so-called inductive sciences. For Whewell, the sciences of astronomy, optics, and mechanics were mixed mathematical ("general") sciences. Electricity, magnetism, and galvanism, like chemistry, were mechanicochemical ("particular") sciences.[39]

Another form of scientific organization that proliferated in the nineteenth century was the large national subscription membership organization, pioneered by the Germans. The British Association for the Advancement of Sciences (BAAS), founded in 1831, initially had six divisions, including Section B for "chemistry" and Section A for "the mathematical and physical sciences" (mathematics, astronomy, theory of tides, meteorology, magnetism, electromagnetism, optics, acoustics, heat, and electricity). Section A of the BAAS and Section B of the American Association for the Advancement of Sciences (AAAS) helped define the parameters of "physics" in the English-speaking scientific community. Before 1899, when the American Physical Society was established in the United States, Section B of the AAAS was the "only place where physicists . . . could get together," according to Ernest Merritt.[40]

The first of the long-lived specialized societies in the physical or chemical sciences was the Chemical Society of London, founded in 1841. There had been earlier chemical groups, for example, Joseph Black's Edinburgh students, who have been called the first chemical club. A chemical society was founded in the Netherlands in 1797 and in Philadelphia by 1798, and there were clubs in London dating from 1781.

A significant policy decision by the Chemical Society was its refusal to publish papers on electricity. It is important to note, too, that many papers published by the Chemical Society in its first decade also were published in the *Philosophical Magazine*. Indeed, midcentury was a period of transition (discussed below) in which the *Philosophical Magazine* gradually was becoming a medium for research papers on energy, electricity, and electromagnetism, helping to define the character of physics for the nineteenth century.[41]

Following the Chemical Society of London, other major chemical societies were established in the next few decades, among them the Société Chimique de Paris in 1857, the Deutsche Chemische Gesellschaft in 1867, and the Amer-

39. William Whewell, *Selected Writings on the History of Science*, ed. Yehuda Elkana (Chicago: University of Chicago Press, 1984): 81–83, 156–158, 186–187.

40. Ernest Merritt, "Early Days of the Physical Society," *The Review of Scientific Instruments* 5 [new series] (1934): 143–148. I am grateful to JoAnn Palmeri for this quotation.

41. See William H. Brock, "The London Chemical Society, 1824," *Ambix* 14 (1967): 133–139; and Robert F. Bud, "The Discipline of Chemistry: The Origins and Early Years of the Chemical Society of London" (Ph.D. dissertation, University of Pennsylvania, 1980). And, more recently, R. F. Bud and G. K. Roberts, *Science versus Practice: Chemistry in Victorian Britain* (Manchester: Manchester University Press, 1984).

ican Chemical Society in 1876. Like the Chemical Society of London, other chemical societies, too, became increasingly oriented toward empiricism and experimentalism, inclining toward a policy of not accepting papers of a purely theoretical nature and inundated with papers in the field of organic chemistry.[42]

Whereas the first long-lived chemical society dates from 1841, the first long-lived "physics" or "physical" society was the Deutsche Physikalische Gesellschaft, founded in 1845. The society's concerns initially included "general physics," acoustics, optics, electricity, heat, and "practical physics," which meant meteorology and physics of the earth.[43] The Société Française de Physique was established in 1873. The Physical Society of London was founded the next year by physics-oriented members of the convivial "B Club," named after Section B (Chemistry) of the BAAS. From 1894 to 1910, the Physical Society met in the Chemical Society's rooms at Burlington House.[44]

Journals for the Physical and Chemical Sciences

The scientific journal is taken to have begun its long history in 1665. There were 2 scientific journals in 1665, 30 in 1700, 750 in 1800, and a few thousand by 1850. Among the oldest journals that have played important roles in the sciences are the *Journal des Sçavans*, founded in 1665 and published by the House of Elsevier in Amsterdam, and the *Philosophical Transactions*, founded a few months later. The *Miscellanea Curiosa Medico-Physica* and *Acta Eruditorum*, Latin counterparts of the French and English scientific journals, began to appear in Leipzig in 1670 and 1682.[45] Just as the *Journal des Sçavans* informed readers about the Paris Academy and the *Philosophical Transactions*

42. See, e.g., David Knight, "Journals," 99–126, in *Sources for the History of Science, 1660–1914* (Cambridge: Cambridge University Press, 1975), on 119. On the exponential rise in organic chemistry, see Jeffrey Johnson, "Academic Chemistry in Imperial Germany,"

43. See Armin Hermann, "Physiker und Physik—anno 1845. 120 Jahre Physikalische Gesellschaft in Deutschland," *Physikalische Blätter* 21 (1969): 399–405. I am grateful to Scott Downie for this reference.

44. See W. H. Brock and A. J. Meadows, *The Lamp of Learning: Taylor and Francis and the Development of Science Publishing* (London and Philadelphia: Taylor and Francis, 1984): 125–130.

45. See A. A. Manten, "The Growth of European Scientific Journal Publishing before 1850," 1–22, in A. J. Meadows, ed., *Development of Science Publishing in Europe* (Amsterdam: Elsevier, 1980). On scientific journals, a valuable reference work is Robert Mortimer Gascoigne, *A Historical Catalogue of Scientific Periodicals, 1665–1900* (New York: Garland, 1985). Also see, D. A. Kronick, *A History of Scientific and Technical Periodicals*, 2d ed. (Metuchen: Scarecrow Press, 1976); and B. Houghton, *Scientific Periodicals: Their Historical Development, Characteristics, and Control* (London: Bingley, 1975).

about the Royal Society of London, so the Leipzig journals kept readers abreast of German as well as broader international scientific developments.[46]

Lavoisier and his colleagues together founded the specialized *Annales de Chimie* in 1789, in part because of their concern to control a publishing outlet for their self-consciously new chemistry. The abbé François Rozier's *Observations et Mémoires sur la physique, sur l'histoire naturelle et sur les arts et métiers* (founded in 1773) strongly supported the phlogiston theory and the old tradition of physique générale.[47] In the first volume of the new chemistry journal, the sciences under review included not only chemistry, mineralogy, medical chemistry, and agriculture but also physique particulière.[48] The *Annales* was renamed the *Annales de Chimie et de Physique* in 1816, with Joseph Louis Gay-Lussac the principal editor for chemistry and D. F. J. Arago the principal editor for physics. Only in 1914 did the journal split into the *Annales de Chimie* and the *Annales de Physique*.[49]

Liebig's *Annalen der Chemie* commenced its long life as the *Annalen der Pharmacie* in 1832, becoming *Justus Liebigs Annalen der Chemie und Pharmacie* in 1840, and taking its present title in 1874. As a German journal for "pharmacie," Liebig's journal was preceded by J. B. Trommsdorff's *Journal der Pharmacie*, founded in 1794. As a German journal for "chemie," Liebig's journal was preceded by several others, including Lorentz Crell's *Chemisches Journal* (1778), F. A. C. Gren's *Annalen der Physik und Chemie* (1790), and Alexander Scherer's *Allgemeines Journal der Chemie* (1798).

Scherer's journal branched out to include "Physik" but reverted to chemistry alone under O. L. Erdmann (the *Journal für praktische Chemie*, 1834).[50] L. W. Gilbert at first eliminated "Chemie" from Gren's *Annalen der Physik und Chemie* but restored it in the form of "Physikalischen Chemie" during 1819–1823, before Poggendorff restored "Chemie" without qualification in 1824, on the grounds that physics and chemistry in their present state could not be separated. Poggendorff excluded pure mathematics from the *Annalen* and included electricity. Many highly mathematical articles by Georg Ohm,

46. See May F. Katzen, "The Changing Appearance of Research Journals in Science and Technology: An Analysis and Case Study," 177–214, in A. J. Meadows, ed., *Development of Science Publishing*. And Holmes, *Eighteenth-Century Chemistry*, 34.

47. See Douglas McKie, " 'Observations' of the Abbé François Rozier (1734–93)," *Annals of Science* 13 (1958): 73–89; and S. Court, "The *Annales de Chimie*, 1789–1815," *Ambix* 19 (1972): 113–128. From 1794 to 1823, Rozier's journal was retitled *Journal de Physique, de Chimie, de Histoire Naturelle et des Arts*. See Aaron J. Ihde, *The Development of Modern Chemistry* (New York: Harper and Row, 1964): 273.

48. *Annales de Chimie* I (1789): 4. The first issues contained memoirs by Guyton de Morveau, Lavoisier, Gaspard Monge, Berthollet, Fourcroy, Pierre Adet and J. H. Hassenfratz, and the Baron P. Dietrich.

49. From 1924 to 1933, the editors of the *Annales de Physique* were Marcel Brillouin, Aimé Cotton, and Jean Perrin; and the editors of the *Annales de Chimie* were Charles Moureu, Georges Urbain, and Marcel Délépine. It is interesting, of course, that the "physical chemist" Perrin here considered himself a physicist.

50. See Ihde, *The Development of Chemistry*, 270–273.

Wilhelm Weber, Franz Neumann, and Thomas Seebeck filled its pages; and following the founding of the "new" physical chemistry in the 1880s, the *Annalen* dropped "Chemie" from its title in 1900.

With the Chemical Society's inauguration of its *Journal* in 1847, English chemists now had a specialized journal of record, which supplemented their previous outlets in the *Philosophical Magazine* (1798) and the *Chemical Gazette* (1843), the latter the predecessor of William Crookes's *Chemical News* (founded 1859). In the case of the *Philosophical Magazine* and the *Chemical Gazette*, we have a well-documented illustration of how the history of scientific journals and the history of the formation of disciplinary specialties overlap with business entrepreneurship, in this case, that of the publishing firm of Taylor and Francis.

Founded in 1798 by Alexander Tilloch, the *Philosophical Magazine* was printed by Richard Taylor, who became its co-editor and co-owner in 1822. The *Philosophical Magazine* absorbed one journal after another: William Nicholson's *Journal of Natural Philosophy, Chemistry and the Arts* (founded 1797, absorbed 1814), Thomas Thomson's *Annals of Philosophy* (founded 1813, absorbed 1827), and David Brewster's *Edinburgh Journal of Science* (founded 1824, absorbed 1832).[51] Strongly encouraged by his son William Francis, Taylor began publishing the *Annals and Magazine of Natural History* in 1841, self-consciously including "geology" in its subtitle to attract contributions from Charles Lyell and Richard Owen.[52]

In the late 1830s and early 1840s, Francis was studying natural history, physics, and chemistry in Germany, France, and London. He and his friends provided news and translations of German chemistry for the *Philosophical Magazine* while Francis finished a doctorate under Liebig in Giessen in 1842. As the number of chemical articles in the *Philosophical Magazine* became more and more numerous, Taylor and Francis agreed on the need for a journal devoted exclusively to chemistry, one more comprehensive than *The Chemist*, established in 1840 by Charles Watt and oriented toward practical chemistry and pharmacy. Thus, the *Chemical Gazette* was launched in 1842. The founding by Taylor and Francis of their new natural history and chemistry journals in 1841 and 1842 quickly changed the character of the *Philosophical Magazine*, a trend accelerated by the appointment of William Francis's friends John Tyndall (1854–1863) and then William Thomson, Lord Kelvin (1871–1907) as its co-editors. The siphoning off of papers on biology and chemistry

> left the former journal with a pool of papers on mathematical and experimental physics. . . . Since Faraday had already chosen to use the *Philosophical Maga-*

51. W. H. Brock, "The Development of Commercial Science Journals in Victorian Britain," 95–122, in Meadows, *Development of Science Publishing*; and W. H. Brock and A. J. Meadows, *The Lamp of Learning*.

52. Brock and Meadows, *Lamp of Learning*, 110.

zine to announce findings in the field of electricity and magnetism, the journal helped to define a separate discipline of physics organized around the themes of thermodynamics, the kinetic theory of gases, and the electromagnetic theory of light, and their associated experimental phenomena.[53]

Thus, the Taylor and Francis publishing house not only provided the wherewithal for the publication of the proceedings, transactions, and journals of major scientific societies throughout Great Britain, including the *Philosophical Transactions of the Royal Society*, but also played a substantial role in refining the definitions of disciplinary specialties like physics and chemistry throughout the nineteenth century.

Lecture Traditions in Chemistry and Physics

Two teaching chairs in chemistry were installed at the Jardin du Roi in Paris during the seventeenth century, and the lectures of Nicolas Lemery and Nicolas Le Fèvre became standard texts for the next fifty years.[54] The teaching of chemistry and pharmacy was widely practiced in eighteenth-century medical faculties, exemplified at a very high standard in the lectures of Hermann Boerhaave at Leiden, Bergmann at Uppsala, and Black at Edinburgh.

Communities of chemists thrived in the late seventeenth and early eighteenth century, as Holmes has discussed in *Eighteenth-Century Chemistry as an Investigative Enterprise.* Several generations of Parisian chemists congregated for lectures at the Jardin des Plantes, and they systematically studied mineral salts and vegetable substances under the auspices of the Paris Academy. In the 1730s, Rouelle gave private courses in pharmacy and chemistry, in addition to public lectures and demonstrations at the Jardin du Roi where Lavoisier was one of his avid listeners.

François Venel, an early student of Rouelle, wrote the article on chemistry in 1753 for the *Encyclopédie.* It reflected the views of Parisian chemists like Rouelle who were concerned to distinguish chemists and their subject matter from rival lecturers and topics in physics.

> Chemistry by its visible operations resolves bodies into certain gross and palpable principles, salts, and sulfurs, etc; but Physics, by delicate speculations, acts on principles as chemistry has done on bodies; it resolves them into simpler principles, into small bodies propelled and shaped in an infinity of fashions.[55]

53. Ibid., 122. For example, translations in the 1850s of German articles by Clausius and Helmholtz on thermodynamics led directly to contributions in response from Thomson and Maxwell.

54. Holmes, *Eighteenth-Century Chemistry*, 11.

55. F. Venel, ''Chymie,'' 408–437, in 1753, in facsimile edition of *Encyclopédie ou Dictionnaire Raisonné des Arts et Métiers*, 3 (Stuttgart: Frederich Frommann, 1966), quoting from *Mem. Acad. Sci.*, 1699, on 409.

Thus, according to Venel, chemists operate, while physicists speculate, on the underlying principles of matter. Venel, Rouelle, and Macquer—like Stahl in Germany—resisted attempts to make chemistry a branch of mechanical philosophy, and they ridiculed the pretensions of the university natural philosopher or physicien who spun out theories at amphitheater rostrums, while failing to exchange the academic robe for a laboratory apron.[56] There is here an interesting and instructive parallel with the earlier history of medical education, in which the "new" anatomists of the seventeenth century claimed authority not just from mastery of the text but from prowess with the instruments of dissection.

Yet, while chemists claimed a direct, empirical, verifiable, and useful knowledge of material substances, they also became increasingly adamant that theirs was a science not only complementary to physics but building on it and thereby equally worthy of the title "philosophy" or "science." For the teaching of chemistry, with its roots in alchemy and industry, this was an important point. Indeed, Meinel has argued that the prominence of Newtonian ideas in chemical discourse is to be explained partially by chemists' campaign for scientific legitimacy. The practice of chemistry as an academic discipline was justified on the grounds that its aim was the understanding of nature (*physis*).[57]

In nineteenth-century German universities, unlike in France, there were no sciences faculties in which both physics and chemistry were taught under the same roof. At the beginning of the nineteenth century, Physik was taught in the German university Philosophical Faculty, Chemie in the Medical Faculty. As late as 1840, both the medical and philosophical faculties at Berlin rejected Liebig's claim that chemistry should be taught in the Philosophical Faculty, arguing against Liebig that a science dependent on laboratory instruction had no place in the Philosophical Faculty. Chemistry, it was said, was not a science of causes; it was not a theoretical science; and only theoretical sciences like physics or physiology had a place in philosophy.[58] Thus, we see placed into action the classification scheme of the philosophy professor Immanuel Kant.

One way of dealing with the Berlin Philosophical Faculty's point of view was to develop a distinction between "pure" and "applied" science so as to distinguish general chemistry from its particular uses in pharmacy, agriculture, manufactures, brewing, and wine making. This distinction allowed chemists to teach as academicians in the philosophical tradition but to continue to advise municipal committees on sanitation measures and perform assays for local industries.

Another path lay in persisting in the line of reasoning that physique precedes

56. J. B. Dumas quoted Nicolas LeFèvre on this point (on 55) in his *Leçons sur la philosophie chimique*.

57. Meinel, "Theory of Practice?"

58. See Steven Turner, "Justus Liebig versus Prussian Chemistry: Reflections on Early Institute-Building in Germany," *HSPS* 13 (1982): 129–162.

chimie and that chemical physics is part of the introductory chemical curriculum. Thus, as the Connecticut chemist Thomas Pynchon wrote in his *Introduction to Chemical Physics*,

> Chemistry is usually divided into two portions. The first treats of the Chemical Agents, Heat, Light, and Electricity, and is commonly called Chemical Physics; the second, of the chemical properties and relations of the various kinds of matter, inorganic and organic.[59]

In Josiah P. Cooke's projected three-volume course on chemical philosophy at Harvard University, the first volume treated chemical physics, the second was to cover electricity, and the third was to deal with stoiochiometry and the principles of chemical classification.[60] Similarly, in Wurtz's *Introduction à l'étude de la chimie* (1885), it was not until page 206 that Wurtz began discussing the law of definite proportions, following a long section on the physical properties of bodies.[61]

This format in chemistry textbooks was regarded as "no longer justifiable" by a book reviewer for the *Philosophical Magazine* in 1869, at a time "when physics is beginning to be taught (even in elementary schools) by a distinct official, or as a distinct subject from chemistry."[62] In 1873, the Natural Sciences Tripos Board concluded that is was no longer possible to "treat physics as a mere appendage to chemistry,"[63] and physics became available as an area of concentration at Cambridge University in 1881.[64]

It is important to note that while the chemical physics part of natural philosophy often was taught by chemists, the mathematical physics part of natural philosophy frequently was pursued by mathematicians in the nineteenth century as a humanistic discipline exercising the faculty of reason and nurturing aesthetic sensibility.[65] Indeed, Kuhn has argued that the disciplinary formation of modern physics along the lines of the early twentieth-century "physics" curriculum was due, in part, to mathematicians' losing interest in applied math-

59. Thomas R. Pynchon, *Introduction to Chemical Physics*, 3d rev. ed. (Philadelphia: Van Nostrand, 1881).

60. Josiah P. Cooke, Jr., *Elements of Chemical Physics* (Boston: Little, Brown and Co., 1860): vi. In his textbook section, "Chemical Physics," William Allen Miller treated elasticity, cohesion, adhesion, light, heat, magnetism, static electricity, galvanic electricity, and thermoelectricity. In W. A. Miller, *Elements of Chemistry: Theoretical and Practical* (London: John Parker, 1855).

61. Adolphe Wurtz, *Introduction à l'étude de la chimie* (Paris: Masson, 1885).

62. Quoted in Nier, "Emergence of Physics," p. 240. Hermann Kopp lectured at Giessen on "theoretical chemistry," which A. W. Hofmann referred to as "chemical physics."

63. Quoted in Nier, from Cambridge University records.

64. Romualdas Sviedrys, "The Rise of Physics Laboratories in Britain," *HSPS* 7 (1976): 405–436, on 430, n. 57.

65. On this, see Fritz Ringer, *Education and Society in Modern Europe* (Bloomington: Indiana University Press, 1978).

ematical problems.[66] During the eighteenth century and the first half of the nineteenth century, elementary "mixed mathematics" included exercises and proofs in Newtonian propositions, hydrostatics, optics, and parts of astronomy.[67] At its higher level, mixed mathematics was the focus of the Cambridge Mathematical Sciences Tripos or, on the Continent, of university-level lectures and examinations in rational mechanics and celestial mechanics.

But these topics, which had been taught by mathematicians, increasingly became the domain of mathematical physicists. By midcentury, mathematicians were abandoning the real world as the basis for mathematics in favor of establishing abstract, algebraic foundations for mathematical analysis and exploring nonphysical, that is, non-Euclidean geometries. In short, they began to indulge in the pursuit of mathematics for its own sake.[68] In Kuhn's interpretation, just as the constricting disciplinary definition of chemistry determined what would be taught as physics in the nineteenth century, so, too, the disciplinary formation of pure mathematics exerted an influence on the meaning of "physics."

Laboratories for Chemistry and Physics

The creation of the physics curriculum at Cambridge developed hand in glove with the creation of the Cavendish Physical Laboratory. No single characteristic more fully distinguished chemistry from physics before the mid-nineteenth century than the role of the laboratory. The laboratory, after all, was the reason given for excluding Liebig's chemistry from the Philosophical Faculty at Berlin. In establishing the laboratory character of modern "physical science," which resulted in the late-nineteenth-century physics institute resembling a factory more than a *palais,* chemists led the way.

To be sure, a "Laboratorium physicum" existed as early as the 1670s in Leiden, where Buchardus De Volder and his successor, Wolferdus Senguerdius, were in the forefront of teaching and demonstrating experimental philosophy based on instruments like the air pump.[69] However, the laboratorium

66. Thomas Kuhn, "Mathematical versus Experimental Traditions in the Development of Physical Science," 31–65, in *The Essential Tension: Selected Studies in Scientific Tradition and Change* (Chicago: University of Chicago Press, 1977): 60–61.

67. Joan Richards, *Mathematical Visions: The Pursuit of Geometry in Victorian England* (San Diego: Academic Press, 1988): 40–41.

68. Ibid., 240; Elizabeth Garber, "Siméon-Denis Poisson: Mathematics versus Physics in Early Nineteenth-Century France," in *Beyond History of Science: Essays in Honor of Robert E. Schofield* (Lehigh University Press, 1990). Also, J. M. Bos, "Mathematics and Rational Mechanics," 327–355, in Rousseau and Porter, *The Ferment of Knowledge*, esp. pp. 329, 334–335, 348.

69. See Willem Hackman, "Experimental Philosophy and the Dutch Republic," 171–178, in Robert P. Maccubbin and Martha Hamilton-Phillips, eds., *The Age of William III and Mary II: Power, Politics and Patronage, 1688–1702* (Williamsburg: College of William and Mary, 1989): 175.

was essentially a "theatrum" (its original name) for purposes of display. Indeed, Schaffer and others have argued that there is a great distance in this early form of natural philosophy from "science as it was later constituted."[70]

Systematic laboratory instruction in exact science, first as a program of collaboration and apprenticeship, then as a program of school training and systematic chemical investigation, was the contribution preeminently of chemists to modern science. The well-organized chemical laboratory existed from at least the 1680s, when the Swedish Board of Mines operated a "laboratorium chymicum."[71] In the early nineteenth century, the *meaning* of "laboratory" was irrefutably chemical. Thus, in 1819, the English *Cyclopedia* defined the laboratory as "a place furnished with chemical apparatus and entirely devoted to the different operations of chemistry whether on the scale of chemical manufacture, or for the purpose of experimental research."[72]

The chemical laboratory became a locus for a quantitative chemistry. In the mid-eighteenth century, Venel had called for a new Paracelsus, who would put chemistry at the side of "la Physique calculée."[73] While Lavoisier aspired to make chemistry a discipline as logically systematic as geometry,[74] his success lay in introducing precise numbers into the chemical laboratory.[75] Thus, Lavoisier, a close collaborator with the mathematician and experimentalist Laplace, was confident that chemistry could attain "results as certain as one can hope for in physics"[76] and that in the future its subject matter could be expressed algebraically.[77]

In teaching the origins of their discipline, many nineteenth-century chemists emphasized this *quantitative* aspect of Lavoisier's role in chemistry, as in Cooke's characteristic historical definition of chemistry: "The History of chemistry *as an exact science* [my emphasis] may be said to date from Lavoisier."[78] In his laboratory work, Lavoisier reported experimental results to

70. See Simon Schaffer, "History of Physical Science," 285–314, in Pietro Corsi and Paul Weindling, eds., *Information Sources in the History of Science and Medicine* (London: Butterworth, 1983): 293; also see, John Heilbron, "Experimental Natural Philosophy," 357–388, in Rousseau and Porter, *The Ferment of Knowledge*.

71. See Lundgren, "The Changing Role of Numbers," 251; and Sten Lindroth, "Urban Hiärne and the *Laboratorium Chymicum*," *Lynchos* (1946–47): 51–112.

72. "Laboratory," in *Cyclopedia* (London, 1819), quoted in Graeme Gooday, "Precision Measurement and the Genesis of Physics Teaching Laboratories in Victorian Britain," *BJHS* 23 (1990): 27.

73. Venel, "Chymie," 409–410.

74. On this theme, see Bensaude-Vincent, "A View of the Chemical Revolution."

75. Lundgren, "The Changing Role of Numbers," 256–258; Holmes, *Eighteenth-Century Chemistry*, 106–107; and F. L. Holmes, *Lavoisier and the Chemistry of Life: An Exploration of Scientific Creativity* (Madison: University of Wisconsin Press, 1985).

76. Lavoisier, *Opuscules physiques et chimiques*, in *Oeuvres de Lavoisier* (Paris: Imprimerie Impériale, 1864), I: 446. On this, too, see Arthur Donovan, "Lavoisier and the Origins of Modern Chemistry," *Osiris*, 2d ser., 4 (1988): 214–231, esp. 219–228.

77. See A. Lavoisier, "Considérations générales sur la dissolution des métaux dans les acides" (1782), in *Oeuvres de Lavoisier* (Paris: Imprimerie Impériale, 1862), II: 509–527.

78. Josiah Cooke, Jr., *Elements of Chemical Physics*, iv.

six, seven, even eight figures, unaware or insensitive to the problem of significant figures. Other chemists took to heart this example of "exact science." Berzelius reported his weight results to four or five decimal places, and he expressed scathing criticism of Humphry Davy for Davy's use of the word "about." Davy's *Elements of Chemical Philosophy* was a work too philosophical and insufficiently exact for Berzelius's tastes.[79]

In addition, Lavoisier and his colleagues introduced programmatically into the chemical laboratory apparatus other than the furnace, the crucible, and the retort, describing and illustrating the new instruments' construction and their use in texts like Lavoisier's *Traité élémentaire de chimie*. Lavoisier employed not only the balance and the thermometer but pneumatic apparatus, the electrical machine, the burning lens, and the calorimeter.[80] As the instruments of the chemical laboratory proliferated, so, too, did the problems chemists dreamed of posing and resolving.

Provision of the instruments of the laboratory became a well-established commerce in the next few decades. Just as entrepreneurs like Taylor and Francis helped create the specialized literatures of distinct scientific disciplines, so other entrepreneurs created the discipline's individualized tools. An example is John James Griffin, who, following chemical studies under Leopold Gmelin, began exhibiting chemical apparatus at BAAS meetings in the 1830s and became a founding member of the Chemical Society of London in 1841. His supplies came largely from Germany, but his designs won him a prize medal at the Great Exhibition of 1851. The Griffin and George firm remains one of the largest purveyors of chemical instruments in Great Britain.[81]

In his work with colleagues at the Arsenal, Lavoisier helped define chemistry as a collaborative, laboratory enterprise. The whole of his work gained disciplinary authority from its being the result, as Lavoisier put it, of a "community of opinions."[82] Natural philosophy had largely been the work of isolated thinkers; chemistry and physics now were to be the endeavor of a laboratory community.[83]

Chemistry and physics flourished together at the Arsenal laboratory, first under Lavoisier and then under Gay-Lussac. Laboratories were set up at the

79. On Lavoisier, see Holmes, *Lavoisier*, 499–500; on Berzelius, see Alan J. Rocke, "Atoms and Equivalents: The Early Development of the Chemical Atomic Theory," *HSPS* 9 (1978): 225–263, on 249–250.

80. Henry Roscoe and C. Schorlemmer, *A Treatise on Chemistry*, Vol. II, *Metals*, Pt. II (London: Macmillan, 1880): 464.

81. William Brock, paper given at colloquium, the University of Oklahoma, Norman, October 1990.

82. Antoine L. Lavoisier, *Elements of Chemistry*, trans. Robert Kerr (New York: Dover reprint, 1965; from 1790 Edinburgh ed., based on 1789 1st ed.): xxxiii–xxxiv. Lavoisier named Berthollet, Fourcroy, [De] Laplace, Monge, and "others."

83. On Macquer's earlier rhetorical consolidations of chemistry as an authoritative and philosophical discipline, see Wilda Anderson, *Between the Library and the Laboratory: The Language of Chemistry in Eighteenth-Century France* (Baltimore: Johns Hopkins University Press, 1984).

Ecole Polytechnique, the Institute of Egypt, and the suburban Arcueil estates of Laplace and Berthollet. Like Gay-Lussac, the chemists Louis N. Vauquelin, Michel Eugène Chevreul, and Louis Jacques Thenard admitted well-recommended students to their private chemical laboratories. By the 1830s, Dumas and Victor Regnault were training students in larger numbers as part of the expected chemical curriculum.[84]

Similarly, by 1820, Thomson began practical chemical training at Glasgow University, which in 1829 formally established a chemical laboratory. Practical chemistry teaching began at the University of Edinburgh in 1823 under Charles Hope and at University College, London, in 1829.[85]

In Germany, Liebig's well-known laboratory at Giessen was preceded by several other chemical laboratories, notably, those of Friedrich Stromeyer at Göttingen and Johann Döbereiner at Jena.[86] It is striking, as Russell McCormmach has noted, that in Germany and elsewhere the first directors of laboratories in experimental physics tended to be men trained in chemistry.[87]

It likely is no accident that it was at Giessen that Heinrich Buff, who had studied chemistry with Gay-Lussac and Liebig, created a university-recognized "institute" for physics in 1838, with space for faculty and students to do research and laboratory exercises. Buff was one of the earliest appointments in Germany to an ordinary professorship defined exclusively for Physik.[88] Weber began systematic laboratory instruction in physics in the 1830s at Göttingen.[89] Gustav Magnus, also trained as a chemist, opened the first university-funded physical laboratory in Germany in Berlin in the 1840s.[90]

Influenced by his experiences in Regnault's Parisian laboratory of chem-

84. Dumas was the first chemist to give a practical laboratory course as a routine part of instruction to students, beginning in 1832 at the École Polytechnique. See Leo J. Klosterman, "A Research School of Chemistry." On the earlier period, see Maurice Crosland, *The Society of Arcueil: A View of French Science at the Time of Napoleon I* (Cambridge, Harvard University Press, 1967); and Crosland, *Gay-Lussac. Scientist and Bourgeois* (Cambridge: Cambridge University Press, 1978).

85. Romualdas Sviedrys, "The Rise of Physics Laboratories," 407, n. 3.

86. Döbereiner was the first chemist in Germany to have an ordinary chemistry professorship in the philosophical, rather than the medical, faculty. He was appointed at Jena in 1811, and he opened a chemistry laboratory for twenty students with state funding in 1820. See Christa Jungnickel, "Teaching and Research in the Physical Sciences and Mathematics in Saxony, 1820–1850," *HSPS* 10 (1979): 3–47, on 25–26. On Liebig, see Morrell, "The Chemist Breeders," Geison, "Scientific Change," and Holmes, "Complementarity," 121–164, esp. 123–124. The first chemical laboratories in Germany were opened by F. Stromeyer at Göttingen (1806), J. N. Von Fuchs at Landshut (1807), J. F. Döbereiner at Jena, and N. W. Fischer at Breslau (1820), according to Ihde, *The Development of Modern Chemistry*, 264.

87. See Russell McCormmach, Editor's Foreword, *HSPS* 3 (1971): xi–xxiv, on xi.

88. Christa Jungnickel and Russell McCormmach, *Intellectual Mastery of Nature* (Chicago: University of Chicago Press, 1986), I: 218.

89. Jungnickel, "Teaching and Research," 23–24.

90. See Nier, "Emergence of Physics," 279. William Thomson set up the first physics laboratory of its kind in Britain at the University of Glasgow. Alexander Wood, *The Cavendish Laboratory* (Cambridge: Cambridge University Press, 1946): 1.

istry, William Thomson (later Lord Kelvin) organized a physical laboratory near his lecture room at Glasgow in 1850, which remained the only academic site for theoretical and practical instruction in electricity in Great Britain until the 1860s, when George Carey Foster established a laboratory at University College, London. Carey had been a student of the chemists Thomas Graham, Alexander Williamson, and Auguste Kekulé and of the physicists Jules Jamin and F. Quincke. The number of academic physics laboratories grew in Britain from ten in 1874 to twenty-four by 1885, a period in which the setting up of engineering professorships and laboratories further contributed to a differentiation between physics and engineering in Great Britain.[91]

If chemistry was characterized in the nineteenth century by the precise measurement of the products of chemical combustion and combination, as well as by the precise calculation of elementary combining proportions or atomic weights, physics, too, came increasingly to be identified not just with experimentalism but with precise measurement and the "last decimal place." As Maxwell put it shortly before his death in 1879,

> This characteristic of modern experiments—that they consist principally of measurements—is so prominent, that the opinion seems to have got abroad, that in a few years all the great physical constants will have been approximately estimated and that the only occupation which will then be left to men of science will be to carry on these measurements to another place of decimals.[92]

Or, as John Trowbridge, director of Harvard University's Jefferson Physical Laboratory in 1895, put it,

> The subject of physics may be characterized as that Branch of Philosophy to which men look for *exact information* [my emphasis]; . . . the difficulty of physical investigation can be realized when we reflect that an accurate determination, for instance, of the mechanical equivalent of heat would take all the time of the most competent physicist for at least a year.[93]

Conclusion: History and Hierarchy

Eighteenth- and nineteenth-century chemists helped create the university and laboratory discipline of modern physics by defining the chemical domain

91. See Graeme Gooday's table for the founding dates of laboratories in Great Britain, on p. 29, in Gooday, "Precision Measurement." Also, M. Phillips, "Laboratories and the Rise of the Physics Profession in the Nineteenth Century," *American Journal of Physics* 51 (1983): 497–503. And Sviedrys, "The Rise of Physics Laboratories," 410, 431.

92. Quoted in J. G. Crowther, *The Cavendish Laboratory, 1874–1974*, (New York: Science History Publications, 1974): 38. Also see; Lawrence Badash, "The Completeness of Nineteenth-Century Science," *Isis* 63 (1972): 48–58.

93. Quoted in Jun Fudano, "Early X-ray Research at Physical Laboratories in the United States of America, circa 1900: A Reappraisal of American Physics" (Ph.D. dissertation, University of Oklahoma, 1990): 317.

against the older, more inclusive domain of natural philosophy, physique générale, and physique particulière. Chemists established specialized academy sections, societies, and journals, and they separated out what chemists called ''physique particulière'' or ''chemical physics'' as prefatory, to their chemical discipline, not its core. Chemists established disciplinary identity and professional prestige within the academies and the universities. They also successfully integrated laboratory practice, that is, the manual arts, not only into the professional medical and pharmacy schools of the university but into the ''scientific'' and ''philosophical'' heart of the university as well.

In his preface to the 1830 edition of the *System of Chemistry*, Thomas Thomson wrote that a new chair should be set up in every university for someone ''to explain the principles of heat, light, electricity, and magnetism'' so that ''this important branch of science'' would be adequately treated. In France, he commented, ''physique'' designated this branch of study.[94] In France, Dumas told his students at the Collège de France that it is to ''physics'' that chemistry must leave the problem of particles and forces, while chemistry moves along its own different path.[95]

By 1830, most chemists were distancing themselves farther from ''physics'' than had been the case for Lavoisier and the middle generation. Institutionally, they did this through the founding of academy sections, specialized societies, chemical journals, an educational curriculum, and chemical laboratories. Conceptually, as we will see in the next chapters, Lavoisier's chemistry built on the eighteenth-century chemistry of Bergmann and others so as to construct a systematic chemistry based in generic groups of combining substances. Dumas and Laurent developed biological analogies of ''types, ''nuclei,'' and ''roots'' to explain the nature and activities of chemical substances. Methods of organic synthesis required relatively little in the way of physical apparatus, and in the enlarging field of organic chemistry, a no-man's-land between the boundaries of chemistry and physics could be great indeed. Willstätter commented in the late nineteenth century on the absence of physical instruments, even of electricity, in von Baeyer's Munich laboratory.[96] While some chemists pursued the study of relations between physical quantities like boiling points, melting points, vapor pressures, or characteristic spectral lines and chemical identity, the majority did not.[97]

By 1873, strong disciplinary boundaries had been mapped out between physics and chemistry in most institutions of higher learning in Europe and

94. Quoted in Nier, ''Emergence of Physics,'' 219.

95. Dumas, *Leçons sur la philosophie chimique*, 3.

96. Quoted in Johnson, ''Academic Chemistry in Imperial Germany,'' 510.

97. Arthur Hantzsch at Würzburg had difficulty getting his papers published because organic chemists editing the *Berichte* of the German Chemical Society insisted on the use of classical chemical methods for establishing the identity of a compound rather than methods of cryoscopy or spectroscopy. See Jeffrey Johnson, ''Hierarchy and Creativity in Chemistry, 1871–1914,'' *Osiris*, 2d ser., 5 (1989): 214–240, on 234.

Great Britain. The mutual cooperation of 1800 had largely been replaced by incomprehension of chemistry on the part of the physicists and suspicion of physicists on the part of the chemists. Chemists were incensed when the physicist Rudolf Clausius thought he was original in proposing the notion of the diatomic molecule in 1857.[98] They were indignant when Lord Kelvin optimistically claimed in the 1860s that his physical vortex atom could explain chemical affinities.[99]

Lavoisier's dictum that physics should precede chemistry became a logicohistorical interpretation, as he meant it to be, instead of a statement of pedagogical or disciplinary strategy. Paradoxically, the contemporary prestige of physics is associated with this logicohistorical tradition and with the classical and aesthetic appeal of abstract mathematics, rather than with the precision laboratory tradition on which much of modern physics, like chemistry, is based. The founder myth of Lavoisier has been perpetuated in the hagiography of the disciplinary clan of chemistry because of his role not only in the conceptual and linguistic foundations of nineteenth-century chemistry but also in a community of practitioners who refined the social definition of the chemical discipline: its formal distinction from ''physique'' in the Paris Academy, its autonomous status as the subject of the *Annales de Chimie*, its Janus-faced position astride the abyss that previously divided the philosophical science of the university from the technical practice of the laboratory.

Yet, as we have seen suggested in the barest outline by the examples of entrepreneurs like Taylor and Francis and Griffin, distinctions between chemistry and physics were carved out not only by leading chemists like Lavoisier but by a larger community, which recognized conceptual and institutional changes taking place in the scientific community and helped further them. We have looked at some social aspects of this development, and we turn now to important conceptual dimensions of nineteenth-century chemistry.

98. See Edward Daub, ''Rudolf Clausius,'' 303–311, in *Dictionary of Scientific Biography* (New York: Charles Scribners, 1971), III: 307.

99. William Thomson [Lord Kelvin], ''On Vortex Atoms,'' *Philosophical Magazine* 34 (1867): 15–24.

3

Philosophy of Chemistry and Chemical Philosophy

Epistemological Values in the Nineteenth Century

The Epistemology of Chemistry

The term "chemical philosophy" was often used in the eighteenth and nineteenth centuries, but what did it mean, and how did its meaning change during this period? Clearly, there were parallels of meaning conceptually between the terms "natural philosophy" and "chemical philosophy," but there also were important differences, particularly as the two distinctive identities of chemistry and physics began to emerge during the first decades of the nineteenth century. We focus now on the changing aims and methods of chemistry, or on the epistemology of chemistry, during much of the nineteenth century.

Philosophers and historians interested in science mostly have focused on physics, biology (especially evolutionary biology), and the social sciences, with only occasional mention of atomic spheres, benzene hexagons, and the periodic table.[1] Attention to chemical epistemology largely has centered on a

1. Of course, important exceptions to disinterest in chemistry among philosophers and historians can be found, especially in the French tradition. See Gaston Bachelard, *La pluralisme cohérent de la chimie moderne* (Paris: Vrin, 1932); François Dagognet, *Tableaux et langages de la chimie* (Paris: Seuil, 1969), and *Ecriture et iconographie* (Paris: Vrin, 1973); Ilya Prigogine, *From Being to Becoming—Time and Complexity in the Physical Sciences* (San Francisco: Freeman, 1980); Ilya Prigogine and Isabelle Stengers, *Order Out of Chaos* (London: New Science Library, 1984; first appeared as *La nouvelle alliance*, 1979; Isabelle Stengers and Judith Schlanger, *Les concepts scientifiques: Invention et pouvoir* (Paris: Editions La Découverte, 1988); and Bernadette Bensaude-Vincent, *A propos de 'méthode de nomenclature chimique': Esquisse historique suivie du texte de 1787* (Paris: Centre de Documentation Sciences Humaines, 1983).

few isolated figures like Paracelsus, Priestley, and Lavoisier and on the chemical hypothesis of the atom.[2]

By and large, a pejorative view of the methodological sophistication of chemical science has prevailed, notably, in comparison to physics. The structure of scientific explanation in chemistry often has been deemed child's play, or kitchen work. Chemistry frequently is characterized as a handmaiden, "like the maid occupied with daily civilization: she is busy with fertilizers, medicines, glass, [and] insecticides . . . for which she dispenses the recipes."[3] The Toulouse physicist Henri Bouasse enraged his colleague Paul Sabatier, who was awarded a Nobel Prize in chemistry in 1912, by jibing that chemists only aim to "faire la cuisine."[4]

In his *Confessions d'un chimiste ordinaire* (1981), Jacques mused that this state of affairs exists partly because chemists rarely write about their discipline for the general public like their biologist and physicist colleagues. And chemists have contributed to the view of their métier as a descriptive, empirical science. Recalling a late-nineteenth-century course he attended at the Sorbonne, Lespieau remarked in 1913, "Four and a half hours was all the time devoted to generalities; if one had doubled this time, it would not have been detrimental to the seventeenth property of chlorous anhydride."[5]

Nor, while flirting with the Scylla of naive empiricism, have practitioners of chemistry escaped the Charybdis of naive realism. Dalton's atomism was in fact unrelentingly and naively realist. His illustrative plates in the *New System of Chemistry* (1808) are claimed there to exhibit "the mode of combination of some of the more simple cases" of ultimate particles forming bigger ones. At midcentury, Williamson, Dalton's compatriot, received some notoriety for his defense of realism, albeit a more sophisticated variety, particularly in a London Chemical Society debate in 1869.[6]

2. On the atom, see Alan J. Rocke, *Chemical Atomism in the Nineteenth Century: From Dalton to Cannizzaro* (Columbus: Ohio State University Press, 1984); Mary Jo Nye, "The Nineteenth-Century Atomic Debates and the Dilemma of an 'Indifferent Hypothesis,' " *SHPS* 7 (1976): 245–268; and Mi Gyung Kim, "The Layers of Chemical Language II: Stabilizing Atoms and Molecules in the Practice of Organic Chemistry," *History of Science* 30 (1992): 397–437. On other issues than atomism, see John H. Brooke, "Laurent, Gerhardt, and the Philosophy of Chemistry," *HSPS* 6 (1975): 405–429, and "Methods and Methodology in the Development of Organic Chemistry," *Ambix* 34 (1987): 147–155; Rocke, "Kekulé's Benzene Theory and the Appraisal of Scientific Theories," 45–161, in A. Donovan et al., eds., *Scrutinizing Science* (Dordrecht: Kluwer Academic Publishers, 1988). On Kuhnian vs. Popperian models, see H. W. Schütt, "Guglielmo Körner (1839–1925) und sein Beitrag zur Chemie isomerer Benzolderivate," *Physis* 17 (1975): 113–125.

3. Jean Jacques, *Confessions d'un chimiste ordinaire* (Paris: Seuil, 1981): 5.

4. See Nye, *Science in the Provinces*, 286, n. 100.

5. Robert Lespieau, "Sur les notations chimiques," *Revue du Mois* 16 (1913): 257–278, on 259. Also, on Deville and Berthelot, see Mary Jo Nye, "Berthelot's Anti-Atomism: A 'Matter of Taste'?" *Annals of Science* 38 (1981): 585–590.

6. John Dalton, *A New System of Chemical Philosophy* (Manchester, 1808). For Williamson,

One result was that a century later, the philosopher John Bradley could comfortably claim that most nineteenth-century chemists were naive realists:

> To Avogadro and Cannizzaro, as to Couper and Kekulé, the molecules and atoms considered in this great theory were real objects: they were thought of the same way as one thinks of tables and chairs.[7]

This view of chemical philosophy has been shared by many physicists, for example, Henry Margenau, professor of physics and natural philosophy at Yale University in 1950, who wrote,

> Twenty years ago many chemists would have defended the theory of bond arms as a satisfactory *explanation* because they had become accustomed to thinking of it as unique and as ultimate.[8]

As we will see, this exaggerated view of naive realism in chemical philosophy does not stand up to historical scrutiny.[9]

Some of the founding documents of modern chemistry do explicitly address questions of epistemology. Where they are not explicit, we can reconstruct epistemological views by looking carefully at lectures, textbooks, and journal articles from the nineteenth and early twentieth centuries. What we find is a plurality of methodologies in the practice of chemistry that belies simplistic stereotypes of a single chemical epistemology. Nineteenth-century chemists were not dyed-in-the-wool instrumentalists or radical skeptics, any more than they were naive empiricists or naive realists. And chemists have not all shared the same views about the aims and methods of their discipline.

We can safely say that chemical explanation from the late eighteenth century onward has been consistently characterized, like physics, by one or more of three kinds of theoretical approaches that are well known to analysts of scientific method: (1) "realist," (2) "positive," and (3) "conventional." In this schema, realist theories can be characterized as hypotheticodeductive, with varying degrees of truth probability for their "picture" or "system" of the

---

see A. W. Williamson, "On the Atomic Theory," *JCS* 22 (1869): 328–365; followed by "Discussion of Dr. Williamson's Lecture on the Atomic Theory," ibid., 433–441.

7. John Bradley, "On the Operational Interaction of Classical Chemistry," *BJPS* 6 (1955–56): 32–42, on 32.

8. Henry Margenau, *The Nature of Physical Reality: A Philosophy of Modern Physics* (New York: McGraw Hill, 1950), 99, n. 1.

9. It is interesting to conjecture that if chemists have spent less effort than physicists reflecting on the philosophical foundations of their subject, the reason may be illuminated by Bas Van Frassen's claim that the closer contemporary *physicists* are to experimental work, the less interested they are in fundamental questions. Chemists have always been closer to experimental work, more thoroughly involved in the laboratory than natural philosophers, and less interested in idealizations of phenomena.

See Bas Van Frassen, "Interpretation in Science and the Arts," paper given at CCACC Conference, "Realism and Representation," Rutgers University, November 10–12, 1989.

world. Realist theories are often characterized as "philosophical" because they are fundamentally concerned with causality in the classical ontological tradition. In contrast, positive theories are largely nonhypothetical, descriptive, and inductive. They aim at quantification and mathematical representation in the sense of "exact" science. Finally, conventional theory is neither "exact" nor "probably true," and especially as used in chemistry, it has overlapping, complementary, but frequently incommensurate parts. Conventional theory is instrumental, and, above all, it is pragmatic.

In this chapter, we begin by examining how eighteenth-century "chemical philosophy" was identified with the aims of "natural philosophy" for establishing a "probably true" or realist picture of mechanical causes and effects in the world. These aims were recognized by Lavoisier, though not by all his colleagues, to be premature. By 1830, "philosophical chemistry" was in decline, as the philosophy or epistemology of chemistry became increasingly positive and conventional. In France, the decline of philosophical chemistry coincided with the decline of the Newtonian program in several areas of physique particulière, and it coincided with vigorous debates in the Academy of Sciences between Georges Cuvier and Etienne Geoffroy Saint-Hilaire about the relative virtues of functional and morphological explanation in a natural system of zoological classification.[10]

We focus in this chapter on a shared epistemology, rooted in the natural history tradition, that helped create a strong sense of discrete disciplinary identity independent of physics for most chemists, particularly as problems of organic chemistry came to dominate their interests.[11] A focus on structure and function replaced the earlier preoccupation with cause and mechanism, defining a set of problems and methods of solution that demarcated chemistry ever more clearly from physics. It was only toward the end of the nineteenth century, following the successful development of organic chemistry and "structure chemistry," as well as initial investigations in early chemical thermodynamics, that theoretically oriented chemists returned to the eighteenth-century project of a chemical philosophy directed to understanding the mechanisms that underlie chemical functions.

Early Chemical Philosophy: Aims and Methods

In the late sixteenth century, Andreas Libavius warned a former student not to associate with chemists who were not philosophers. Libavius was the author

10. On Laplace, see Robert Fox, "The Rise and Fall of Laplacian Physics," *HSPS* 4 (1974): 89–136; and on Cuvier, see Toby A. Appel, *The Cuvier-Geoffroy Debate: French Biology in the Decades Before Darwin* (Oxford: Oxford University Press, 1987).

11. On the natural history tradition as an example for chemistry, see David Knight, *The Transcendental Part of Chemistry* (Folkestone, Kent: Dawson, 1978); and, more recently, Mi Gyung Kim, "The Layers of Chemical Language, I: Constitution of Bodies v. Structure of Matter," *History of Science* 30 (1992): 69–96.

of *Alchemia* (1597) and the founder of modern chemistry, according to the historian of chemistry Owen Hannaway.[12] In the philosophical pursuit of chemistry, Libavius followed the humanist tradition of Lutheran intellectualism embodied in the teachings of the Lutheran theologian Philipp Melanchthon. Knowledge was to be gained by sense experience and reason, but it would be rash to claim absolute conviction for naturally acquired knowledge. Progress in science (*scientia*) would come from collective endeavor in which individual contributions are subject to the scrutiny of one's peers and measured against the collective wisdom of the past. The nature of chemical reasoning was to be analogical, invoking the juxtaposition of natures or ratios, and therefore, it was by nature probabilistic. Following Petrus Ramus, Libavius indicated the origin of knowledge in practice but stressed that chemistry must be a demonstrative science, abstracted from its applications.[13]

This distinction between ''theoretical'' and ''practical'' chemistry was one observed in textbooks throughout the eighteenth and nineteenth centuries. A tradition of ''philosophical chemistry'' answered Libavius's challenge for chemistry to abandon alchemical magic and Paracelsian iatrochemistry in favor of newly philosophic principles in chemistry. Jacob Barner's seventeenth-century work, *Chymia philosophica,* is an early example; later, more famous texts in chemical philosophy are those of John Dalton (1808), Davy (1812), and Dumas (1837).[14] But texts called chemical philosophy were fewer than those in ''natural philosophy,'' and very few texts in chemical philosophy were written after 1840.[15] Why was this the case?

12. Owen Hannaway, *The Chemist and the Word* (Baltimore: Johns Hopkins University Press, 1975): 75–79. H. W. Schütt claims for Robertus Vallensis the status of author of the first chemical text and history of chemistry, *De veritate et antiquitate artis chemicae* (1561); in Hans-Werner Schütt, ''Chemiegeschichtsschreibung—'Zu welchem Ende'?'' *Chemie in Unserer Zeit* 22 (1988): 139–145, on 140.

13. Hannaway, *The Chemist and the Word*, 95–96, 109, 121–123, 141–142.

14. Jacob Barner, *Chymia philosophica* (Norbergae: Sumtibus Andreae Ottonis, 1689), a system of acids and alkalis, written by the physician to the king of Poland. Antoine F. de Fourcroy, *Philosophie chimique, ou, vérités fondamentales de la chimie moderne disposés dans un nouvel ordre* (Paris: Imprimerie de Cl. Simon, 1792); this is a reprint of the ''Axiomes'' from Vol. II of the *Encyclopédie méthodique* article on ''Chymie.'' John Dalton, *A New System of Chemical Philosophy* (Manchester: S. Russel, 1808); Humphry Davy, *Elements of Chemical Philosophy* (London: Printed for J. Johnson by W. Bulmer, 1812); Jean-Baptiste Dumas, *Leçons sur la philosophie chimique* (Paris: Ebrard, 1837).

On the ''chemical philosophy'' of the Paracelsian tradition, see Allen G. Debus, *The Chemical Philosophy: Paracelsian Science and Medicine in the 16th and 17th Centuries*, 2 vols. (New York: Science History Publications, 1977). Also see Arthur Donovan, *Philosophical Chemistry in the Scottish Enlightenment: The Doctrines and Discoveries of William Cullen and Joseph Black* (Edinburgh: Edinburgh University Press, 1975).

15. The most famous work of ''chemical philosophy'' in the second half of the nineteenth century is Stanislao Cannizzaro, ''Sketch of a Course of Chemical Philosophy'' (1858; Edinburgh: Alembic Club reprint 18, 1947), in which he argues the identity of the chemical and physical atom.

Pt. II of Benjamin Silliman, Jr.'s *First Principles of Chemistry* (Philadelphia: Loomis and Peck, 1847) is entitled ''Chemical Philosophy,'' encompassing laws of combination, nomencla-

In the mid-eighteenth century in Paris, Macquer and Antoine Baumé self-consciously began their chemistry course with a set of general principles rather than facts to "provide links between facts and make it easier for students to learn." They proposed at the same time to "indicate where principles are only suppositions, probabilities, and matter for further research."[16]

The word "principle" had two meanings for mid-eighteenth-century chemists—the ideal, abstract, and hypothetical and the material and empirical. The aim of the chemistry of Macquer and Baumé was the explanation of chemical change through laws of affinity that describe power relations among the fundamental chemical principles. Macquer vacillated between the hypothetical and material meanings of "principle," that is, the Newtonian notion of hypothetical massy particle and the Stahlian notion of material elementary substance.[17]

If chemical "principles" were understood by philosophers to be invisible or visible matter, "cause" was understood to be invisible force. Like so many other chemists in the eighteenth century, Macquer and Baumé assumed that the forces of chemical affinity are simply instances of physical forces.[18] They took Newton to be a student of these affinity forces, and they found congenial Peter Shaw's view that "it was by means of chemistry that Sir Isaac Newton has made a great part of his discoveries in natural philosophy."[19] The notion of the identity of chemical and physical force was based in the principle of economy, expressed in the standard form that similar effects "must arise from the same law, if we are not to multiply causes."[20]

What were the characteristics of chemical philosophy in the early nineteenth century? For Davy, the popular lecturer at the Royal Institution in London, the aim of chemical philosophy was to "ascertain the causes of all [chemical changes] and to discover the laws by which they are governed." In the British tradition of utility and natural theology, Davy taught that chemical philosophy has "ends" as well as "aims," namely, its applications and uses "for increasing the comforts and enjoyments of man, and the demonstration of the order, harmony, and intelligent design of the system of the earth." Regarding meth-

---

ture, affinity, crystallization, and chemical effects of electricity. Josiah P. Cooke published in 1860 the *Elements of Chemical Physics*, which was intended to be the first volume of an extended work on the "Philosophy of Chemistry" (see vi).

16. A. Baumé and P. J. Macquer, *Plan d'un cours de chymie expérimentale et raisonnée avec un discours historique sur la chymie* (Paris: Herissant, 1757): 1–7.

17. For an analysis of the language of "matter," substance," and "body" in natural philosophy, chemistry, and natural history, see Mi Gyung Kim, "Layers of Chemical Language," 71–80. Kim similarly analyzes the terms "attraction," "affinity," and "relationship" and "aggregation," "composition," and "constitution."

18. Ibid., 8.

19. Peter Shaw, *A New Method of Chemistry*, 2d ed., 2 vols. (London: Longman, 1741): 173, n.; quoted in Schaffer, "Natural Philosophy," 65.

20. Adam Walker, A *System of Familiar Philosophy in Twelve Lectures* (London: Printed for the Author, 1799): 144. In Newton's *Principia*, the wording is similar.

odology, "[in] chemical philosophy . . . observation, guided by analogy, leads to experiment, and analogy, confirmed by experiment, becomes scientific truth."[21]

Dumas was more explicit about principles and causes in his lectures on chemical philosophy to students at the Sorbonne in 1837:

> Chemical philosophy . . . has for its aim to reveal the general principles of the science . . . [to give its history], to give an explanation of the most general of chemical phenomena, to establish the link between facts and the cause of the facts. Chemical philosophy makes abstract the special properties of bodies; it is composed of the general study of the material particles chemists call atoms and the forces to which these particles are submitted.[22]

In short, Dumas's chemical philosophy aimed at general abstract principles, which were identified with chemical atoms and chemical forces; and it taught the history of chemistry as a guide to the progress of philosophical truth.

In marked contrast to Dumas's philosophical chemistry is the mode of presentation and the tradition of Lavoisier's *Traité élémentaire de chimie* (1789). Lavoisier self consciously began his text with observations, not first principles, stating that "chemistry is an incomplete science, not like geometry." The new nomenclature that he and his colleagues had developed is claimed to be based in a "natural order of ideas," not in metaphysics.[23]

What are the aims of chemistry? For Lavoisier, they are identical to chemical operations:

> Chemistry, in submitting different natural bodies to experiment, has for its goal decomposing them and putting them in a state to allow examining separately the different substances which enter into combination. . . . We cannot be assured that what we think simple today really is so; all that we can say is that some substance is the present end at which chemical analysis arrives.[24]

The laboratory method of chemistry is analysis/synthesis, and the conceptual method of chemistry is analogy expressed in systematic language, for, as Etienne Bonnot de Condillac had demonstrated, "languages are true analytical

21. Davy, *Elements of Chemical Philosophy*, 1. For natural theology and chemical philosophy, see William Prout's Bridgewater Treatise, *Chemistry, Meteorology and the Function of Digestion Considered with Reference to Natural Theology* (London: Pickering, 1834).

22. J. B. Dumas, *Leçons sur la philosophie chimique*, 1–2.

23. Antoine Lavoisier, *Elements of Chemistry* [1789], trans. Robert Kerr (Edinburgh: William Creech, 1790; Dover reprint ed., 1965): xix–xx, xxvi. And see, Antoine Fourcroy, *Philosophie chimique*. J. A. Deluc notes that Fourcroy's "chemical philosophy" is better understood to mean the "new chemistry" of Lavoisier. See Deluc, *Introduction à la physique terrestre*, 167.

24. Antoine Lavoisier, 193–194, in Vol. I, *Traité élémentaire de chimie*; in Dover edition, 176–177.

methods. . . . The art of reasoning is nothing more than a language well arranged.''[25]

On the face of it, nothing could seem more diametrically opposed than the views of Lavoisier and Dumas on the aims of chemistry, at least as expressed in these programmatic statements of 1789 and 1837. Yet, ironically, Dumas was at a crossroads in the mid-1830s, and his interpretation of the explanatory aims of chemistry was shifting from an atom-and-force program of explanation to a structure-and-function one. In so shifting, he appears to move in the direction of Lavoisier's definition of chemistry and away from the notion of a chemical philosophy of mechanical forces that he and Davy had espoused. In short, Dumas was shifting to positivist and conventional methods of chemistry from the more mechanical, realist, and ''philosophical'' method.

What must be remembered is that while the twentieth-century definition of ''philosophy'' is broad and flexible, eighteenth- and early-nineteenth-century notions of philosophy more narrowly identified philosophy with unified systems of causal demonstration. Natural philosophy was nothing if not causal, and it was oriented to strongly probable, or certain, knowledge.[26]

In this sense, Lavoisier's *Elementary Treatise* is self-consciously nonphilosophical from the start. It begins with facts; it is an open-ended enterprise with no appearance of the certainty of geometric demonstration; its symbols may look like algebra, but they are only ''simple annotations, of which the object is to ease the labors of the mind.''[27] Its method is one of analysis, namely, a material or operational analysis accomplished by the instruments of the laboratory and a conceptual analysis achieved by a method of classification. The method is one of a compositional grammar of ''principles'' defined by physical parameters like volume, weight, melting points, and boiling points and by chemical parameters like acid, alkali, and neutral (or saturated or unsaturated) orders of substances. It is a generic system.

To the question, what is the *cause* of a chemical reaction? Lavoisier's answer was descriptive and concrete, not geometric and abstract. Lavoisier did not rule out the possibility of an abstract and geometric chemical philosophy in the future but thought it was premature at the time. An avid investigator of the chemistry of organisms as well as of mineral substances, Lavoisier for the present shared the view of his contemporary, the British natural philosopher John Robison, that the phenomena of fermentation, nutrition, secretion, and crystallization are not susceptible to simple mechanical reasoning.

25. Lavoisier, quoting Condillac, in the Dover edition, xiii–xiv; on analysis and synthesis, 33; on analogy, 16–17, 61, 146, 156, 168.

26. See Stephen E. Toulmin, *Cosmopolis: The Hidden Agenda of Modernity* (New York: Free Press, 1990), for an analysis of this seventeenth-century program.

27. See C. C. Gillispie, *The Edge of Objectivity* (Princeton: Princeton University Press, 1970): 242–245; and A. L. Lavoisier, ''Considérations générales sur la dissolution des métaux dans les acides'' (1782), in *Oeuvres de Lavoisier*, Vol. II, *Mémoires de chimie et de physique* (Paris: Imprimerie Impériale, 1862): 509–527.

> Whenever we see an author attempting to explain these hidden operations by invisible fluids, by aethers, by collisions, and vibrations, and particularly if we see him introducing mathematical reasonings into such explanations—the best thing we can do is to shut the book, and take to some other subject.[28]

In the early 1800s, there was a decided difference of opinion among chemists about the merits of pursuing philosophical chemistry. For Dalton, like the young Dumas, the *cause* of chemical reactions and behavior lay in elementary atoms and forces of attraction and repulsion lying in atoms and in the clouds of caloric surrounding them.[29] Similarly, for Berzelius,

> . . . the cause of chemical proportions [in reacting compounds] must be that bodies are composed of particles which should be mechanically indivisible; this explains all phenomena, . . . especially multiple proportions. . . . Probabilities lead to imagining these elementary bodies in spherical form.[30]

Liebig, too, saw the aim of chemistry as the search not only to consolidate the truth of chemical proportions but to study the causes of the regularity and constancy of these proportions. Liebig took the cause of chemical action to lie in Newtonian type atoms and forces of the Berzelian variety, that is, spherical atoms and electrical affinity forces.[31] This is a problem-solving tradition focused on atoms and powers, or mechanism and materialism.[32]

Dumas was to be the pivotal figure in shifting the study of causes in chemistry from atoms and forces to structure and function. This shift away from force as causal agent undermined the ''philosophical'' or ''positive'' status of chemistry, which some chemical philosophers already feared was becoming increasingly ''unphilosophical'' on other grounds, namely, that chemistry was becoming a science of too many conventions.

One issue that concerned Berzelius arose from the experimental variability of the equivalent combining values of some chemical elements. The combining proportion of hydrogen with nitrogen, for example, varies depending on circumstances. Was nitrogen to be assigned an equivalent combining value of 3 or 5? The fundamental concept of chemical ''equivalent,'' Berzelius wrote Liebig, ''was no longer positive, i.e., empirical, but merely conventional.''[33]

28. John Robison, quoted in Nier, ''The Emergence of Physics in Nineteenth-Century Britain,'' 88.

29. Dalton, *A New System of Chemical Philosophy*, 141.

30. Jöns Jakob Berzelius, *Essai sur la théorie des proportions chimiques et sur l'influence chimique de l'électricité* [1819], intro. Colin Russell (Johnson reprint, 1972): 21, 23.

31. See Justus von Liebig, *Introduction à l'étude de la chimie* (Paris: L. Mathias, 1837): 110.

32. See Arnold Thackray, *Atoms and Powers: An Essay on Newtonian Matter Theory and the Development of Chemistry* (Cambridge: Harvard University Press, 1970), and Robert E. Schofield, *Mechanism and Materialism: British Natural Philosophy in an Age of Reason* (Princeton: Princeton University Press, 1970).

33. Berzelius to Liebig, in Justus Carrière, ed., *Berzelius und Liebig: Ihre Briefe von 1831–1845*, 2d ed. (München: J. F. Lehmann, 1898): 206.

Further, in 1823, Döbereiner, Pierre Dulong, and Thenard recognized the operation of a new kind of chemical "affinity," in which a substance can affect a chemical reaction without appearing itself to change. Referring to this discovery of chemical "catalysis," Graham concluded that it was unphilosophical to simply refer the phenomenon to an unexplained force : "The doctrine of catalysis must be viewed in no other light than as a convenient fiction, by which we are able to class together a number of decompositions."[34] Fictional forces were to give way to new modes of chemical explanation around 1830.

Natural History and Chemical Explanation in the Nineteenth Century

Dumas told students in his 1836 course of lectures that the principle of an indivisible atom was an indifferent hypothesis and therefore a matter of convention. "What difference does it make for the facts of chemistry if [elementary chemical masses] were capable of being cut up infinitely by forces independent from chemistry?"[35]

What led Dumas to this conclusion was a failed program of chemical philosophy. For some years, he had developed and employed the technique of vapor-density analysis to calculate relative atomic weights of elements that are easily vaporized, using as working hypotheses Dalton's atomic hypothesis and Avogadro's hypothesis (that equal volumes of vapors under identical conditions contain equal numbers of particles). Dumas further assumed that all elemental gas particles are composed of two halves (diatomic molecules in modern terminology), but this assumption got him into difficulty with the vapors of mercury (Hg), sulfur [S_6], and phosphorus [P_4]. He began losing confidence in the general atomic theory.

In addition to pursuing the experimental program of studying vapor densities, Dumas also kept returning in the early 1830s to implications of a practical problem he had been presented following a court soirée in 1827 when bleached wax candles ruined the evening by emitting noxious chlorine vapors. In 1838, Dumas published a paper on the chlorination of organic substances in which he presented a new theory not of chemical *forces* but of chemical *types*.

> [C]hlorinated vinegar is still an acid, like ordinary vinegar; its acid power has not changed. It saturates the same quantity of base as before, . . . So here is a

34. Thomas Graham, *Elements of Chemistry: Including the Application of the Science in the Arts*, 2d ed., 2 vols., (London: Baillière, 1850). I: 234.

35. Dumas, *Leçons sur la philosophie chimique*, 233–234. See later, Henri Poincaré, in *Science and Hypothesis*, trans. W. J. Greenstreet (New York: Dover, 1952): 152.

> new organic acid in which a very considerable quantity of chlorine has entered, and which exhibits none of the reactions of chlorine, in which the hydrogen has disappeared, replaced by chlorine, but which experiences by this remarkable substitution only a gradual change in its physical properties.[36]

Dumas's argument stressed the malleability of the chemical "type" (in this case, the organic acid) that maintains its fundamental properties even when new and very different elements enter into its constitution. In chapter 4, more details will be given about the images and language employed in this new explanatory strategy in chemistry. Here we stress that Dumas now began to emphasize *differences* in properties and laws between physics and chemistry, rather than their similarity or complementarity, and he fairly quickly won converts to his view.

The conversion came at a time when the Newtonian program of explanation had lost ground in several fields of laboratory studies, including physical optics, electricity, and heat. Intellectually, this loss of influence was epitomized by the publication in 1826 of Augustin Fresnel's 1819 prize memoir on the diffraction of light, in which he abandoned the Newtonian corpuscular theory. Institutionally, the decline was registered by the 1822 election of Fourier to the office of permanent secretary of the Academy of Sciences, despite the opposition of Laplace, who along with Berthollet had earlier personified the Newtonian tradition in France.[37]

Further, the summer of 1830 was the culminating period of a long dispute between Cuvier and Etienne Geoffroy Saint-Hilaire over the basis for a natural system of classification in zoology. Pursuing a set of doctrines that he called "philosophical" or "transcendental" anatomy, Geoffroy Saint-Hilaire challenged Cuvier's basic premise that an animal's needs and functions determine its structure. Geoffroy Saint-Hilaire believed, in contrast, that one could work out a generalized plan for both vertebrate and invertebrate anatomy by studying the connections between parts and determining homological relationships. As Toby Appel puts it in her masterful study of these debates, "For the philosophical anatomists, animal organization appeared to have a constancy in the number and arrangement of parts that was independent of the form of the parts and the uses to which they were put."[38]

Before coming to Paris as a young man, Dumas studied in Geneva under the botanist Augustin P. de Candolle, who had published a botanical classification based on the concept of "type" in 1813. One of Dumas's good friends in Paris was the biologist Henry Milne Edwards, who developed a type theory

36. J. B. Dumas, "Mémoire sur la constitution de quelques corps organiques et sur la théorie des substitutions," *Comptes Rendus* (1839): 609–622.

37. See Fox, "The Rise and Fall of Laplacian Physics," 110–113, 126.

38. Appel, *The Cuvier-Geoffroy Debate*, 4; and Knight, *The Transcendental Part of Chemistry*.

in the 1830s and who attempted to work out a systematic compromise between the extreme emphases of Cuvier on function and Geoffroy Saint-Hilaire on structure.[39] Dumas, in turn, became perhaps the most significant figure to advocate a new explanatory system, or philosophy, for chemistry. A theory of molecular or constitutional type could meet the needs of chemical experimentation and explanation. Dumas simply borrowed this from the natural history of living organisms.

Friedrich Wöhler, Berzelius, and Liebig were among those who initially opposed and even ridiculed Dumas's new approach. The distinguished physical chemist and historian of chemistry J. R. Partington has given us the following vignette:

> Wöhler, who met Dumas on a visit to Paris (''Babylon'') in 1833, although calling him a ''windbag'' and ''Jesuit'' (Dumas was a Catholic), said he was a ''very industrious fellow'' (ein sehr fleissiger Kerl) and had ''a good heart.'' . . . Berzelius on repeating Dumas' experiments, could only confirm them. Liebig and Dumas were soon reconciled . . . and at a banquet in Paris in 1867, over which Dumas presided, Liebig said he had given up organic chemistry ''since with the theory of substitution as a foundation,'' organic chemistry needs only labourers.[40]

In Liebig's banquet comments, he was disingenuous about his reasons for giving up organic chemistry. But he did give credit to the transformation of chemical philosophy into what we might call a paradigmatic ''normal'' science. The substitution theory and then the type and structure theories *were* extraordinarily successful in producing and predicting new chemical results, including the synthesis of tens of thousands of compounds in the next decades. As Berthelot noted, chemistry became the first science to create its own objects of study.[41] Its characteristic theories gave chemistry an identity independent from both natural philosophy and natural history, while borrowing from both. Its practical successes gave chemistry a convincing warrant for believing that its theories were very much on the right track.

Substitution of elements or groups of elements (''radicals'') into the molecular type became the explanation for the production of compounds that then could be understood as, say, ammonia or methane ''derivatives'' (''degenerates''). The theory spawned further explanatory schemes. The carbon ''chain'' was a ''backbone'' or ''vertebral column'' for organic molecules. Why does carbon give birth to so many compounds? The explanation combines the early nineteenth-century equivalency theory and the mid-nineteenth-century carbon-

39. See Appel, *The Cuvier-Geoffroy Debate;* and Mi Gyung Kim, ''Practice and Representation: Investigative Programs of Chemical Affinity in the Nineteenth Century'' (Ph.D. dissertation, University of California, Los Angeles, 1990): 75, n. 37.

40. J. R. Partington, *A History of Chemistry* (London: Macmillan, 1964), IV: 339.

41. Marcellin Berthelot, *La synthèse chimique* (Paris: Baillière, 1876): 275.

backbone theory: "carbon is a quadrivalent element and it can combine with itself."[42] None of these theories makes use of the indivisibility of the atom, rectilinear or curvilear motion, or centers of force.

In addition, as is discussed further in chapter 4, molecules' structures, like organisms' structures, were found to have many overlapping functions. The empirical formula for a substance might be expressed $X_aY_bZ_c$, but further representations of the substance needed to set out parallel or incommensurate functions of alcohol or aldehyde, acid or base, dependent on environmental circumstances. During the nineteenth century, the abandonment of force causality, on the one hand, and the adoption of multiple-explanation and conventional methodology, on the other hand, led many chemists to forsake the language of earlier "chemical philosophy" and to stress the differences between physics and chemistry. The distinction between the two became all the more marked as mathematical physics became more thoroughly based in abstract analytic methods.

In this vein, Liebig wrote his former student August Wilhelm von Hofmann, shortly after the latter arrived in London in 1849,

> The new compounds of aniline are very interesting, namely the bases corresponding to ethylamide or amidethyl. You will certainly not regret having spent so much time working with aniline, because *the history and constitution* of such noteworthy compounds is *so valuable for theoretical chemistry*; every single new compound is a point of departure for *a series of homologous compounds and the idea out of which the compounds originated* is a kernel of a seed which bears its fruit in the mind by similar work.[43]

For Liebig and Hofmann, theoretical explanation meant understanding what *has happened* as well as what would happen. It is this temporal emphasis of chemistry that aligns it with biology as much as with physics.[44]

Atomism and Chemical Explanation in the Nineteenth Century

Citing their successes, many chemists tended to emphasize the empirical and descriptive characteristics of their work, distancing themselves from spec-

42. Charles Friedel, *Cours de chimie organique, professé à la Faculté des Sciences, Paris* (Paris, [1886–87] 1887): 2.

43. My emphasis. Quoted by Michael Keas, "The Structure and Philosophy of Group Research," 111; who also quotes a letter from Hofmann to Hermann Kolbe (from a collection at the Deutsches Museum), in which Hofmann says that the entire history [*Geschichte*] of the ammonia type was completed in six weeks. In Keas, "The Structure and Philosophy of Group Research," 114.

44. On chemistry and temporality, see D. W. Theobald, "Some Contributions on the Philosophy of Chemistry," *Chemical Society Reviews* 5 (1976): 203–213, on 209. Also, O. T. Benfey, "Concepts of Time in Chemistry," *JChem.Ed.* 40 (1963): 574–577, esp. 577.

ulative theory, especially the use of hypothesis, and finding algebraic representation irrelevant. Perhaps the most extreme statement of this view was that of Berthelot, who exerted considerable authority over chemical education and chemical careers in France for the last four decades of the nineteenth century.[45]

An antiatomist and powerful member of the French academic establishment, Berthelot dissuaded students from writing their examinations in the notation of atomic weights and symbols that were used almost everywhere outside France after 1860. Berthelot argued that the notation of equivalent weights is based on *chemical* analysis and analogies, whereas atomic weights are based on *physical* hypotheses; that chemistry is a science in the tradition of natural history, employing a system of classification founded on analogies, types, and functions. Chemistry, he argued, is a positive science, whereas physics relies too much on speculative and metaphysical hypotheses.[46] In this spirit, too, Hermann Kolbe attacked van't Hoff and Johannes Wislicenus, as he had criticized Charles Gerhardt and Laurent for their abstract "pencil and paper chemistry."[47]

Chemists attached to this pencil and paper chemistry were usually careful to make limited claims for its correspondence with reality. Berzelius himself wrote that theory

> is only a manner of imagining [se répresenter] the interior of phenomena, although in a certain period of development of science, it serves it altogether like a true theory. . . . [With new facts accumulated through the centuries] one will probably change the modes of imagining phenomena in science, without perhaps ever finding the truth. . . . [I]t sometimes happens that two different explanations can equally take place: it becomes necessary to study them both. . . . The true *savant* . . . studies the probabilities, and gives preference to no opinion unless it is founded on decisive proofs. . . . If we change theory it must fit the known facts better.[48]

On these grounds, Berzelius then introduced his theory of the identity of electricity and chemical affinity.

On the controversial question of the atomic hypothesis, Edward Frankland was more circumspect:

> I neither believe in atoms themselves, nor do I believe in the existence of centres

45. For an excellent, if unsympathetic, biography of Berthelot, see Jean Jacques's, *Berthelot: Autopsie d'un mythe.*

46. Marcellin Berthelot, *Leçons sur les méthodes générales de synthèse en chimie organique* (Paris: Gauthier-Villars, 1864): 453–454, n. 5, and 521–523.

47. See H. Kolbe, "Zeichen du Zeit," *Journal für praktische Chemie* 122 (1876): 268–278; and "Zeichen du Zeit. II.," ibid., 123 (1877): 473–477, cited in J. R. Partington, *A History of Chemistry*, 503; and Rocke, *Chemical Atomism*, 235.

48. J. J. Berzelius, *Essai sur la théorie des proportions chimiques et sur l'influence chimique de l'électricité*, 18–19.

> of forces, so that I do not think I can be fairly charged with this very crude notion.[49]

Kekulé, who also suffered Kolbe's attacks on pencil and paper chemistry, defended the use of theoretical hypotheses on the grounds of utility, rather than probability:

> The question whether atoms exist or not . . . belongs rather to metaphysics. In chemistry we have only to decide whether the assumption of atoms is an hypothesis adapted to the explanation of chemical phenomena . . . [and] whether a further development of the atomic hypothesis promises to advance our knowledge of the mechanisms of chemical phenomena. . . . I rather expect that we shall some day find, for what we now call atoms, a mathematico-mechanical explanation, which will render an account of atomic weight, of atomicity, and of numerous other properties of the so-called atoms.[50]

Alan Rocke claims that the advocacy of the method of hypothesis, which had been familiar to physicists for a generation, began to replace inductivist rhetoric in chemical circles just about the time that Kekulé was formulating his benzene theory. But the method of hypothesis was a familiar one to chemical leaders like Berzelius and Dumas, and Dalton hardly avoids the fact of his hypothetical reasoning by his, indeed, inductivist rhetoric.[51]

Rocke suggests that, like the atomic theory, the benzene theory was *both* extremely popular—because it was successful—*and* disbelieved in realist terms.[52] In both cases, and in the case of the carbon tetrahedron, there were two compelling problems that deterred most chemists from believing they had a real, probably true theory.

First, neither the chemical atom nor the benzene ring had its own mechanics, that is, a set of laws of motion for what was going on inside the chemical species or molecule. Second, neither the chemical theory of the atom nor the chemical theory of the benzene ring corresponded to a single explanatory device but rather to overlapping, partially incommensurate explanations of chemical behavior. The chemists' atoms constituted a molecule that acted more like an organism responding to its environment than like a billiard ball or smoke ring, and rigorous deductions from one or two simple causes did not work.

49. Edward Frankland, in ''Discussion,'' 302–305, following Benjamin Brodie's paper, ''On the Mode of Representation afforded by the Chemical Calculus, as Contrasted with the Atomic Theory,'' *Chemical News* 15 (1867): 295–302; both reprinted in David Knight, ed., *Classical Scientific Papers* (New York: American Elsevier, 1968): 250 in original text, 302 in Knight.

50. Auguste Kekulé, ''On Some Points of Chemical Philosophy,'' *Laboratory* 1 (1867): 303–306; reprinted in Knight, 255–258, on 304 [257].

51. Rocke, ''Kekulé's Benzene Theory,'' 156–157.

52. Ibid.

Mechanics, the Elusive Dream

Chemistry thrived in the middle decades of the nineteenth century as a distinctive scientific discipline with a pragmatic epistemology that suited well the practical needs of everyday laboratory life. The idea of a chemical philosophy rooted in causal mechanisms of atoms and forces languished because it presented few useful strategies in the chemical laboratory. The cumulative successes of organic chemistry were rooted in positive and conventionalist epistemologies that provided schemes for collecting information, interpreting behavior, and inventing new substances. This does not mean that the mechanist or "philosophical" dream was relinquished; rather, it was repressed.

Chemistry is more than a descriptive science, said Thomas Graham, the dean of English chemistry following Dalton's death and England's most physics-minded chemistry professor at midcentury. Chemistry is not just classification, because it embraces the action of bodies on each other.[53] "To suppose that rest, rather than motion, is the normal state of the particles of matter, is at variance with all that we know of the effects of heat, light, and electricity."[54] Thus, chemical causality still was understood by Graham, and by his younger German colleague, Lothar Meyer, to encompass mechanics but perhaps not a mechanics of "force."

> If we look for an explanation of this astonishing property by which some atom can be united only to one other atom, while another kind of atom can be united to two, three, or even four, five, six atoms . . . we find ourselves before the door to which chemistry has been knocking for a hundred years without finding an answer.[55]

To be "scientific," continued Meyer, who is often identified as a "theoretical" chemist, we must find out the cause of variable valence and how it is that an atom changes its nature according to its immediate environment.

The answer, Meyer thought, lies in the kinetic theory of heat and matter. This physical theory had been given explicit chemical meaning by Williamson's inference from studies of the synthesis of diethyl ether that atoms in chemical compounds must be continually changing places.[56] Molecules are not empty boxes in translation or rotation but little Pandora-like boxes filled with active entities. The goal of chemistry must be the understanding of chemical phenomena using theories of motion, not just theories of species or types.

Frankland, former Manchester chemist and professor at the Royal College

53. Thomas Graham, *Elements of Chemistry*, 2d rev. ed. (New York: Baillière, 1850) I. 217.

54. Graham, *Elements of Chemistry*, 2d rev. ed., (New York: Baillière, 1857), II: 600–603.

55. Lothar Meyer, *Die Modernen Theorien der Chemie und ihre Bedeutung für die Chemische Statik*, 3d ed. (Breslau, Maruschke und Berendt, 1877), pp. 158–159.

56. Meyer, *Les théories modernes de la chimie et leur application à la mécanique chimique*, trans. from 5th German ed., Albert Bloch (Paris: Carré, 1887), I. viii.

of Chemistry, similarly characterized atoms within the molecule as having their own proper motions, vibratory or otherwise, in addition to motion common to the whole molecule. But, he stressed, there also are bounds beyond which no atom can pass without causing rupture and decomposition of the molecule. We must understand the nature of these bounds that define the species or type as well as the motion against them.[57]

Chemical "affinity" remained part of the tool kit of the chemist, however badly defined and understood. Affinity cannot simply be explained away as heat, insisted Wurtz, a leading advocate of chemical and physical atomism in France in the generation following Dumas.[58] As we will see in chapter 5, "energy" replaced "affinity" in the late 1800s as the driving force of chemical reactions. In addition, the concepts of spontaneity and irreversibility entered the domain of physics, undermining the classical mechanics of matter and force in which processes are, in principle, reversible. Conceptually, the notions of spontaneity and irreversibility were more closely allied with experimental results in classical chemistry than in classical physics.

During the nineteenth century, then, chemistry became a powerful discipline with roots as strongly entrenched in the philosophical tradition of the academy and the university as in the practical work place of industry and agriculture. In defining their aims and methods, chemists developed a plurality of strategies that constituted an epistemology of chemistry fully as sophisticated as the epistemology of physics. This epistemology helped constitute the independent disciplinary identity of chemistry well demarcated from the old chemical philosophy.

In conclusion, it should be reiterated that realism, positivism, and conventionalism all were aspects of chemical epistemology, as they were of physical epistemology.

More radically, it can be argued that chemists recognized *before* most physicists the conventional character of the basic definitions and premises of scientific explanation systems, an argument usually identified in physics with Heinrich Hertz, Henri Poincaré, and Edouard LeRoy at the end of the nineteenth century. And finally, chemists recognized early on that multiple explanations are superior to a simple but wrong explanation. In short, chemistry had a principle of complementarity long before physics did.

57. Edward Frankland, "Contributions to the Notation of Organic and Inorganic Compounds," 4–25, in Frankland, *Experimental Researches in Pure, Applied, and Physical Chemistry* (London: J. Van Voorst, 1877): 4–5.

58. Adolphe Wurtz, *Introduction à l'étude de la chimie*, 8–9.

4

Language and Image in Nineteenth-Century Chemistry

Signs and Meanings

Charles Peirce distinguished three kinds of signs: the icon, the index, and the symbol. According to Peirce, the icon conserves the element, or prototype, represented. The index recalls the prototype by indirect means of empirical connections. The symbol evokes the prototype by pure conventions and a system of relations.[1] These categories are a useful starting point in thinking about the language and imagery of the chemical discipline in the nineteenth and twentieth centuries.

''Icon'' originally referred to paintings, sculpture, or other artwork depicting angels or saints in the Eastern Orthodox church. The modern semiotic meaning of ''icon'' is broader; the icon stands for its object because it looks like, or triggers an association with, the thing it represents.[2] There is a broader meaning still. Just as Rockefeller Center has become an ''American icon,'' so ''icons of science'' popularly have come to mean physical objects that are revered artifacts or instruments of scientific knowledge; among these we might include the Cavendish Laboratory or Galileo's telescope.[3] But more in keeping

1. See François Dagognet on Peirce, in *Ecriture et iconographie* (Paris: Vrin, 1973): 43. Charles Peirce, *The Collected Papers of Charles Sanders Peirce*, Vols. I-VI, eds. Charles Hartshorne and Paul Weiss (Cambridge: Harvard University Press, 1931–1935); Vols. VII-VIII, ed. Arthur Burks (Cambridge: Harvard University Press, 1958), II:249.

2. On Macintosh and other icons, see William Safire, ''I Like Icon,'' *New York Times Magazine*, February 4, 1990, 12–14.

3. The Rockefeller Tower was referred to as an American icon after its purchase by Japanese business interests. The *International Herald Tribune* reported in October 1989, ''Communist protesters, showing no respect for capitalist icons, hurled eggs, flour and tomato sauce at Walt Disney Company Chairman Michael D. Eisner.'' Ibid., 12.

with the original meaning of ''icon,'' we may identify scientific icons with scientific ''models.'' But what is the meaning of ''model''?

Dalton already was constructing wooden models of chemical molecules in the early years of the 1800s. The molecular world could be modeled in wood and brass just like the earth and the heavens.[4] That meaning of ''model'' corresponds to the Greek Orthodox meaning of ''icon.'' In contrast, as Suzanne Bachelard and others have noted, the word ''model'' appears with a new and different meaning in the late nineteenth century in the writings of Lord Kelvin, James Clerk Maxwell, and Oliver Lodge. Their method of modeling attracted vituperative attacks from the French physicist Pierre Duhem, who characterized the British electromagnetic-ether models as quintessentially English, reeking of the engineer's world and the Victorian factory.[5]

About this time, at the turn of the twentieth century, Ludwig Boltzmann wrote an article entitled ''Model'' for the eleventh edition of the *Encyclopedia Britannica* in which he discussed different kinds of models, including representations of thermodynamic relations by surfaces, demonstrations of the origin of surfaces from lines, and representations of bodies as motion of particles. He discussed the purposes of modeling in physical science, in particular Maxwell's aim, which Boltzmann characterized as mechanical analogy and dynamical illustration. Boltzmann noted that whereas physicists used to assume the probability of the actual existence of modeled mechanisms, now ''philosophers postulate no more than a partial resemblance between the phenomena visible in such mechanics and those which appear in nature.'' Boltzmann excluded maps, charts, musical notes, and figures from the category of model on the grounds that models ''always involve a concrete spatial analogy in three dimensions.''[6]

The meaning of ''model'' was to become far less concrete in the next couple of decades. In 1929, Irving Langmuir criticized mechanical models, like those of Lord Kelvin and Maxwell, on the grounds that the relationships of their parts are restricted to what is already known in mechanics, electricity, or magnetism, limiting the possibility of new insights into new phenomena. ''Mathematical relationships are far more flexible,'' he claimed, and ''the mathematical theory is a far better model of the atom than any of the mechanical

4. See O. Bertrand Ramsey, *Stereochemistry* (London: Heyden, 1981), for detailed descriptions of wooden, brass, etc., models.

5. See Suzanne Bachelard, ''Quelques aspects des notions de modèle et de justification des modèles,'' in P. Delattre and M. Thullier, eds., *Elaboration et justification des modèles: Applications en biologie*, 2 vols. (Paris: Maloine, 1979), quoted in Terry Shinn, ''Géometrie et langage: La structure des modèles en sciences sociales et en sciences physiques.'' ms., 2. And W. H. Leatherdale, *The Role of Analogy, Model and Metaphor in Science* (Amsterdam: North Holland, 1974).

6. Ludwig Boltzmann, ''Model,'' *Encyclopedia Britannica* (New York, 1911), XVII: 638–640. As discussed further, there is a problem with the distinction between ''model'' and ''analogy.'' Georges Canguilhem has claimed to demonstrate that the distinction is perfectly impossible; in Canguilhem, *Etudes d'histoire et de philosophie des sciences* (Paris: Vrin, 1979).

models which are possible.''[7] Thus, the idea had arisen that a mathematical equation is a ''model.''

More recently, the educational psychologist Jerome Bruner has classified models into three categories: (1) iconic (perception and imagery), (2) enactive (manipulation and action), and (3) descriptive or analytic/mathematical (symbolic apparatus).[8] In a book on modeling in chemistry, Colin Suckling, Keith Suckling, and Charles Suckling agree that ''models'' include mathematical systems, and they classify models in chemistry into four types: (1) quantum mechanics, (2) functional groups, (3) effects, and (4) reacting systems.[9] The philosopher Mary Hesse describes a model as any system ''whether buildable, picturable, imaginable or none of these, which has the characteristic of making a theory *predictive*,'' where theories are (1) formal theories, (2) conceptual models, or (3) material analogue models.

There are some scientists and philosophers who still claim that a model by definition ''furnishes a concrete image'' and ''does not constitute a theory.''[10] But if the model is the mathematical description, then the question of whether the model is the theory appears to become moot, since most people accept the view that rigorous mathematical deduction constitutes theory. For others, like Hesse and Kuhn, even if the model is a concrete image leading to the mathematical description, it still has explanatory or theoretical meaning, for, as Kuhn put it, ''it is to Bohr's model, not to nature, that the various terms of the Schrödinger equation refer.''[11] Indeed, as is especially clear from a consideration of mathematical models in social science, where social forces are modeled by functional relations or sets of mathematical entities, the mathematical model turns out to be so much simpler than the original that one immediately sees the gap between a ''best theory'' and the ''real world.''[12]

The real world of the chemical molecule was understood by late-eighteenth-century chemical philosophers to be sufficiently complicated that it was not likely to be reducible to the ''geometric certainty'' aimed at by natural phi-

7. Irving Langmuir, ''Modern Concepts in Physics and Their Relation to Chemistry,'' *Science* 70 (October 25, 1929): 385–396, on 390, 394.

8. Jerome Bruner, *Toward a Theory of Instruction* (Cambridge: Harvard University Press, 1967): 28, quoted in Colin J. Suckling, Keith E. Suckling, and Charles W. Suckling, *Chemistry through Models. Concepts and Applications of Modelling in Chemical Science, Technology, and Industry* (Cambridge: Cambridge University Press, 1978): 10.

9. Suckling et al., *Chemistry through Models*, 65.

10. Jean-Louis Detouches, ''Sur la notion de modèle en microphysique,'' *Synthese* 12 (1960): 176–181, on 180. On models in microphysics: a model 'is a mechanical system obeying the laws of classical mechanics, either of point mechanics or the mechanics of continuous media. A model does not constitute a theory but furnishes a concrete image.''

11. Mary B. Hesse, *Models and Analogies in Science* (South Bend, Ind.: Notre Dame University Press, 1966): 19, also 129; Thomas Kuhn, ''Metaphor in Science,'' 409–419, in Andrew Ortony, ed., *Metaphor and Thought* (Cambridge: Cambridge University Press, 1979): 415.

12. Kenneth J. Arrow, ''On Mathematical Models in the Social Sciences,'' 1951, cited and discussed in Max Black, *Models and Metaphors. Studies in Language and Philosophy* (Ithaca: Cornell University Press, 1962): 223–225.

losophers or physicists. Nor had the real world of the molecule been much enlightened by alchemical traditions of the last millennium. Eighteenth-century chemists owed some debts to alchemy and especially to the practical arts for laboratory apparatus and methods of preparation for the metals and other useful substances.

From the mid-eighteenth century, chemists adopted a communal strategy of setting up conventions that were negotiated as socially accepted constructions for the objects of their joint inquiry. To this end, they used information networks and personal contacts, chemical societies and journals, and conferences like the international congress convened at Karlsruhe in 1860.[13] The most pressing problem was nomenclature. Once Dalton's and Gay-Lussac's ideas of integral relative combining weights (chemical equivalents, elementary atoms, or proportional numbers) proved useful, agreement had to be reached on the standard elementary substance on which to base a quantitative system of proportional numbers and on formulas for standard compounds from which to calculate relative atomic weights or equivalents. Further, conventional symbols for the elements and their combination, as well as the operations that could legitimately be carried out on the symbols, had to be negotiated. The meanings of words used in chemical descriptions had to be standardized, since these words usually had well-established literal meanings in the ordinary world of human objects and behavior. Since the words we use and the symbols we manipulate are instrumental in producing new knowledge as well as in expressing what we already know, the course of chemical research in the nineteenth century and early twentieth century became as closely linked to the conventions of language and imagery as to the instruments chemists built and the network of social circumstances that influenced the problems they chose to study.[14] All these factors were mutually determinative.

We will look first at the descriptive language of chemistry and the signs adopted in that language. We will see how metaphors became definitions by convention during the course of the nineteenth century and how systems of nomenclature were expressed in ordinary language and displayed in tableaus that acquired the power of explanatory model. We then will consider symbolic language, first, the development of symbolic, nonvisual formulas for the objects of chemistry, including pseudo-algebraic representations. We then will

13. See M. J. Nye, ed. and intro. xixxx–xxxi, *The Question of the Atom: From the Karlsruhe Congress to the First Solvay Conference, 1860–1911* (San Francisco: Tomash, 1984).

14. See Ian Hacking, *Representing and Intervening* (Cambridge: Cambridge University Press, 1983), and Peter Galison, *How Experiments End* (Chicago: University of Chicago Press, 1987), on instruments; and M. J. Nye, *Science in the Provinces*, on aspects of social and economic conditions determing scientific problem choice. Humphry Davy in *Elements of Chemical Philosophy*: "Nothing tends so much to the advancement of knowledge as the advancement of a new instrument" (28); Charles Brunold: "Chemistry and physics differ more by their experimental technique than by the object of their researches," in *Le problème de l'affinité chimique et l'atomistique: Etude du rapprochement actuel de la physique et de la chimie* (Paris: Masson, 1930): 2.

turn to schematic, visual formulas, superficially suggestive of iconic modeling, that, in contrast to appearance, were concerned with chemical function, not concrete mechanisms or physical positions of atoms or groups in three-dimensional space. Combinations of these "constitutional" (later, "structural" and stereochemical) formulas modeled the multiple functions of a chemical molecule and reinforced most chemists' view that theirs was not a subject expressible by any single "rational" or "true" language or image. Thus, the detailed scrutiny of language and image in nineteenth-century chemistry illustrates the history of chemical philosophy and philosophy of chemistry delineated in chapter 3, as well as the process by which the signs and representations of chemistry diverged from natural philosophy and physics.

Metaphors and Definitions

In the *Topics*, Aristotle urged a distinction between genuine definitions and metaphors, and in the early days of the European scientific academies, their members attempted to proscribe the flowery language of literature in favor of a "mathematical plainnesse" thought appropriate for the new science.[15] Why should the metaphor be ruled out? Because it is a "rule violation" of logical grammar, and it is meaningless to assign truth or falsity to metaphors. Michel Clôitre and Terry Shinn, for example, argue that metaphors are agents of "polysemy," that is, a confusion resulting from the collision of two disjoint elements. Thus, metaphor is not a mode of reasoning that is, or can be, structured in linear fashion, and therefore it does not meet the needs of hypothetico-deductive explanation.[16]

For these reasons, Max Black claimed that metaphors are, or should be, restricted in scientific reasoning to pretheoretical stages of a discipline, or to the realm of heuristics, pedagogy, or informal exegesis. The roles of heuristic "aid to discovery" and pedagogical "aid to teaching" are familiar roles played by hypothesis.[17] But Richard Boyd and others have argued that metaphor performs another important function in introducing theoretical terminology where none previously existed, "providing epistemic access to a natural phenomenon."[18] That metaphor in this way influences the way scientists investigate and treat their objects is Hesse's view.

15. On Aristotle, Ortony, *Metaphor and Thought*, 3; on "plainnesse" in the new scientific language, Thomas Sprat, *History of the Royal Society*, 111, 113; and, more recently, Steven Shapin and Simon Schaffer, *Leviathan and the Air-Pump*, 65–69.

16. See Max Black, "More about Metaphor," 19–43, in Ortony, *Metaphor and Thought*, 40–41; and Michel Clôitre and Terry Shinn, "Enclavement et diffusion du savoir," 34-p. ms., 12–13; see published version in *Information sur les Sciences Sociales* 25, 1 (1986): 161–187.

17. Though not by Clôitre and Shinn. Black, *Models and Metaphors*, 37; quoted by Richard Boyd, "Metaphor and Theory Change: What Is 'Metaphor' a Metaphor For?" 356–408, in Ortony, *Metaphor and Thought*, 357. And see Alan P. Lightman, "Magic on the Mind: Physicists' Use of Metaphor," *American Scholar* 58 (1989): 97–101: "Metaphor is not just a pedagogical device, but an aid to discovery" (101).

18. Boyd, "Metaphor and Theory Change," 357–358.

> Men are seen to be more like wolves after the wolf metaphor is used, and wolves seem to be more human. Nature becomes more like a machine in the mechanical philosophy, and concrete machines themselves are seen as if stripped down to their essential qualities of mass in motion.[19]

The role of metaphor in defining a scientific object and suggesting a method of investigation is demonstrated in the history of the chemical discipline, both in the development of conventional *definitions* of the causes of chemical effects and in the working out of a *system*, which, by describing substances in the language of natural history, encouraged chemists to think about these objects along genealogical and morphological lines.

It is the profligacy of metaphor in chemical language, along with the liberality of pictorial imagery, that accounts for the philosophers' sometimes naive view of the naiveté of the chemists. Consider some of the enduring and legion examples of metaphor in the mainstream of modern chemistry. In the mid-nineteenth century chemistry of Laurent, atoms or particles were active in the "chase," in "copulation," "conjugation," "birth," and "marriages of convenience."[20] Hofmann's aldehydes absorbed oxygen not indifferently but with "greed."[21] For Henry Armstrong, molecules were keys fitting locks; ducks on a ramrod; and curls, like a judge's wig.[22] In the 1830s, Dumas revived Leucippus' trope that particles of matter are "like letters transposed, which thus form altogether different words."[23] For Laurent, particles form structures like Gothic cathedrals:

> Some are ornated with two towers, others have only one, some are topped with steeples, with or without lateral galleries, but the fundamental plan of the edifice is always the same, it is the Greek cross.[24]

Chemical form became biological in Laurent's metaphor of the chemical "tree": "I have searched if there does not exist in all parts of a same chemical tree, something analogous to this mother cell, in a word a nucleus common to all compounds of the same series."[25] Like so many chemists, Armstrong could not resist the military metaphor, for example, in describing the chemistry of camphor: "Whatever the agent, the attack is always delivered from the oxygen center and . . . the direction in which the attack becomes effective depends on

19. Hesse, *Models and Analogies in Science*, 163.
20. Auguste Laurent, *Méthode de chimie* (Paris: Mallet-Bachelier, 1854): 29.
21. A. W. Hofmann, "On the Combining Power of Atoms," *Chemical News* 12 (1865): 166–169, 175–179, 187–190, on 166.
22. Henry Edward Armstrong, "Presidential Address of the Chemical Section," *BAAS Rep.*, Winnipeg, 1909, 443–446.
23. Jean Baptiste Dumas, *Leçons sur la philosophie chimique*, 316.
24. Laurent, *Méthode de chimie*, 176.
25. Ibid., 397–398.

the position which the agent can take up relative to the various sections of the molecule.''[26]

But this metaphorical, descriptive language should not be misleading. These examples originate both in texts aimed at fellow chemists and in texts aimed at students or popular audiences. The metaphors are intended to be illuminating but not rigorous. The language is playful and whimsical, or simply picturesque and striking to the memory. That some of it could be changed without harm to fundamental ideas has been recognized. Commenting on metaphorical language in a recent American organic chemistry manual (''The attack of the electrophile proceeds from the less hindered side of the molecular plane of the double bond''), a Canadian chemist proposed substituting the vocabulary of electrophilic addition and nucleophilic substitution for the vocabulary of ''attack.'' ''Why, in conclusion, don't we abandon an 'agressive' chemistry in favor of a non-violent chemistry?''[27]

However, in contrast to such whimsical and playful description, some metaphors developed into serious definitions that became essential to rigorous chemical classification and explanation. Indeed, the electrophilic/nucleophilic language is an example from the development of chemical theory which is the subject of later chapters of this book. Let us consider three other examples of metaphor-turned-convention that dominated eighteenth- and nineteenth-century chemistry.

The term ''affinity'' has its roots in very old ideas to the effect that like attracts like and that bodies combine with other bodies because of mutual affection or *affinitas.* This meaning is employed in Etienne François Geoffroy's *Table des différents rapports observés entre différentes substances* (1718) for replacement reactions.[28] However, in the middle of the eighteenth century, Boerhaave spoke of the affinity of a substance for others *unlike* it, giving the word ''affinity'' a new meaning. Boerhaave interpreted Geoffroy's table as a representation of Newtonian-type forces of gravitational attraction or electrical attraction and repulsion.[29]

But if affinity was *like* gravity, electricity, or, to a lesser extent, magnetism, it could not be identical to them. An important difference was that affinity was selective or ''elective.'' Elementary chemical substances chose friends and foes on the basis of kind, not just the quantity of a thing and its distance. Defining affinity in the nineteenth century, Wurtz called it the ''force which

26. Armstrong, ''Presidential Address,'' 440.

27. Claude Benezra, ''La chimie des flèches ou la notion d'agresseur et d'agressé en chimie organique,'' *L'Actualité Chimique* (May 1975): 21.

28. A still valuable treatment of the topic of ''chemical affinity'' is by M. M. Pattison Muir, *A History of Chemical Theories and Laws* (New York: Wiley, 1907), Chap. XIV, ''Chemical Affinity,'' 379–430.

29. See Arnold Thackray, *Atoms and Powers*, 90, 213.

presides over chemical combination,''[30] a metaphor of governance similar to ''Gravity ruling'' or ''Nature selecting.''

> We have thought of affinity or chemical force as an attractive force, like a form of universal gravitational attraction exerted between celestial bodies. It is a hypothesis in both cases, or if one wants, a useful representation. It is figurative language. It is a way of expressing a fact rather than giving an explanation.[31]

In his article on affinity for Wurtz's *Dictionnaire de chimie*, Georges Salet in 1869 noted that chemists were less and less preoccupied with affinity and that Gay-Lussac, Dumas, and Gerhardt had not treated it as a topic in their work from the 1820s to the 1860s. In Salet's view, the theory of electricity, once established, would be a key to understanding what had been called affinity.[32] Two decades later, Meyer referred to the ''fiction'' of a force of affinity. ''I personally prefer the application of concepts of potential and kinetic energy . . . to seeing affinity as an attractive force. . . . We are on the verge of a kinetic theory of affinity if one can hope to consider the very essence of electricity as a particular form of movement. Chemical statics will receive a very different form from what Berthollet envisioned.''[33]

In a second example of conventionalized metaphor, consider the word ''atomicity.'' Wurtz in 1867 defined atomicity as the ''force or power of combination which resides in atoms, and which is exercised in different manner according to the nature of the atoms.''[34] However, the existence of indivisible atomic particles corresponding to the chemical elements was itself in dispute among chemists. The term ''atom'' combined the property of concrete object with the opposite property of infinitesimal point, a perfect instance of polysemy in metaphor. Many chemists used the word ''atomicity,'' indeed ''atom,'' purely as a convention, without commitment to its etymological meaning. Thus, Hofmann wrote in his *Introduction to Modern Chemistry* (1865) that ''atomicity'' was a barbarous expression because it suggested atomic structure. He preferred the observational word ''quantivalence,'' a word shortened to valence (*Valenz*) by Kekulé and by Hermann Wichelhaus.[35] The valence, or value, of an element varies according to the number of hydrogen atoms with which it can combine.

Another word expressing this idea of combining power was ''saturation'' capacity. Using an electrochemical theory of chemical reaction, Berzelius hid the assumption of electrical neutralization by the material analogy or metaphor

30. Adolphe Wurtz, *Leçons élémentaires de chimie moderne* (Paris: Masson, 1867): 6.
31. Wurtz, *Introduction à l'étude de la chimie*, 8–9.
32. Georges Salet, ''Affinité,'' 69–83, in Adolphe Wurtz, ed., *Dictionnaire de Chimie Pure et Appliquée*, 3 vols. (Paris: Hachette, 1869–1874), I (1869): 78.
33. Meyer, *Les théories modernes de la chimie*, 12, 17, 303.
34. Wurtz, *Leçons élémentaires*, 219.
35. See Colin A. Russell, *A History of Valency* (New York: Humanities Press, 1971): 85–86.

of saturation.[36] "Valence" was meant to be an ontologically neutral word, originating from the term "equivalence" without implying atoms. It expressed exact combining proportions of weights of substances with one another, that is, 1 to 1, 1 to 2, 1 to 3, and so on. Thus, using hydrogen or oxygen as a standard substance and assigning the number 1, 10, or 100 as the standard dimensionless unit for hydrogen or oxygen, tables of equivalents could be constructed, that is, 1 gram of hydrogen is equivalent to 16 grams of oxygen or to 12 grams of carbon in chemical combining power.

However, as already noted, chemists quickly recognized that both the atoms system and the equivalent system, which were intended to be "philosophic" or "positive" (and not metaphysical or metaphorical), were based in convention.[37] In 1813, Thomson submitted his table of atomic weights "to the chemical world as more convenient" than any other way of talking about combining units, and he also noted that the choice of oxygen, rather than hydrogen, for the base value 1 (or 100) was rooted in the fact of oxygen's easy facility for chemical combinations, making it among elements "the most convenient."[38] Typically, when Dumas explained the system to his students in 1837, he used the words "convention" and "convenir," as did Liebig in his introductory lectures in chemistry.[39] The whole system of equivalents, intended to be empirical and nonhypothetical, is "purely conventional, strongly arbitrary, and cannot have any pretense to a scientific value," wrote Charles Marignac in the 1870s.[40] Conventions, then, were at the base of nineteenth-century chemistry and widely recognized to be so. This fact contributed to the reticence of chemists to make confident claims about the correspondence of their systems with reality.

Nomenclature and Taxonomy

The language of chemistry before Lavoisier was full of mystery and ambiguity. Flowers of zinc, diaphoretic antimony, butter of antimony, powder of algaroth, vitriolated tartar, and ethiops per se were among the names no longer familiar to modern chemists. Although some chemists have thought of Lavoisier's rules as the "true principles of chemical nomenclature,"[41] most chem-

36. Ibid., 23.

37. Recall the letter from Berzelius to Liebig noting that the system of equivalents was "no longer positive but merely conventional."

38. Thomas Thomson, "On the Daltonian Theory of Definite Proportions in Chemical Combinations," *Annals of Philosophy* 2 (1813): 32–43, on 41–42.

39. Dumas, *Leçons sur la philosophie chimique*, 225–228; Justus von Liebig, *Introduction à l'étude de la chimie*, trans. Charles Gerhardt (Paris: L. Mathias, 1837): 85.

40. Quoted in Edouard Grimaux, *Introduction à l'étude de la chimie*, 40. However, it is important to note that correctness of systems using conventions *can* be demonstrated, e.g., water is H_2O, not HO.

41. Muir, *A History of Chemical Theories and Laws*, 190.

ists, particularly those who first opposed the use of these principles, recognized the rules and the names as conventional and, above all, *French.*

The practice of binomial nomenclature in chemistry was already carried out in some seventeenth- and eighteenth-century works, including Oswald Croll's 1609 *Basilica chymica.* Indeed, in the mid-eighteenth century, Bergmann embarked on a project of reform before Louis Bernard Guyton de Morveau, Fourcroy, Berthollet, and Lavoisier. Bergmann argued that a standard nomenclature should be in Latin, like the natural history nomenclature of his Swedish colleague Linnaeus.[42]

Similarly, well before Lavoisier, in London in the mid-eighteenth century, the Royal College of Physicians convened the Pharmacopoeia Committee, which adopted the principle that names of compounds should be related to their constituents rather than to their observable properties or to analogies with things like butter and flowers. The physicians' 1746 dictionary of names influenced Macquer's 1766 dictionary of chemistry.[43]

The new French nomenclature appeared in 1787 under the authorship of Fourcroy, Berthollet, Lavoisier, and Guyton de Morveau. Fourcroy taught the new naming system at the Jardin du Roi, the Lycée des Arts, and later the Ecole Polytechnique. Lavoisier incorporated it with much fanfare into his *Traité élémentaire de chimie* (1789). It met such hostility from Jean Claude de la Métherie, the editor of *Observations sur la physique*, that Lavoisier and his colleagues had to establish their own journal, the *Annales de chimie*, in 1789. Davy claimed that the nomenclature was not a "philosophical language": "metals, earths, alkalies are appropriate names for the bodies they represent, and independent of all speculative views; whereas oxides, sulphurets, and muriates are terms founded upon opinions."[44] Priestley wrote a scathing attack on the new system, but, like Davy, he had to learn it to keep up with the new chemical literature.[45]

Chemists working on the same problems could only know they were studying the same substances if they spoke the same language. The chemical language, like chemical instruments, defined the discipline.[46] The instruments and the nomenclature were illustrated in elaborate diagrams and "tableaus." Lavoisier's "Tableau des substances simples" is one of the most famous. (See fig. 1.) Here he organizes thirty-three simple substances into four categories:

42. Maurice Crosland, *Historical Studies in the Language of Chemistry* (Cambridge: Harvard University Press, 1962): 134, 148, 151.

43. Ibid., 95–96, 120–122.

44. Davy, *Elements of Chemical Philosophy*, 25.

45. Crosland, *Historical Studies*, 189, 198.

46. A recent writer on medical research noted that bacteriologists have had a different language from geneticists, e.g., the vocabulary of "bacterial dissociation" rather than "mutation." As a consequence, geneticists were not prepared to accept information from studies of the pneumoccus as having any bearing on the genetics of higher organisms. See Maclyn McCarty, *The Transforming Principle* (New York: Norton, 1985): 214–215.

DES SUBSTANCES SIMPLES.

TABLEAU DES SUBSTANCES SIMPLES.

	Noms nouveaux.	*Noms anciens correſpondans.*
Subſtances ſimples qui appartiennent aux trois règnes & qu'on peut regarder comme les élémens des corps.	Lumière	Lumière.
	Calorique	Chaleur.
		Principe de la chaleur.
		Fluide igné.
		Feu.
		Matière du feu & de la chaleur.
	Oxygène	Air déphlogiſtiqué.
		Air empiréal.
		Air vital.
		Baſe de l'air vital.
	Azote	Gaz phlogiſtiqué.
		Mofete.
		Baſe de la mofete.
	Hydrogène	Gaz inflammable.
		Baſe du gaz inflammable.
Subſtances ſimples non métalliques oxidables & acidifiables.	Soufre	Soufre.
	Phoſphore	Phoſphore.
	Carbone	Charbon pur.
	Radical muriatique	Inconnu.
	Radical fluorique	Inconnu.
	Radical boracique	Inconnu.
Subſtances ſimples métalliques oxidables & acidifiables.	Antimoine	Antimoine.
	Argent	Argent.
	Arſenic	Arſenic.
	Biſmuth	Biſmuth.
	Cobolt	Cobolt.
	Cuivre	Cuivre.
	Etain	Etain.
	Fer	Fer.
	Manganèſe	Manganèſe.
	Mercure	Mercure.
	Molybdene	Molybdène.
	Nickel	Nickel.
	Or	Or.
	Platine	Platine.
	Plomb	Plomb.
	Tungſtène	Tungſtène.
	Zinc	Zinc.
Subſtances ſimples ſalifiables terreuſes.	Chaux	Terre calcaire, chaux.
	Magnéſie	Magnéſie, baſe du ſel d'Epſom.
	Baryte	Barote, terre peſante.
	Alumine	Argile, terre de l'alun, baſe de l'alun.
	Silice	Terre ſiliceuſe, terre vitrifiable.

Figure 1. Lavoisier's "Tableau des Substances Simples" from his Traité élémentaire de chimie *(Paris: Cuchet, 1789), p. 192. Courtesy of History of Science Collections, University of Oklahoma.*

the five known elementary bodies (light, caloric, oxygen, nitrogen, and hydrogen); nonmetallic simple substances that are oxidizable and acidifiable; metallic simple substances that are oxidizable and acidifiable; and the earthy simple substances. Other of his tables include a list of oxidizable and acidifiable ''radicals or bases'' that are compound substances acting like simple substances, for example, the acetic and tartaric ''radicals.'' And the *Traité* includes a large foldout table showing the old and new names for compounds of simple substances and oxygen, where oxidation occurs to different degrees. This table illustrates the still-practiced ''-ide,'' ''-ous,'' ''-ic'' rules of nomenclature; thus, the compounds of oxygen and phosphorus are ''oxide de phosphore,'' ''acide phosphoreux,'' ''acide phosphorique,'' and ''acide phosphorique oxigéné.''[47]

Adoption of the French system committed one to the theory of oxidation as a new starting point of chemical investigation. Davy and others soon demonstrated some of the weaknesses of the system, for example, that the so-called muriatic acid radical does not in fact contain oxygen but is instead the simple elementary substance Davy now called chlorine. The nomenclature also seemed to commit its users to a theoretical position about the constitution of compound substances that many did not accept. Graham, for example, registered the complaint that names like sulfate of soda imply the continued existence of sulfuric acid and soda in the salt but that we should not take the names to mean ''actual formation of subordinate compounds, or anything more than what are considered to be the predominating set of attractions among all the possible attractions which the elements have for each other.''[48] That is, Graham wanted to be clear that chemical nomenclature indicates the *origin* or *genealogy* of the salt, not its *real constitution.*

By the mid-nineteenth century, approximately sixty-five elements were known. When Graham planned a chemistry course in the mid-nineteenth century, he divided the elements into ''groups or natural families,'' based on their properties; he divided the metals into nine ''orders'' distributed among three ''classes'' (alkalis and alkali-earths; metals of earths; and metals proper, divided according to affinity for oxygen).[49] The classes of elements are not abruptly separated, he stated, but shade ''into each other in their characters, like the classes created by the naturalists for the objects of the organic world.''[50]

Such systems of classification were becoming common; alphabetical organization typical of dictionaries was unilluminating, and really simple schemes became more difficult as the number of elements increased. Whereas this classification mimicks the classification schemes of the naturalists, Graham

47. See Antoine Lavoisier, *Traité élémentaire de chimie*, 192, 196, 203.
48. Graham, *Elements of Chemistry*, I:208.
49. Graham, I:168, 517; and II:1.
50. Ibid., I:168.

CH_3-
methyl

C_2H_5-
ethyl

C_6H_5-
phenyl

C_6H_5CO-
benzoyl

-OH
hydroxyl

$-NH_2$
amino

$-NO_2$
nitro

$-OCH_3$
methoxy

Figure 2. Radicals. Drawings courtesy of Glenn Dryhurst.

himself was unsympathetic to chemists aligning themselves with biology against physics.

A second ''natural history'' scheme of order was suggested in the rationale of the substitution and type theories that took their origins in Liebig's and Wöhler's demonstration that some atoms form groupings that remain units in many different reactions, for example, the benzoyl ''radical'' or root, composed of carbon, hydrogen, and oxygen (now, C_7H_5O).[51] (See fig. 2.) Hofmann was an important contributor to these theories, proving, as Liebig put it, ''that the chemical character of a compound does not depend, as the electrochemical theory supposes, upon the nature of the elements it contains, but solely on the manner of their grouping.''[52]

For Laurent, who used metaphors of the chemical tree and the biological nucleus, the type was an ordering principle of extraordinary power. Berzelius was suspicious of this orientation because it was at odds with his theory of electrical polarities between elementary substances.[53]

The water, ammonia, hydrogen chloride, and methane ''types'' (fig. 3.)

51. See O. Theodor Benfey, *From Vital Force to Structural Formulas* (Washington, D.C.: American Chemical Society, 1975; 1st ed., 1964): 34. As noted above, Lavoisier's 1789 tableaus include a table of ''radicals'' in which he surmised elementary substances stayed together as a group.

52. Liebig's footnote, p. 1, to A. W. Hofmann, ''Erzeugung organischer Basen, welche Chlor und Brom enthalten,''*Liebig's Annalen* 53 (1845): 1–57; quoted in Keas, ''The Structure and Philosophy of Group Research,'' 55.

53. J. J. Berzelius (1846), quoted in Knight, *Transcendental*, 669.

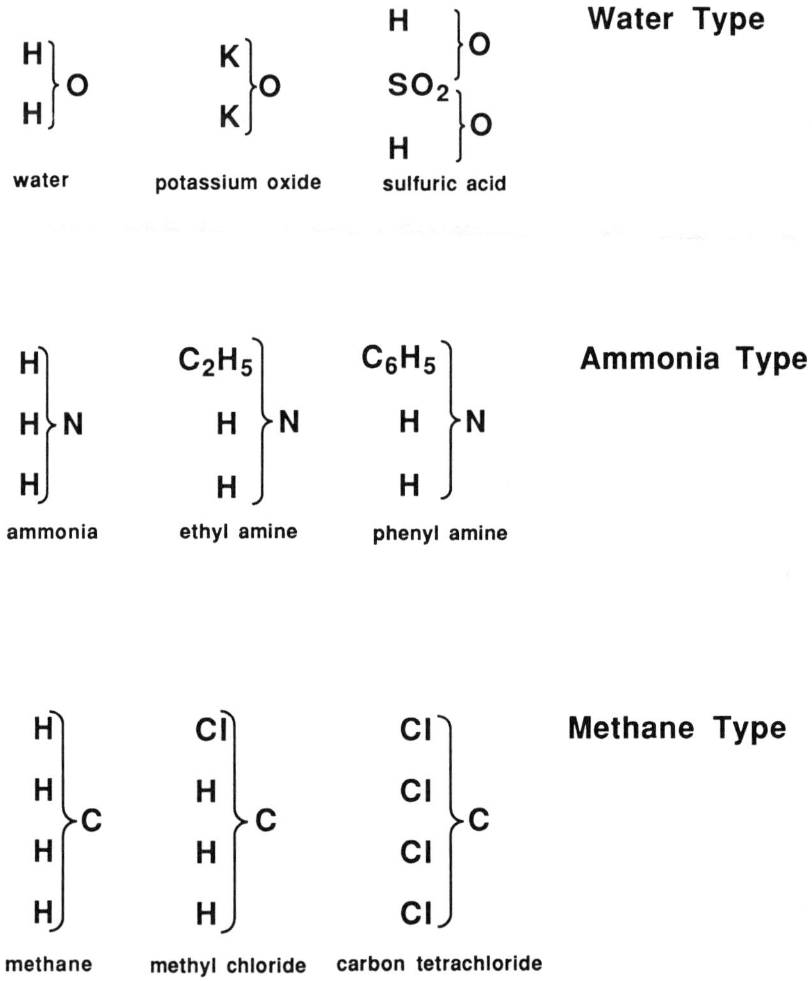

Water Type

$\left.\begin{matrix} H \\ H \end{matrix}\right\} O$ — water

$\left.\begin{matrix} K \\ K \end{matrix}\right\} O$ — potassium oxide

$\left.\begin{matrix} H \\ SO_2 \end{matrix}\right\} O$, $\left.\begin{matrix} SO_2 \\ H \end{matrix}\right\} O$ — sulfuric acid

Ammonia Type

$\left.\begin{matrix} H \\ H \\ H \end{matrix}\right\} N$ — ammonia

$\left.\begin{matrix} C_2H_5 \\ H \\ H \end{matrix}\right\} N$ — ethyl amine

$\left.\begin{matrix} C_6H_5 \\ H \\ H \end{matrix}\right\} N$ — phenyl amine

Methane Type

$\left.\begin{matrix} H \\ H \\ H \\ H \end{matrix}\right\} C$ — methane

$\left.\begin{matrix} Cl \\ H \\ H \\ H \end{matrix}\right\} C$ — methyl chloride

$\left.\begin{matrix} Cl \\ Cl \\ Cl \\ Cl \end{matrix}\right\} C$ — carbon tetrachloride

Figure 3. Types. Drawings courtesy of Glenn Dryhurst.

carried mental associations with biological forms, ordered morphologically rather than physically. The "homologous" hydrocarbons had a basic vertebral column of carbon atoms, or carbon equivalents, to which new "radical" or "residue" groups might add themselves, increasing the sites and kinds of active functions of the column or carbon "chain." As the column lengthens in the simplest possible way, for example, from CH_3Cl to C_2H_5Cl to C_3H_7Cl, the properties of the substance change in a systematic way, so that a gas, condensing at 23 degrees, becomes a liquid, boiling at 26.5 degrees, and then a liquid, boiling at 46.5 degrees. Relating changes in morphological structure to physical properties and chemical behavior or *habits* of the compound is a fundamental aim in chemistry.

Dumas played the role in French chemistry that Cuvier played in French zoology, although not to uniform praise and certainly not to praise from the younger Laurent.

> Having treated this subject more in the manner of a poet than a scientist [*savant*], the result is that it is difficult to know exactly his [Dumas's] thought. . . . Whether this statue should be in bronze, in marble, or in ivory? the material is of little importance, says M. Dumas; it is always the same statue, the same type. In effect, it little matters that this sulfate is a salt of copper, or of lead or of iron; it always belongs to the sulfate type. But where do we stop? ammonium sulfate, sulfuric ether, alum, do they belong to the sulfate type?[54]

What was necessary, continued Laurent, was to take into account the many properties of bodies, like chemical function, atomic constitution, gas volume, density, boiling point, etc. and weigh these characteristics against each other, "as the naturalist organizes hierarchically the number, situation and function or organs to each other" (393–394).

The explanatory system which dramatically combined the classification methods of natural history with the quantitative methods of physical laws was the periodic system worked out by Dmitri Mendeleev (and independently, although less successfully, by Meyer). What is the great tableau that is the periodic table? Is it icon, index, or symbol? It is not metaphor. Is it a model? There is no chemical laboratory in the world where Mendeleev's table does not hang on the wall, despite the fact that the original version is well over one hundred years old. Its center remains untouched. Give a chemist a choice between the periodic table (fig. 4) and Schrödinger's equation. Which would she take?

For an eloquent description of the periodic table, François Dagognet's cannot be matched. The table relies on two principles. The atom, or element, is put at the intersection of a square formed by its four immediate neighbors. The element's atomic mass is the mean of those surrounding it. All the attributes of the element only confirm or explain its relative atomic weight. In the table, the horizontal period or series includes substances with very different chemical properties but similar atomic weight and physical properties. Vertical groups include families of bodies with very different atomic weights but the same chemical properties. A cosmic dichotomy is expressed in the table. Electronegative elements are on the right, and electropositive elements are on the left. The earth and sky, metals and gases are in two centers, the lower left and upper right. Here the true opposition is not median but diagonal. A "ladder" of metalloids, sorts of hybrids, includes substances that have the form or aspect

54. Laurent, *Méthode de chimie*, 362–363.

The periodic table of the elements

Groups

Periods	IA	IIA	IIIB	IVB	VB	VIB	VIIB	VIIIB	VIIIB	VIIIB	IB	IIB	IIIA	IVA	VA	VIA	VIIA	VIIIA
1	1 H																	2 He
2	3 Li	4 Be											5 B	6 C	7 N	8 O	9 F	10 Ne
3	11 Na	12 Mg											13 Al	14 Si	15 P	16 S	17 Cl	18 Ar
4	19 K	20 Ca	21 Sc	22 Ti	23 V	24 Cr	25 Mn	26 Fe	27 Co	28 Ni	29 Cu	30 Zn	31 Ga	32 Ge	33 As	34 Se	35 Br	36 Kr
5	37 Rb	38 Sr	39 Y	40 Zr	41 Nb	42 Mo	43 Tc	44 Ru	45 Rh	46 Pd	47 Ag	48 Cd	49 In	50 Sn	51 Sb	52 Te	53 I	54 Xe
6	55 Cs	56 Ba		72 Hf	73 Ta	74 W	75 Re	76 Os	77 Ir	78 Pt	79 Au	80 Hg	81 Tl	82 Pb	83 Bi	84 Po	85 At	86 Rn
7	87 Fr	88 Ra																

57 La	58 Ce	59 Pr	60 Nd	61 Pm	62 Sm	63 Eu	64 Gd	65 Tb	66 Dy	67 Ho	68 Er	69 Tm	70 Yb	71 Lu
89 Ac	90 Th	91 Pa	92 U	93 Np	94 Pu	95 Am	96 Cm	97 Bk	98 Cf	99 Es	100 Fm	101 Md	102 No	103 Lr

Figure 4. The Periodic Table. From Harry N. Holmes, General Chemistry *(New York: Macmillan, 1921).*

of metals but are not metals. Between the real metals and the authentic non-metals, then, is a *région marécageuse* inhabited by sorts of "flying fish." The table simultaneously saves perfect continuity (the slow transition and passage across a band) and a discrete discontinuity (the two antagonistic poles).[55]

The power of the tableau was illustrated in 1875, only six years after Mendeleev laid out its eight fundamental principles to the Russian Physical Chemical Society: the first that "the elements, if arranged according to their atomic weights, exhibit an evident *periodicity* of properties," the sixth that "we must expect the discovery of many yet *unknown elements*, for example, elements analogous to aluminium and silicon, whose atomic weight would be between 65 and 75."[56] One of the vacant spaces in the table was filled in by Paul E. Lecoq de Boisbaudran in 1875 on the discovery of what Mendeleev had called *eka aluminium* and the Frenchman now named gallium. Predictions and verifications were striking: atomic weight (predicted 68/observed 69.9), density (5.9/5.93); melting point (low/30.1° C.); oxide formula (Ea_2O_3/Ga_2O_3).[57] Discoveries of *eka boron* and *eka silicon* soon followed.

The table uses a system of symbols for the elements and a system of conventions for atomic weights; it employs a classification, or a visual array, that groups the symbols so that their relations and properties are immediately suggested to the viewer who knows the principles of classification and a few facts. Deductions can be made both to the facts that established the table and to the facts that were unknown when the table was first set out. Here is a scheme that is an explanatory and predictive model and an icon in both the semiotic and the popular senses of the word. But its power comes from visual display, from image, not the principles and facts that can be recorded in ordinary or conventional language.

Symbols and Algebraic Formulas

Let us take up in some detail the symbolic language of chemistry. Alchemical symbols, for example, those associating the metals with the planets, persisted as chemical signs into the eighteenth century. The notion of representing all the elements of the world by their true signs and generating all possible combinations of signs for all possible compounds in the universe is a very old idea. Ian Hacking cites the alchemist Raymond Lull (1234–1315) as the founder of this theory of combinations; the theory of combinations had precedents in the Spanish cabalist practice of meditating on combinations of the sacred

55. François Dagognet, *Tableaux et langages de la chimie*, 100–158.

56. D. I. Mendeleev, "The Periodic Law of the Chemical Elements," 160–182, in C. S. Gibson and A. J. Greenaway, eds., *Faraday Lectures 1869–1928* (London: Chemical Society, 1928). Appeared originally in *JCS* 55 (1889): 634–656, on 635–636.

57. See Ihde, *The Development of Modern Chemistry*, 248.

Hebrew alphabet.[58] Modern symbolic combinations in chemistry are extraordinarily powerful. As Charles Friedel told his students in 1887, the fact that a combination of atoms can arrange themselves in different ways (isomerism) within a single molecule means that it is possible, for example, that there are 799 isomers, or different arrangements, corresponding to the formula $C_{13}H_{28}$. It is up to the chemist to investigate which of these isomers exist and what their properties really are.[59]

Lavoisier still used some of the old alchemical or newer pictograph symbols for the elementary substances. He and his colleagues Pierre Auguste Adet and Jean Henri Hassenfratz also used pseudo-algebraic symbols on occasion, but Lavoisier explicitly warned that these symbols were not in fact algebraic but shortcuts to "ease the workings of the mind." One day, he hoped, it would be possible to "know the energy of all these forces, to succeed in giving them a numerical value, to calculate them—this is the aim which chemistry must have."[60]

Dalton's table of atomic weights included geometrical symbols like inscribed circles for elements and compounds, and Berzelius, like Dalton, imagined elemental atoms as spheres. But Berzelius broke with pictorial or geometrical symbols in 1814 as a formal means of communication. Instead, he used letters of the alphabet, index numbers, dots, bars, and the sign of addition to represent so-called double-decomposition reactions (electrochemical dualism).[61] These alphabetical and indexed linear formulas included grouped letters within the linear formulas, and Berzelius introduced a distinction between "empirical" and "rational" formulas, the latter indicating how atoms were thought to be grouped within a molecule or compound, held together by opposite electrical polarities.[62] Thus, using modern notation, the empirical formula is NH_4Cl or KNO_3; the rational formula is

$$NH_3 \cdot HCl \text{ or } KO \cdot NO_2.$$

It is not surprising that there were objections to Berzelius's new symbolic system, Dalton among them, on grounds of philosophical preference for pic-

58. See Ian Hacking, *The Emergence of Probability* (Cambridge: Cambridge University Press, 1975): 18; and Jamie C. Kassler, "The Emergence of Probability Reconsidered," *AIHS* 36 (1986): 17–44, on 18, n. 4.

59. Charles Friedel, *Cours de chimie organique*, 6.

60. Quoted from Lavoisier, *Oeuvres de Lavoisier*, (Paris, 1862), II: 525; in Maurice Crosland, "The Development of Chemistry in the Eighteenth Century," *Studies on Voltaire and the Eighteenth Century* 24 (1963): 369–441, on 407. Included in the 1787 *Nomenclature* are six folding plates demonstrating a symbolic scheme for chemical representation devised by Hassenfratz and Adet.

61. On Dalton's symbols, see W. W. Haldane Gee, Hubert Frank Coward, and Arthur Harden, "John Dalton's Lectures and Lecture Illustrations," Memoir No. XII in *Mem.Manchester LPS* 59 (1914–15), esp. 41–66. J. J. Berzelius,"Essay on the Cause of Chemical Proportions, and on some Circumstances relating to Them: Together with a Short and Easy Method of Expressing Them," *Annals of Philosophy* 2 (1813): 443–454, and 3 (1814): 43–52. Also see, Crosland, 268–275.

62. Crosland, *Historical Studies*, 323.

tograph symbols and xenophobic opposition to foreign chemistry.[63] There also were objections to the notion of the "rational" formula, particularly since more than one rational formula was possible for most compounds. We will return to this objection later.

Whewell criticized Berzelius's original superscript symbols on the grounds that they contravened the most elementary rules of algebra. NH^4Cl means $(x \cdot y^4 \cdot Cl)$, not, as Berzelius intends, $(N + 4H + Cl)$.[64] Whewell complained of the new notation,

> Its formulae are merely unconnected records of inferences which are in some degree arbitrary; the analysis itself, the fundamental and certain fact from which inferences are made, is not recorded in the symbol; and the connexion between different formulae, the identity of which is a necessary and important circumstance, can be recognized only by an entire perversion of all algebraical rules.[65]

In 1833, Edward Turner became the first author of a British chemistry textbook to use Berzelius's symbols, explaining in the preface that he was introducing them halfway through the text after confronting the difficulty of using ordinary language to give an account of Liebig's and Wöhler's work on cyanogen. The symbols, he said, "are not only fitted to be a convenient abbreviation among educated chemists, but may be made a powerful instrument of instruction by teachers of chemistry."[66]

Graham similarly adopted the new notation "in such cases where it may be expedient," for example, in expressing his results for the acids and salts of phosphorus.[67] The British Association for the Advancement of Science constituted a committee that reported in 1835 that a majority of the sixteen members approved the new "continental" notation, with the proviso that "it is desirable not to deviate . . . from algebraic usage except so far as convenience requires."[68]

Benjamin Brodie, a professor of chemistry at Oxford University, renewed the project for algebraic reform of chemical formulas in a paper presented to the Chemical Society of London in 1867. Chemists recognized the aim of the project to be a "strictly philosophical system of chemical notation by means of actual formulae instead of mere symbols."[69] But again, there was little

63. On this see, Timothy L. Alborn, "Negotiating Notation: Chemical Symbols and British Society, 1831–1835," *Annals of Science* 46 (1989): 437–460, esp. 441.

64. See William Whewell, "On the Employment of Notation in Chemistry," *J. Royal Inst.* 1 (1831): 437–438; Alborn, "Negotiating Notation"; W. H. Brock, "The British Association Committee on Chemical Symbols 1834: Edward Turner's Letter to British Chemists and a Reply by William Prout," *Ambix* 33 (1986): 33–37.

65. Whewell, "On the Employment of Notation."

66. Edward Turner, *Elements of Chemistry*, 4th ed. (London: Printed for J. Taylor, Bookseller, 1833): vii–viii.

67. T. Graham, in *Philosophical Magazine* ser. 3, 4 (1834): 404; quoted in Crosland, *Historical Studies*, 281.

68. *BAAS Rep.* (1835), 207; quoted in Alborn, "Negotiating Notation," 458.

69. William Odling, "Presidential Address," *BAAS Rep.* (1864): 21–24, on 23.

interest.[70] Chemists did not find Brodie's algebraic representation with Greek symbols useful, in comparison to their own nonalgebraic, *chemical* formulas.

Berthollet, Laplace, and J. B. Biot were among those who attempted in the early nineteenth century to devise a chemical mechanics that could be expressed algebraically or analytically. These early attempts failed to account for chemical facts. Many scientists and philosophers subsequently held (as we saw in chap. 3) that chemistry could not be considered a ''philosophical'' or true science, insofar as it proved unamenable to geometric or analytic mathematics. Yet chemistry was a *quantitative* and *exact* science, and it was a precise science in which laboratory investigators drove themselves hard to calculate vapor densities, combining weights, and relative atomic weights to the farthest possible decimal place, in part to test the hypothesis that every elementary substance is made up of hydrogen or submultiples of hydrogen. The triumph of Mendeleev's table toward the end of the century was rooted in all these numbers, not just in qualitatively described properties of bodies. Mendeleev's principle was the quantitative one of periodicity, and periodic functions were a mainstay of mathematical physics.

Symbols and Structural Formulas

As the theory of organic ''radicals'' became increasingly important in the 1830s, it was easy to represent these ''radicals'' by letters and subscripts, or superscripts, separated spatially from other letters representing elements in a compound. However, from the start, chemists debated whether the ''radicals'' were groups ''convenient for description'' or real organic equivalents of inorganic elements. Dumas claimed radicals really exist, that NH_3, for example, is ''preformed'' in ammonium chloride. Laurent disagreed, criticized the inclusion of over one hundred hypothetical radicals in one of Liebig's treatises, and argued that preformed groups could not be inferred from reactions because atoms may be arranged differently in a nonreacting compound than in a reacting one.[71] Kekulé shared Laurent's view, as did Gerhardt, Laurent's colleague and ally, who rechristened the radicals ''residues'' and insisted that the units were arbitrarily chosen for the convenience of the chemist.[72] Whereas Berzelius had seemed to have a Platonic notion of one ''true'' rational formula, Gerhardt's motto was ''autant de réactions autant de formules rationelles.''[73]

The formulas began to take on a nonlinear presentation both in the electro-

70. Benjamin Brodie, ''On the Mode of Representation Afforded by the Chemical Calculus, as contrasted with the Atomic Theory,'' *Chemical News* 15 (1867): 295–305, esp. 298–300. See W. H. Brock, ed., *The Atomic Debates: Brodie and the Rejection of the Atomic Theory* (Leicester University Press, 1967).

71. For an analysis of Laurent's views, see John Hedley Brooke, ''Laurent, Gerhardt, and the Philosophy of Chemistry,'' *HSPS* 6 (1975): 405–429, 411–412, 428.

72. Benfey, *From Vital Force*, 58.

73. See Crosland, *Historical Studies*, 331.

chemical and the radical theories. In 1850, Graham suggested writing a formula in two lines, placing the negatively charged, or polar, constituents on the upper line and the positive on the lower line. Such formulae, he suggested, would "exhibit" the polar attractive forces of the elements.[74] Thus:

$$\text{potash} \quad \frac{\mathrm{O}}{\mathrm{K}} \qquad \text{ammonia} \quad \frac{\mathrm{N}}{\mathrm{H_3}}$$

$$\frac{\mathrm{N}}{\mathrm{H_3}} \ \text{and} \ \frac{\mathrm{Cl}}{\mathrm{H}} \ \text{give} \ \frac{\mathrm{NCl}}{\mathrm{H_2H_2}} \ .$$

The new "type" theory similarly used nonlinear formulas, reminding the reader of the crystallographic and biological forms that were their inspiration. Gerhardt used elongated brackets in 1856, followed by Edward Frankland and William Odling, although Williamson would not employ them. Gerhardt insisted that the arrangement of brackets, letters, and other marks on paper did not imply actual two-dimensional, physical arrangement of parts.[75] Typical symbolism included the water "type" and its varieties:

$$\left.\begin{matrix}\mathrm{H}\\\mathrm{H}\end{matrix}\right\}0 \qquad \left.\begin{matrix}\mathrm{K}\\\mathrm{K}\end{matrix}\right\}0 \qquad \begin{matrix}\left.\begin{matrix}\mathrm{H}\\\mathrm{SO_2}\end{matrix}\right\}0\\\left.\mathrm{H}\ \right\}0\end{matrix}$$

And, from Wurtz[76] (note the grappling hook):

$$\left.\begin{matrix}\boxed{\mathrm{Cl}}\\\mathrm{Cl}\end{matrix}\right\}0 + \left.\begin{matrix}\boxed{\mathrm{H}}\\\mathrm{H}\end{matrix}\right\}0 = \left.\begin{matrix}\mathrm{H}\\\mathrm{Cl}\end{matrix}\right\}0 + \left.\begin{matrix}\mathrm{Cl}\\\mathrm{H}\end{matrix}\right\}0 \ .$$

With the exception of some unique symbols of William Higgins in 1789, generally, straight lines appeared in published chemical formulas only when Archibald Couper introduced them in 1858 to indicate valences (units of atomicity, saturation capacity, or quantivalence).[77] Whereas innocent accent marks or superscript dashes had been used at midcentury to indicate valence or value, straight lines now suggested a less abstract meaning, despite disclaimers like Alexander Crum Brown's that the lines indicated the "chemical," not "physical," positions of atoms.[78]

74. Thomas Graham, *Elements of Chemistry*, I:205–207.

75. See C. Russell, *History of Valency*, 94–95. Liebig used brackets in representations of affinity relations, in *Introduction*, 59–60.

76. Wurtz, *Leçons élémentaires*, 77.

77. See Russell, *History of Valency*, 100–101.

78. Ibid., 102.

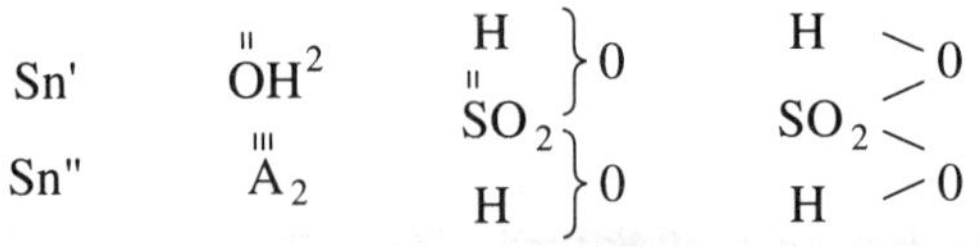

Sn' $\overset{\shortparallel}{O}H^2$ H ⎫ 0 ... H > 0

Sn" $\overset{\shortparallel\shortmid}{A}_2$ SO_2 ... SO_2 ... H ⎭ 0 ... H > 0

As Colin Russell has noted in his *History of Valency* (1971), a straight line is a much "more tangible symbol, . . . like rods, tie-bars, and so on."[79] With the "bond," the pictorial or structural formula had made its debut.

The so-called unsaturated properties of benzene and other hydrocarbons, their ability to "soak up" more hydrogen, for example, resulted in cartoonlike graphic representations for the valence bond.[80] Thus:

Alexander Crum Brown (1864) —Ⓒ—Ⓒ— —Ⓒ—Ⓒ—

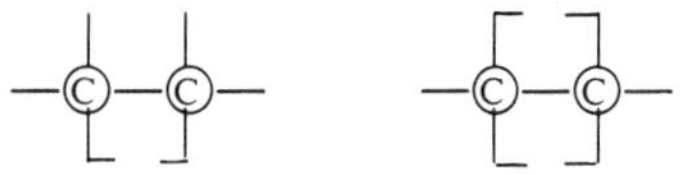

Josef Willbrand (1865) C |||| |||| C C |||| |||| C C |||| |||| C

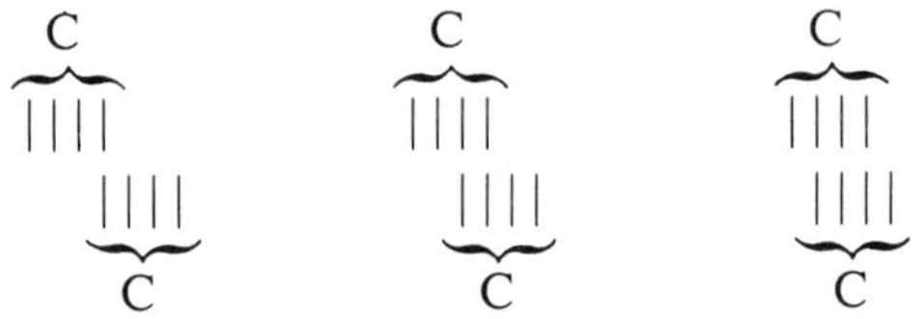

The puzzling behavior of benzene, its unreactivity, and the inference that all six carbon atoms must be linked in the same way could not be described by an open carbon chain (or vertebrate), nor was it satisfactory to write one true "rational" formula for "empirical" C_6H_6 benzene. Albert Ladenburg's nonplanar prism formula was the best single representation,

79. Ibid., 100.
80. Ibid., 234; also see, Benfey, *From Vital Force*, 105–107.

but Ladenburg himself used the prism formula side by side with Kekulé's "ring" or cyclohexatriene formulas:[81]

Rival schemes to the prism and hexagon included figures in which Josef Loschmidt used a large circle to signify neither atoms nor a ring structure for benzene but a structurally indeterminate benzene nucleus:[82]

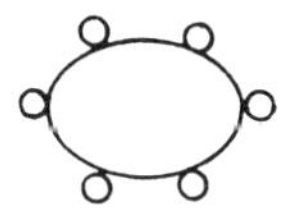

A truly impressionistic, or pointillist, scheme appears in the work of Johannes Thiele, to which we will return in chapter 5. Thiele attempted to explain anomalies in the behavior of butadiene, as well as benzene, by "floating" or "indeterminate" valences:[83]

81. See Alan J. Rocke, "Kekulé's Benzene Theory," 155–156; A. Ladenburg, *Theorie der aromatischen Verbindungen* (Brunswick, 1876).

82. Alan J. Rocke, "Hypothesis and Experiment in the Early Development of Kekulé's Benzene Theory," *Annals of Science* 42 (1985): 355–381, on 366.

83. F. K. Johannes Thiele, "Zur Kenntnis der unsättigten Verbindungen, Pt. I," *Liebig's Ann.* 306 (1899): 87–142, on 89, 126.

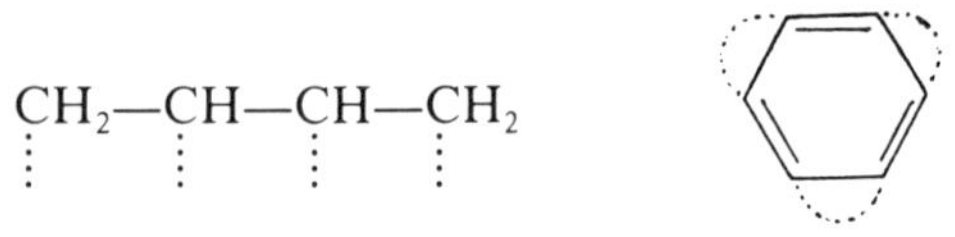

$$CH_2—CH—CH—CH_2$$

After rejection of alchemical and geometrical symbols in favor of the linear symbolic formulas of the early nineteenth century, chemical formulas had become full of visual imagery by the late nineteenth century. This imagery was of a graphic and peculiarly cartoonlike nature, more nearly like so-called primitive art than like realist or photographic art by way of stressing or exaggerating special aspects of functionality in the chemical species.[84]

In a scathing attack on molecular imagery at the Chemical Society in 1867, Brodie amused his audience by describing a commercially available set of molecular models:

> There are seventy balls in all for the representation of atoms—monads, dyads, triads, tetrads, pentads, and hexads [valence] being distinguished by the number of holes pierced in the balls. To connect these into rational formulae . . . brass balls, straight or bent, and occasionally flexible bands are employed.[85]

On this occasion, Brodie clearly interchanged the roles in chemical explanation of the rational or constitutional formula and the molecular model, because the rest of his lecture is a critique of the formula.

But what was the meaning of the pictorial formulas? What did they explain? Were they theoretical models? What was the aim of Hofmann's croquet balls, Dewar's brass-and-rod models, Kekulé's wooden sausages, and van't Hoff's cardboard tetrahedrons?[86]

When Couper used lines or bars in graphic formulas, he stated that he had in mind a representation of the forces of attraction between the elements in organic molecules.[87] Meyer and Kekulé were among those who quickly but circumspectly adopted the notation. But Frankland tried to be very careful in using the word ''bond'' and its bar representation:

> By the term bond, I intend merely to give a more concrete expression to what has received various names from different chemists, such as atomicity, an atomic power, and an equivalent. . . . It is scarcely necessary to remark that by this term I do not intend to convey the idea of any material connection between the ele-

84. See Roald Hoffmann and Pierre Laszlo, ''La representation en chimie,'' *Diogène* 147 (1989): 24–54. Also see, Guy Ourisson, *L'Actualité Chimique*, cited by Hoffmann and Laszlo.

85. Brodie, ''On the Mode of Representation,'' 296.

86. On these, see Rocke, ''Hypothesis and Experiment''; Crosland, *Historical Studies*, 336; and especially, Bert Ramsay, ''Molecules in Three Dimensions,'' *Chemistry* 47, no. 1 (1974): 6–9, and no. 2 (1974): 6–11; and Ramsay, *Stereochemistry*.

87. See Crosland, *Historical Studies*, 332–332.

ments of a compound, the bonds actually holding the atoms of a chemical compound being, as regard their nature, much more like those which connect the members of our solar system.[88]

Logically, more bonds between atoms should mean a greater force of attraction and increased carbon density. But doubly bonded alkenes are more reactive, not less reactive, than singly bonded alkanes or triply bonded acetylenic compounds. Further, how could the force associated with a polyvalent atom be divided a priori to react to the atoms encountered, sometimes dividing itself in half, other times by thirds? To assume that valences act across empty space, free of atoms, some chemists noted, "is only possible on paper or in a model where there are lines or wires but not forces."[89]

Chemical language is full of spatial language and the imagery of concrete particles and geometrical shapes, even in so simple a description as one of chemical solution (by "prolonging the operation, and multiplying the points of contact, the salt redissolves and enters into combination with the liquid"[90]). At midcentury, Laurent invoked the image of the prism as a device for imagining the "fundamental form or fundamental nucleus" of organic molecules, imagining eight atoms of carbon to occupy the peaks and twelve atoms of hydrogen to occupy the midpoints of the edges of the prism.[91] References to the molecular "edifice" or "structure" are ordinary in all chemical language.[92]

But what is the meaning of this edifice? It was easy to deny that constitutional formulas in two dimensions represented chemical "reality." After all, molecules could hardly exist in just two dimensions.[93] The French school of chemistry was clear on this point, beginning with Laurent, who wrote, "The formula represents the functions of the compound,"[94] a point of view shared by Wurtz and Edouard Grimaux. Wurtz claimed that "these formulas . . . do not give any indication on the form of the molecule in space."[95] Similarly,

88. Edward Frankland, *Lecture Notes for Chemical Students: Embracing Mineral and Organic Chemistry* (London: John Van Voorst, 1866): 25. On Frankland, see Colin Russell, *Lancastrian Chemist: The Early Years of Sir Edward Frankland* (Milton Keynes: Open University Press, 1986).

89. Quoting R. Demuth and Victor Meyer, in Russell, *History of Valency*, 166.

90. C. L. Berthollet, *Researches into the Laws of Chemical Affinity*, trans. M. Farrell (Baltimore: Philip Necklin, 1809): 37. Or, "It appears . . . as if the condensation increases to the degree that a larger part of the surface of the elementary atoms is hidden in the interior of the compound atom"; in Berzelius, *Essai sur la théorie des proportions*, 54.

91. Laurent, *Méthode de chimie*, 237. He "supposed" that in ammonia and benzoyl chloride the atoms are arranged in hexagons (408).

92. Wurtz, *Leçons élémentaires*, 226.

93. William Wollaston wrote, "The atomic theory could not rest content with a knowledge of the relative weights of elementary atoms but would have to be completed by a geometrical conception of the arrangement of the elementary particles in all the three dimensions of solid extension." In "On Superacid and Sub-acid Salts," *Philosophical Transactions* (1808): 96.

94. Laurent, *Méthode de chimie*, 421.

95. Wurtz, *Introduction*, 255. "Let us guard against envisioning the preceding formulas as representing really the positions of atoms in space" (*Leçons élémentaires*, 226).

Grimaux said, "None of us imagine, with figures traced in a plane, to determine the real place of atoms in space. . . . These formulas recall to us the reactions themselves."[96]

In addition, there was strong recognition among chemists of the difficulty, perhaps the impossibility, of reconciling the different rational formulas for a single compound into one single "true" rational formula. There were precedents for this idea of incommensurability. In an 1849 essay on John Stuart Mill's logic, Whewell noted, "It is not true that if one theory be true, the other must be false; several different views may all be true."[97] But chemists, earlier than other scientists, had to practice incommensurability because of the necessity for using multiple, apparently mutually exclusive, formulas to represent different aspects of the behavior or multiple functions of a single molecule.

Thus, Graham wrote at midcentury that the formula is an expression "not of the constitution of a body in a state of rest, but of the manner in which the atoms are supposed to arrange themselves when subjected to certain influences. It is no longer the question what the absolute constitution of a substance may be."[98] In Watts's *Dictionary of Chemistry* (1872), George Carey Foster chided critics for not understanding that chemists who use different constitutional formulas for the same substance do this out of the necessity that "is nothing more than what we meet with in common language, where as we know, several distinct partial descriptions of the same object may be each of them quite accurate as far as they go. . . . [Critics misconstrue] the purpose of rational formulae to be symbolic expression of . . . the constitution of compounds, rather than descriptions of their behavior."[99]

Perhaps critics of the chemists were understandably confused, since the formulas frequently were referred to as "constitutional" or "structural" formulas. (See fig. 5.) Frankland's use of the terms "graphic" and "glyptic" formulas is less misleading. This notation, he wrote, expresses the chemical function of atoms, and while some critics counsel the danger that students will regard them as representations of the actual physical position of atoms, Frankland reported that in practice he had not found "this evil to arise."[100]

The difficulty of arriving at one true rational formula, whether glyptic or structural, was reiterated at the 1892 international congress on nomenclature in Geneva. Here arguments were made against establishing one official name for each compound and in favor of establishing rules for arriving at the several

96. Edouard Grimaux, *Introduction à l'étude de la chimie*, 145.

97. William Whewell, "Mr. Mill's Logic" [1849], 336–353 in Whewell, *Selected Writings on the History of Science*, 342.

98. Graham, *Elements of Chemistry*, II:522.

99. George C. Foster, "Formulae, Rational," 695–705, in Henry Watts, ed., *A History of Chemistry and the Allied Branches of Other Sciences*, 2d ed., 5 vols., (London: Longman, Green and Co., 1870), II:699.

100. Edward Frankland, *Lecture Notes for Chemical Students: Embracing Mineral and Organic Chemistry* (London: Van Voorst, 1866): v–vi.

methane CH_4

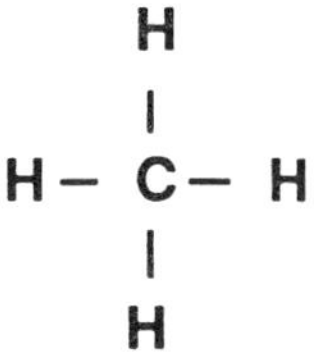

$$H-\overset{\overset{\displaystyle H}{|}}{\underset{\underset{\displaystyle H}{|}}{C}}-H$$

methyl alcohol CH_4O

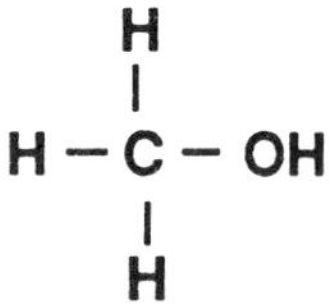

$$H-\overset{\overset{\displaystyle H}{|}}{\underset{\underset{\displaystyle H}{|}}{C}}-OH$$

formaldehyde CH_2O

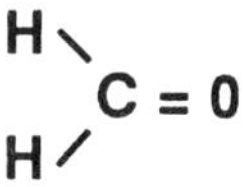

$$\begin{matrix} H \diagdown & \\ & C = 0 \\ H \diagup & \end{matrix}$$

formic acid CH_2O_2

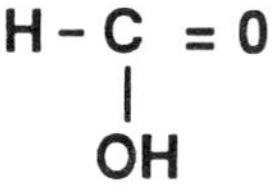

$$H-\underset{\underset{\displaystyle OH}{|}}{C} = 0$$

acetic acid $C_2H_4O_2$

$$CH_3-\underset{\underset{\displaystyle OH}{|}}{C} = 0$$

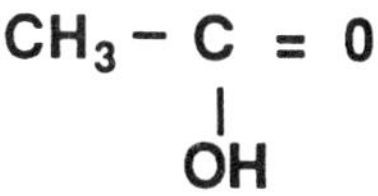

Figure 5. "Constitutional" or "Structural" Formulas. Drawings courtesy of Glenn Dryhurst.

different names that are equally valid to describe each particular function of a substance.[101]

But what about the three-dimensional images or formulations of molecules? What about ''la chimie en l'espace'' introduced by Joseph Achille Le Bel, van't Hoff, and Wislicenus toward the end of the nineteenth century? Were these carbon tetrahedra realistic ''models'' of real molecules in space? Van't Hoff argued in favor of the carbon tetrahedron that if atoms were arranged in a plane, there would be more isomers of the type $CR_1R_2R_3R_4$ predicted in principle than are actually observed. With the tetrahedral structure, only two isomers are possible, related to each other as mirror images.[102]

It often is noted that van't Hoff made the apparently ''positivist'' statement that ''the representations themselves, atom, molecule, their dimensions, and perhaps their shapes, are after all something doubtful, as is the tetrahedron itself.''[103] However, as the theoretical chemist Roald Hoffmann has aptly suggested, this does not necessarily mean that van't Hoff doubted ''chemistry in space'' but perhaps that he did not feel committed to the *exact* geometrical form that is the Platonic tetrahedron, a form that is not only idealized but static.[104]

In the 1930s, the French chemist Georges Urbain argued that the carbon tetrahedron is a valuable mental construct but that it cannot be a ''model,'' because in physical science a model must take mechanism and force into account. The representation of the double bond by one of the sides of a double tetrahedron is a pure symbol.

> It is unconcerned with the forces issuing from the atom considered to be a compact and homogeneous block at the center of the tetrahedron. . . . It hardly conforms to modern scientific thought to confuse purely symbolic figures—which are synoptic signs of writing—with mechanisms which are fictive phenomena substituted for the real phenomena whose laws they share.[105]

Roald Hoffmann and Pierre Laszlo recently have clarified the kind of argument that we find in Urbain's statement, noting that structural formulas in chemistry are ''in-between symbols and models,'' existing in uneasy hybrid states. ''A chemical formula is at one and the same time a metaphor, a model in the sense of a technical diagram, and a theoretical construct.''[106]

Neither the three-dimensional nor the two-dimensional structural formula

101. Crosland, *Historical Studies*, 347–348.

102. See ibid., 337.

103. Quoted, e.g., in Andrew G. Van Melsen, *From Atoms to Atom* (New York: Harper, 1960): 151.

104. In private communication, 5 October 1990.

105. Georges Urbain, *La coordination des atomes dans la molécule: la symbolique chimique*, Actualités Scientifiques et Industrielles, nos. 52 and 53 (Paris: Hermann, 1933): 13.

106. Hoffmann and Laszlo, ''Representation in Chemistry.''

could correspond to molecular reality because the formulas were static representations of what must really be a phenomenon of dynamics. Already in the nineteenth century, many chemists were convinced that not only the molecule but the "atom" itself likely had a substructure capable of dynamic modeling.[107] Gerhardt, Kekulé, and others were convinced that the molecule at rest must have a different internal structure from the molecule in reaction, and Williamson's 1850 paper on etherification had concluded that an exchange of atoms between molecules goes on all the time. "Chemical action results from a constant motion among the ultimate particles of bodies, this same movement likewise giving rise to the phenomena of heat, light, and electricity," suggested Henry Watts in 1872.[108]

This view was given further support by the discovery of "tautomerism," that is, that some compounds behave as if they have two different structures simultaneously. Peter Laar suggested in 1886 that this can best be explained as the result of continual oscillation of a hydrogen atom between two positions within a single molecule,[109] a hypothesis influenced by Kekulé's suggestion that the peculiarities of benzene are the result of an oscillation of atoms in benzene and that "equivalence" or valence is "the relative number of contacts which occur in a unit of time between atoms."[110]

Thus, the inner workings of things would be explained by a mathematical theory, perhaps something like the vortex theory of the atom. In the meantime, the *function* of things had been worked out by chemists' analogical reasoning, expressed in increasingly graphic and pictorial imagery. These images portrayed alternative functions of single substances under different conditions, according to conventions agreed on by most chemists. A set of complementary representations of a hydrocarbon might emphasize first its acid character, now its aldehyde property, then its substitution behavior.[111] This was a great achievement of nineteenth-century organic chemistry.

As Rocke has noted, the structure theory in organic chemistry had no real rivals in the nineteenth century; opponents were forced to retreat into positiv-

107. See M. J. Nye, "Nineteenth-Century Atomic Debates" and "Berthelot's Anti-Atomism," and Alan Rocke, "Subatomic Speculations and the Origin of Structure Theory," *Ambix* 30 (1983): 1–18.

108. Henry Watts, "Chemical Affinity," 850–866 in Watts, ed., *Dictionary*, Vol. I (1872): 866.

109. See Alexander Findlay, *A Hundred Years of Chemistry*, 2d ed. (London: Duckworth, 1948; 1st ed., 1937): 119.

110. Ibid., 125. See Auguste Kekulé, "On Some Condensation Products of Aldehyde," *Liebig's Ann.* 162 (1872): 77–124; abstracted by Henry Armstrong in *JCS, Abstracts* 25 (1872): 612–619, on 614.

111. For a semiological discussion of some of these issues, see Renée Mestrallet Guerre, "Communication, linguistique, et sémiologie. Contribution à l'étude de la sémiologie. Etude sémiologique des systèmes de signes de la chimie" (Universitat Autonoma de Barcelona thèse de doctorat, 1980), especially on multiple functions, 199–207. She concludes that systems of chemical notation are only partially isomorphous with systems of language (476). I am grateful to Roald Hoffmann for directing me to this thesis.

ism.[112] Some chemists, notably Aleksandr M. Butlerov, aimed at a structure theory that would allow writing true rational formulas.[113] But this view was in the minority. It might be said that the structure theory used metaphorical thinking, since it proposed inconsistent images of single substances.

The lack of dynamic models and rigorous mathematics makes nineteenth-century chemistry a different science from physics, but it is no less methodologically sophisticated. Chemists employed varieties of signs, metaphors, and conventions with self-conscious examination and debates among themselves. Nineteenth-century chemists were neither militant empiricists nor naive realists. These chemists were relatively unified in their focus on problems and methods that provided a common core for the chemical discipline, and the language and imagery they used strongly demarcated mid-nineteenth-century chemistry from the field of mid-nineteenth-century physics and natural philosophy.

By the middle of the twentieth century, explanations of chemical *function* had provided a key to molecular *form* and, for that matter, to the physicists' problem of electron distributions in space. Wave mechanics corroborated nearly all the suppositions of the structure theory, many of which had been roughly expressed in graphics and pictorial imagery. This corroboration would neither surprise nor chagrin most nineteenth-century chemists specializing in organic reactions. For if they characterized their knowledge as provisory and contingent, conventional at base, and susceptible to multiple representations, they nonetheless recognized its successes and were confident that their science was progressive and faithful to the phenomena.

112. Rocke, "Kekulé's Benzene Theory," 145.
113. According to Crosland, *Historical Studies*, 331.

Part Two

Chemical Problems and Research Schools

5

Physical Chemistry as Theoretical Chemistry

A Dynamics for Matter at the Turn of the Century

The Disciplinary Origins of Physical Chemistry

The conventional account of the origins of physical chemistry is reiterated by one of the discipline's preeminent twentieth-century practitioners, Henry Eyring, in an essay celebrating the centennial of the American Chemical Society in 1976. ''Unofficially physical chemistry is as old as efforts to improve cooking or to develop the implements of peace or war. Yet its more formal beginning, around 1887 [when the *Zeitschrift für physikalische Chemie* was founded], was only a little later than the birth of the American Chemical Society.'' The founders were Svante Arrhenius, Wilhelm Ostwald, and J. H. van't Hoff. ''These three great men from Sweden, Germany and Holland combined to give physical chemistry as auspicious a send-off as could be imagined.''[1]

Differing with Eyring, some historians have called into question the claim that the field or discipline of physical chemistry suddenly appeared in the late 1880s.[2] Diana Barkan, for example, has analyzed scientists' self-conscious creation of a disciplinary mythology about the founding of physical chemistry in the 1890s and early 1900s. Ostwald reconstructed the history of what he called the electrochemical foundations of physical chemistry in his largely historical treatise, *Electrochemistry*. Walther Nernst, in an opening address at

1. Henry Eyring, ''Physical Chemistry: The Past 100 Years,'' *CENews* 54, no. 15 (April 6, 1976), *Centennial: American Chemical Society, 1876 to 1976*: 88–104, on 88.

2. Robert Scott Root-Bernstein, ''The Ionists: Founding Physical Chemistry, 1872–1890'' (Ph.D. dissertation, Princeton University, 1980); and R. G. A. Dolby, ''The Transmission of Two New Scientific Disciplines from Europe to North America in the Late 19th Century,'' *Annals of Science* 34 (1977): 287–310.

his newly built physical chemistry institute in Göttingen, described the path by which, he claimed, van't Hoff's 1885 paper on dissociation reunited the sciences of chemistry and physics. Duhem lamented the withering away of Lavoisier's embryonic physical chemistry, while welcoming its now-achieved birth. Van't Hoff described physical chemistry as a "new world" that had arisen between the two "continents" of physics and chemistry.[3]

In fact, "physical chemistry" was a phrase much in use before 1887,[4] and as noted in chapter 2, it sometimes was a subject of study, by that very name, among nineteenth-century chemists.[5] An independent chair of physical chemistry was created for Kopp at Heidelberg in 1863, and a section for physical chemistry was established in the physical-mathematical faculty at the University of Kharkow in 1864. The first instructional laboratory at a German university was not Ostwald's but his predecessor's at Leipzig, Gustav Wiedemann, whose physical chemistry laboratory was authorized in 1871 to provide a strong basis for the development of a "theoretical chemistry."[6]

The need for a theoretical chemistry was felt in institutions outside Germany as well. Writing in 1872, Peter Guthrie Tait argued that the "ideal university" should be a combination of the Scottish and English university systems, "to the exclusion of what is manifestly bad in each." Among other improvements,

> let there be, for instance, a Professor of Experimental Physics . . . and a Professor of Applied Mathematics, in the place of the present solitary Professor of the enormous subject Natural Philosophy; let us have a Professor of Chemistry and Medicine, and a Professor of the Theory of Chemistry.[7]

As we have seen, the discipline of chemistry in the nineteenth century was dominated by the expanding achievements and the research agenda of organic

3. See Barkan, "Walther Nernst," 41–60, analyzing Wilhelm Ostwald, *Elektrochemie*; Walther Nernst, *Die Ziele der physikalischen Chemie* (Göttingen: Vandenhoek & Ruprecht, 1896); Pierre Duhem, "Une science nouvelle: La chimie physique"; and Jacobus Henricus van't Hoff, *Physical Chemistry in the Service of the Sciences*, trans. Alexander Smith (Chicago: University of Chicago Press, 1903).

4. The title of the predecessor of Poggendorff (1824–1877) and Wiedemann's (1877–1899) *Annalen der Physik und Chemie* was the *Annalen der Physik und der Physikalische Chemie* (1819–1823).

5. In this connection, Servos mentions, among others, Robert Bunsen at Heidelberg, who invented the carbon-zinc battery and the spectroscope; H. H. Landolt at Bonn, later Berlin, who studied the refractive power of the molecule in relation to the refractivities of its atoms; Heinrich Rose at Berlin, who followed up on Berthollet's theory of mass action; and Cato Guldberg and Peter Waage in Norway, who did so more thoroughly. See John W. Servos, *Physical Chemistry from Ostwald to Pauling*, 11–15.

6. See Barkan, "Walther Nernst," 20–21; who draws on W. Girnus, "Grundzüge der Herausbildung der physikalischen Chemie als Wissenschaftsdiziplin: Eine wissenschaftshistorische Fallstudie zur Disziplinenengenese in der Wissenschaft" (Ph.D. dissertation, Akademie der Wissenschaften der DDR, Berlin, 1982).

7. Quoted in C. G. Knott, *Life and Scientific Work of Peter Guthrie Tait* (Cambridge: Cambridge University Press, 1911): 248. I thank S. S. Schweber for this reference.

chemistry. The aims and methods of chemistry, its language and imagery, were largely those of organic chemistry. The organic chemistry that first emerged as a subfield of general chemistry between 1820 and 1840 largely defined the journal literature and the laboratory work of chemists for the next six decades.[8] The exponential expansion of the *Berichte* of the Deutsche Chemische Gesellschaft, which first appeared in 1868, was largely the result of the publication of synthetic and structural chemistry papers, which far outnumbered other kinds of submissions.[9]

As Russell has noted, there is a curious paradox for sociologists in the founding of the *Journal of Physical Chemistry* (1896) and the *Journal of Organic Chemistry* (1936), since the founding dates represent the reverse order of field specialization. Liebig's *Annalen der Chemie* and Kolbe's *Journal für praktische Chemie*, although given general titles, focused on organic chemistry throughout the nineteenth century. The *Zeitschrift für physikalische Chemie* was founded in 1887 because only then was there sufficient work in this field to fill a volume.[10] Chairs and laboratories of "general," "mineral," and "applied" chemistry existed throughout the nineteenth century, but the output in papers and students was much smaller than in organic chemistry. Most applied chemistry was organic chemistry: chemistry applied to agriculture, physiology, medicine, pharmaceuticals, dyestuffs, brewing, and viticulture.[11]

Physical chemistry began to prosper partly from institutional and industrial causes. Some students who set out to study organic chemistry in the late nineteenth century were dissuaded from their aim by overcrowded conditions in the instructional and research laboratories. One example is Arthur A. Noyes, who was to establish the first physical chemistry research laboratory in America at the Massachusetts Institute of Technology. He set out for Germany in 1888 with his friend Samuel Mulliken, father of the later theoretical and quantum chemist, Robert Mulliken. Noyes's aim was to study in Baeyer's Munich laboratory of organic chemistry, but, deterred by the conditions he found there, Noyes studied with Ostwald at Leipzig.[12]

Noyes, like so many physical chemists, was to work on practical problems

8. The emergence of organic chemistry is the subject of a major study in progress by F. L. Holmes. See F. L. Holmes, "Justus Liebig and the Construction of Organic Chemistry," in Seymour Mauskopf, ed., *Chemical Sciences in the Modern World.*

9. See Johnson, "Academic Chemistry in Imperial Germany," 500–524; Alan J. Rocke, "The Quiet Revolution of the 1850s."

10. Colin A. Russell, "Specialism and Its Hazards," Unit III, in *The Structure of Chemistry* (Walton Hall, Milton Keynes: Open University Press, 1976): 6. On the *Zeitschrift für physikalische Chemie*, see Thomas Hapke, *Die Zeitschrift für physikalische Chemie; Hundert Jahre Wechselwirkung zwischen Fachwissenschaft. Kommunikationsmedium und Gesellschaft* (Herzberg: Verlag Traugott Bautz, 1990).

11. For example, see the discussion of institutes of applied science in France in Harry Paul, *From Knowledge to Power: The Rise and Fall of the French Science Empire* (Cambridge: Cambridge University Press, 1986), and Mary Jo Nye, *Science in the Provinces.*

12. Servos, *Physical Chemistry from Ostwald to Pauling*, 59, 106.

in new areas of industrial development at the end of the nineteenth century, for example, developing a process for recovering alcohol and ether vapors from air that he signed over to the company that became Eastman Kodak.[13] Late-nineteenth-century investigations in physical chemistry and in revived inorganic chemistry often focused on electrolysis, equilibrium conditions at high pressures and high temperatures, catalytic agents, and photochemistry, all areas of research with expanding commercial implications. Expanding enrollments and numbers of graduates in universities and technical schools; increasing international competition in electrical, photochemical, and nonorganic chemical products; and the use of private industrial monies to establish industrial and research laboratories all furthered the practice of physical chemistry by 1900.[14]

In addition, the award of Nobel Prizes in chemistry to van't Hoff, Arrhenius, and Ostwald during 1901–1909 firmly established the legitimacy and success of the new field. It surely was an advantage in arguing for these awards that investigations in physical chemistry were both efforts in fundamental science and bulwarks of modern industry, especially befitting the legacy of Alfred Nobel.[15]

That said, we will focus in this chapter on the principal theoretical claims of the new physical chemistry and the extent to which these were responses to the professed needs of the majority of chemists, namely, the needs of the organic chemists. We also will assess the extent to which the concerns of physicists and chemists overlapped or dovetailed at the close of the nineteenth century. We will keep in mind the claim often made that physical chemistry is a bridge between physics and chemistry (which seems to make it peripheral to both)[16], or that it is a "unifying" and "general" discipline (Ostwald's

13. Ibid., 109.

14. Ibid., esp. 64–69, 91–99, 207–219; Russell, "Specialism and Its Hazards," 32–34; George Dubpernell and T. H. Westbrook, ed., *Selected Topics in the History of Electrochemistry* (Princeton: Electrochemical Society, 1978).

15. On the prizes and physical chemistry, see Elisabeth Crawford, "Arrhenius, the Atomic Hypothesis, and the 1908 Nobel Prizes," 503–522; and *The Beginnings of the Nobel Institution. The Science Prizes, 1901–1915* (Cambridge: Cambridge University Press, 1984).

16. Harry Clary Jones: "It is now generally known that within the last fifteen years a new branch of science has come into existence. This branch, occupying a position between physics and chemistry, is known as physical chemistry. . . . The new physical chemistry really begins with the chapter on solutions." In *The Elements of Physical Chemistry* (New York: Macmillan, 1902): vi.

In 1893, Charles Friedel sought to persuade his Sorbonne colleagues of the need to teach physical chemistry, saying that there are new theories and new needs in the teaching of chemistry each day; "this is especially true for *phenomena on the boundaries of chemistry and physics*, and where considerations of the *chemical molecule* intervene, which appears to be, in a lot of cases at least, identifical with the *physical molecule* [my emphasis]. "Rapport par Charles Friedel," 109–110, in *Pièces-annexes: Procès-Verbaux de la Faculté des Sciences de Paris*, dated June 1893.

allgemeine chemie), which basically entails a physical reductionist approach to chemistry.[17]

Nernst, for example, argued that physics forms the theoretical basis of all sciences, including chemistry.[18] A statement by Emile Dubois-Reymond to the effect that there was a need for a physical chemistry to create a "mathematical mechanics of knowledge" and a "complete picture of molecular processes"[19] was used to preface the first issue of the *Zeitschrift für physikalische Chemie*. For van't Hoff, this was to be achieved through "the application of physical expedients, methods, and instruments to chemical problems" in establishing "comprehensive principles."[20] Less concretely, Jean Perrin wrote that "really, there is no particular method proper to chemical physics [but] rather a physico-chemical *esprit*."[21] Perrin's successors at the Université de Paris understood this remark to mean that physical chemistry at the turn of the century was "whatever interested Jean Perrin, just as physical chemistry at Berkeley was whatever interested Gilbert N. Lewis."[22]

In her recent study of Nernst, Barkan argues the strong thesis that physical chemistry embodies modern scientific pluralism, constituted from classical electrodynamics, electrochemistry, thermodynamics, kinetic theory, and atomism. She further argues that physical chemistry at the turn of the century was essentially a *disunified* practice of science.[23] However, the striking successes of its journals, institutes, curriculum, and myth building suggest sufficient unity to convince many scientific practitioners that their interests, or their self-interest, lay together. A centrally unifying interest that was addressed by a plurality of methods was the problem of chemical affinity.

If Barkan calls into question physical chemistry as unified science, Russell argues the even more striking thesis that physical chemistry was no bridge-building science at all. Not only was there a larger gap between physics and chemistry at the end of the nineteenth century than at its beginning, he suggests, but physical chemistry "became a kind of wedge inserted between" the already too separate fields of inorganic and organic chemistry, which "beyond

17. Wilhelm Ostwald, *Lehrbuch der allgemeine Chemie*, 2 vols. (Leipzig: Engelmann, 1885 and 1887). Nernst's principal text is *Theoretische Chemie vom Standpunkte der Avogadroschen Regel und der Thermodynamik* (Stuttgart: Ferdinand Enke, 1893).

18. Nernst, *Die Ziele der physikalischen Chemie*, 2–3, cited in Barkan, "Walther Nernst," 44–46.

19. Quoted in Root-Bernstein, "The Ionists," 1–2; from Wilhelm Ostwald, "An die Leser," *Zeitschrift für physikalische Chemie* 1 (1887): 1–4, on 1–2.

20. J. H. van't Hoff, *Physical Chemistry*, 16–18, quoted in Barkan, "Walther Nernst," 63–64.

21. Jean Perrin, "La chimie physique," 18–28, in Perrin et al., eds., *L'orientation actuelle des sciences* (Paris: Félix Alcan, 1930): 18.

22. Jules Guéron and Michel Magat, "A History of Physical Chemistry in France," *Ann.Rev.P.Chem.* 22 (1971): 1–25, on 7.

23. Barkan, "Walther Nernst," 17–18.

doubt helped to keep them apart for many more years.'' Physical chemists, Russell suggests, removed problems of common interest from the domains of inorganic and organic chemistry and appropriated them for a new discipline.[24]

In contrast, it is a central thesis of this book that the disciplinary boundary between physics and chemistry became less well defined after 1900 than in the middle decades of the nineteenth century and that physical chemistry played an important role in this development. It is equally a central thesis of this study that a compelling and unifying interest among organic, inorganic, and physical chemists, that is, *of chemists*, was the problem of chemical affinity and the dynamics of chemical activation and reaction.

The practice of physical chemistry came to include many subfields of research: thermochemistry and thermodynamics, solution theory, phase equilibria, surface and transport phenomena, colloids, statistical mechanics, kinetics, spectroscopy, crystallography, photochemistry, and radiation. Here I concentrate only on three approaches within physical chemistry that had some promise for meeting the needs of organic chemists who wanted to explain affinity and reaction dynamics.

I first examine arguments at the century's end that classical organic chemistry had reached conceptual limits that threatened its progress. I then analyze three promising theoretical strategies from physical chemistry for understanding the activation and mechanism of chemical change. These approaches were those of (1) chemical thermodynamics and kinetics, (2) the radiation hypothesis of chemical activation, and (3) the application of ionic, then electronic, theories to organic structure and chemical reaction mechanisms. I explore developments that took place largely among physicists and physical chemists in search of a theoretical chemistry from roughly the 1860s through the 1920s.

The Limits of Classical Organic Chemistry

As we have seen, many nineteenth-century chemists minimized the dependence of their field on hypothetical reasoning and stressed its basis in empirical laws, rational analogy, and systematic classification. Nonetheless, chemists used and debated among themselves the validity and implications of several fundamental chemical theories.

First and foremost among these was the atomic theory. Throughout the century, chemists were divided about the validity of so-called physical atomism, namely, that there exists a unique indivisible particle specific to each chemical element and characterized by an atomic weight that some chemists sought to prove was a multiple of the standard weight of hydrogen (or its subweight).

While divided over the atomic hypothesis, which many chemists deemed metaphysical rather than physical, the entire chemical framework of the nine-

24. Russell, ''Specialism and Its Hazards,'' 27, 31.

teenth century nevertheless rested empirically and conceptually on calculations of the relative combining weights, or so-called equivalent weights, of the elementary substances. At the international Karlsruhe conference in 1860, chemists specifically distinguished and debated the question, as they put it, of whether the physical atom and the chemical atom were one and the same.[25]

In addition to atomism, the principal chemical theories of the nineteenth century included electrochemical dualism, the radical theory, the type theory, and the structure theory, the latter strongly identified with what chemists called the ''law of linking'' of carbon atoms. The valence theory evolved as a way of tying together the notions of chemical equivalence and chemical structure, and it carried along the old problem that some chemical elements (e.g., nitrogen) exhibit different combining values with another element in different circumstances.

Berzelius's theory of electrochemical dualism was rooted experimentally in the invention of the Voltaic pile and was based on the principle that the chemical character of a compound depends entirely on the identity of its elements. The radical theory, argued by Gay-Lussac, Liebig, Wöhler, and Robert Bunsen in the early nineteenth century, had foundations partly in electrochemical conceptions but also in Lavoisier's notion of the ''radical'' that combines with oxygen to form acids. Radical theory became a theory of ''genetic'' relationships, providing an explanation of the root or stock from which variations in types of organic substances were derived. However, as more and more interest was placed on the ''type'' rather than the ''root,'' arguments became increasingly powerful that chemical character depends more on the mode of grouping than on the identity of substituents.

Strong antagonisms developed among proponents of different theories, even though the difference often was a matter of emphasis. Liebig defended the radical theory against Dumas's new theory of substitution; Dumas and Gerhardt argued their meaning, as well as relative priority, for type theory; Kolbe's lingering attachment to electrochemical dualism led him into confrontations with Kekulé on the question of carbon chain formation and the skeletal structure of hydrocarbons. Kekulé's attachment to the principle of constant valency led him into conflict with interpretations of the newly discovered existence of tautomerism.[26]

The astonishing successes of organic chemistry in the nineteenth century were based in the methods associated with the overlapping theories called the

25. On this, for the first half of the nineteenth century, see Rocke, *Chemical Atomism*; and for the later period, M. J. Nye, ed., *The Question of the Atom*, and *Molecular Reality: A Perspective on the Scientific Work of Jean Perrin* (London: Macdonald, New York: American Elsevier, 1972).

26. On ''genetic'' and ''functional'' (or what she calls ''environmental'') relationships, see Mi Gyung Kim, ''Practice and Representation,'' e.g., 65–76, 98, 103–105. On Kolbe, see Alan J. Rocke, ''Kolbe versus the 'Transcendental Chemists': The Emergence of Classical Organic Chemistry,'' *Ambix* 34 (1987): 156–168.

radical, substitution, type, and structure theories. For most of this work, at least with respect to isolating natural substances and synthesizing new ones, methods and theories of early-nineteenth-century electrical studies were superfluous, if not downright misleading. But there was some discontent among chemists about the limits of the overall theoretical framework within which they were working.

Muspratt and Hofmann typically expressed dissatisfaction at midcentury with the failure of organic chemists to understand the mechanisms of chemical change, writing that there still was no understanding of the modus operandi of a decomposition.[27] Van't Hoff was still concerned with this question in the 1870s, writing that the two weak points of modern chemical theory were its failure to consider the relative position that the atoms occupy in the molecule and the nature of their motion.[28] As a young man who had studied organic chemistry, mathematics, and physics, van't Hoff attempted to remedy the first failure by giving three-dimensional structure to the so-called structure formula, thereby helping to create "space" or stereochemistry.[29] Within a few more years, he began a two-volume treatise on organic chemistry in which he dealt with the second problem,[30] but in fact he did not really reconcile the mechanical and spatial approaches within the framework of organic chemistry, because his tetrahedral carbon atom was an essentially static one in which the motion of atoms within the molecular structure is not addressed. His essentially gravitational conception of chemical affinity, or valence, met a dead end.

Alfred W. Stewart, trained in stereochemistry and organic chemistry in the early 1900s under J. N. Collie at University College, London, increasingly incorporated new theories from physical chemistry and radioactivity into his understanding of organic reactions. Ordinary structural formulas, he argued, failed to meet the demands put on them by new discoveries and investigations in the late nineteenth century. What was needed was a "dynamic" rather than a "static" view of the molecule. He argued, however, that the purely "energetic" theories of the physical chemistry of Ostwald did not meet these needs.[31]

Among the problems confronting organic chemists, two especially seemed to test the limits of the classical structure theories, namely, benzene and acetoacetic ester, also known as ethyl acetoacetate. Arthur Lachman, a Munich-

27. Quoted in Michael Keas, "The Structure and Philosophy of Group Research": 50. See Muspratt and Hofmann, "On Certain Processes in which Aniline is Formed," 249.

28. See the discussion of the history of chemical theories in Ferdinand Henrich, *Theories of Organic Chemistry*, trans. Treat B. Johnson and Dorothy A. Hahn (New York: John Wiley, 1922; trans. and enlarged from the revised 4th German ed. of 1921); on van't Hoff, 18.

29. J. H. van't Hoff, "Sur les formules de structure dans l'espace," *Archives néerlandaises des sciences exactes et naturelles* 9 (1874): 445–454.

30. J. H. van't Hoff, *Ansichten über die organische Chemie*, 2 vols. (Braunschweig, 1878–1881).

31. A. W. Stewart, *Recent Advances in Organic Chemistry*, introd. by J. Norman Collie (London: Longman, Green & Co., 1908): 16–17, 262–263.

trained professor of chemistry teaching at the University of Oregon, wrote in 1899 that the problem of the constitution of acetoacetic ester, first prepared by A. Geuther in 1863, was second only to benzene. Like benzene, acetoacetic ester was a valuable reagent. By the proper choice of alkyl halide and conditions of hydrolysis, it could be used for preparing ketones or acids, its structure being represented by Geuther (1864) as the enol

$$CH_3-C(OH) = CH-COOEt$$

and by Edward Frankland and B. F. Duppa (1866) as the ketone

$$CH_3-CO-CH_2-COOEt.$$

The difficulty lay in explaining how this substance comes to possess acidic properties. Wilhelm Wislicenus now suggested the role of "electrochemical polarization" by the influence of the two carbonyl groups on the methylene between them,

$$CH_3-CO-CH_2-CO-OC_2H_5$$

resulting in a shifting of the hydrogen atom during reactions. Others, including Lachman, deemed the notion of polarization unnecessary, since "there is no connection with electrical phenomena as far as we know."[32]

An accessory proposal was Arthur Michael's hypothesis that many reactions proceed by addition, for example, a polymerization of acetaldehyde ($CH_3CH = O$) in the presence of bases (OH^-) to an aldol ($CH_3CHOHCH_2CHO$), with subsequent loss of water to form crotonaldehyde ($CH_3CH = CHCHO$). Michael, educated in America, Germany, and France, made use of Kekulé's idea that two molecules may form a "polymolecule" or molecular compound, which, in turn breaks up to yield the final products.[33] Lachman expressed fairly standard misgivings about this proposal of an intermediary and transition form: "If we are going to explain reactions by means of addition products which we do not or cannot isolate, our explanation loses its definiteness. It becomes simply a *possible* explanation, and its conclusions are by no means binding."[34]

Kekulé's notion of polymolecules or molecular compounds similarly was rejected, indeed ridiculed, when Ira Remsen returned from Germany to teach

32. Arthur Lachman, *The Spirit of Organic Chemistry* (New York: Macmillan, 1899): 71. Also see Russell, "Specialism and Its Hazards," 20–21; and Ihde, *The Development of Modern Chemistry*, who notes that while Wilhelm Wislicenus in 1896 isolated two ethyl formyl phenyl acetates that had *keto* and *enol* properties, it was only in 1911 that each form of acetoacetic ester was isolated free of contamination by the other at temperatures around -78°C (330).

33. Arthur Michael (1853–1942) is a subject of discussion in D. Stanley Tarbell, "Organic Chemistry: The Past 100 Years," *CENews* 54 (1976): 110–123, on 111–112.

34. Lachman, *The Spirit of Organic Chemistry*, 78.

chemistry at Johns Hopkins University. As a chemist primarily interested in mineral substances rather than organic substances, Remsen presented to his students the formula

$$PtCl_4 \cdot 2KCl,$$

observing that "that period [punctuation] has for many years been a full stop to thought. Don't let such devices keep you from trying to find out what lies behind them."[35]

Chemists attempting to understand the mechanism of reactions had to come up with some model or schema for the behavior of the chemical elements within a chemical substance, usually invoking change of position of particles. Typical of early-nineteenth-century efforts of this kind was Dumas's analogy between a molecular system and a planetary system. In the solar system, there are planets that are units within the larger system, and there are subsystems, each composed of a planet and its satellites. So, in a molecule there may be different kinds of bodies held together through systems of mutual attractions, and one body may be replaced by another (chemical substitution) without disturbing the overall "equilibrium" of the entire system if the interchangeable subunits resemble one another sufficiently.[36]

In contrast to van't Hoff, who attempted to interpret the chemical bond by a mathematical theory of gravitational attractions in the *Ansichten*, Kekulé specifically rejected the planetary analogy or model in his well-known oscillation hypothesis for explaining the stability of the benzene ring and the equivalence of the carbon-carbon links in benzene. As quoted earlier, Kekulé proposed the mechanical hypothesis that all atoms in the benzene ring oscillate around equilibrium positions, a hydrogen atom making an oscillation that results in one collision with the adjacent hydrogen atom in the time that a carbon atom makes an oscillation resulting in collision twice with an adjacent carbon atom, once with the other adjacent carbon atom, and once with the adjacent hydrogen atom.[37]

By 1897, Collie was describing benzene as a vibrational system in three

35. Quoted, without citation, in Russell, "Specialism and Its Hazards," 11. To explain the combination of apparently saturated molecules, Alfred Werner introduced the theory of "coordination number" to supersede the old idea of "molecular compounds." Werner intended his theory to apply equally to organic and inorganic chemistry. Ibid., 15, 17.

36. J. B. Dumas, "Premier mémoire sur les types chimiques," *Ann.Chim.Phys.* 73 (1840): 73–104, on 73–74, quoted in Carl Schorlemmer, *The Rise and Development of Organic Chemistry*, rev. ed. (1894): 40.

37. See abstract by Henry Armstrong in *JCS*, 25 (1872), 612–619 of Auguste Kekulé, "Ueber einige Condensationsprodukte des Aldehyds," *Liebig's Ann.* 162 (1872): 77–124, 309–320, on 87–89. Other proponents of oscillating atoms to explain affinity included H. Buff, "Einige Bemerkungen zur Affinitätslehre," *Berichte* 2 (1869): 142–147, on 145; and A. Michaelis, "Ueber die physikalische Möglichkeit der neusten Hypothese Kekulés über das Benzol," *Berichte* 5 (1872): 463–467, cited in Russell, *A History of Valency*, 264.

dimensions, in which six carbon atoms each occupy an apex of a regular octahedron and the hydrogen atoms are arranged in two sets of three each. A vibration of the carbon atoms occurs such that one triplet of atoms moves toward the central cavity of the molecule while the second triplet simultaneously withdraws from the center toward the periphery of the molecule. Using the notion of a transition intermediate, Collie proposed that a loose [poly]molecular complex is formed when a reagent's substituent attaches at the periphery and then is swept in toward the center of the molecule.[38]

While not proposing a specific mechanism, Baeyer was just as willing to invoke a physical, although noncorpuscular, explanation for benzene's unreactivity: "The cause of the passivity of the central bonds we cannot at present explain; whether it depends on the geometrical structure which prevents other atoms from penetrating into the interior, or on a more dense state of ether in the centre of the ring must be left undecided."[39]

Ether chemistry was not unheard of in the nineteenth century, and it provided a palatable escape for those uncomfortable with simple corpuscular theories. Thus, Berthelot's chemical mechanics was consistent with ether mechanics:

> The hypothesis of a matter formed undoubtedly by diverse condensations of the ethereal substance and, therefore, basically identical while multiform in appearances and in each of them characterized by a particular kind of movement . . . is perhaps the most probable of all.[40]

Twenty years later, Isidor Traube, professor of physical chemistry at Berlin's Technische-Hochschule, distinguished between an inner atomic volume corresponding to the material core of the atom and an outer volume that included an atmosphere of bound ether; the whole of the molecule then moved in a larger "co-volume" of free ether.[41] Farther still from mainstream nineteenth-century chemistry, Karl Pearson developed a mathematical theory of "aether squirts," setting up a quantitative measure of chemical affinity in terms of the pulsation periods of the squirts.[42]

38. J. N. Collie, "A Space Formula for Benzene," *Trans.CS* 71 (1897): 1013–1023, discussed in A. W. Stewart, *Recent Advances in Organic Chemistry*, 6th ed., 2 vols. (London: Longmans, Green and Co., 1931), II: 376–377.

39. A. von Baeyer, "Ueber die Constitution des Benzols," *Liebig's Ann.* 245 (1888): 103–190, discussed in Schorlemmer, *The Rise and Development*, 176.

40. Marcellin Berthelot, *Essai de mécanique chimique fondée sur la thermochimie*, 2 vols. (Paris, 1879), I: 455, quoted in Helge Kragh, "The Aether in Late Nineteenth-Century Chemistry," *Ambix* 36 (1989): 49–65, on 59.

41. See Kragh, "Aether," 59, citing Isidor Traube, *Ueber den Raum der Atome* (Stuttgart: F. Enke, 1899), and "Die Eigenschaften der Stoffe als Funktionen der Atom-und Molekularräume und Gedanken über die Systematik der Elemente," *Zeitschrift für anorganische Chemie* 80 (1904): 372–384.

42. Also discussed in Kragh, "Aether," 51. See Karl Pearson's articles "On a Certain Atomic

Most chemists were more comfortable with speculations about movements of atoms than with flows of aether squirts. In particular, the idea of hydrogen atom mobility was to become a leading theme in late-nineteenth-century organic chemistry, based in the work of Williamson at midcentury. Williamson's investigations of etherification led him to a theory of the water "type" as well as to experimental proof that water is H_2O, not HO. Williamson clearly expressed the idea of chemical equilibrium as a balance between two sets of molecules in which some atoms or (uncharged) radicals may exist freely for short periods of time.[43] In addition to its uncontestable central role in the "quiet revolution" of the 1850s,[44] this was a paper that inspired both chemists and physicists to think about the "degree and kind of motion"[45] of atoms within the molecule as well as the motion of the molecule as a whole.

Kekulé's oscillation hypothesis in 1872 was a direct outcome of Williamson's work twenty years earlier, as was Laar's suggestion in 1885 that wandering or oscillating hydrogen atoms account for the phenomenon of tautomerism in acetoacetic ester. Schorlemmer likened hydrogen atoms to "comets" of the chemical universe that, when replaced by more slowly moving elements or radicals, result in stabler molecular forms.[46]

Among physicists, Clausius was directly influenced by Williamson's ideas about motion and equilibrium to argue that small portions of an electrolyte decompose even in the absence of an electric current and that there is a dynamic equilibrium between the decomposed and undecomposed species.[47] Arrhenius took this hypothesis into an even more radical direction, stating that electrolytes exist in solution as independent ions, while van't Hoff used ideas about mobility and kinetics to develop what he called a "chemical dynamics." Just as chemical questions were influential in starting off these developments in what became the new physical chemistry, so the problem of chemical affinity was central to the origins of modern chemical thermodynamics.

From Chemical Affinity to Chemical Thermodynamics

In contemporary physical chemistry, a distinction is made between thermodynamics and kinetics. In nineteenth-century chemistry, the terms "me-

Hypothesis," *Transactions of the Cambridge Philosophical Society* 14 (1887): 71–120; "On a Certain Atomic Hypothesis," *Proceedings of the London Mathematical Society* 20 (1889): 38–63; and "Ether Squirts. Being an Attempt to Specialize the Form of Ether Motion which Forms an Atom in a Theory Propounded in Former Papers," *American Journal of Mathematics* 13 (1891): 309–362.

43. A. W. Williamson, "Results of a Research on Etherification" and "Suggestions for a Dynamics of Chemistry Derived from the Theory of Etherification," reprinted in *Papers on Etherification and on the Constitution of Salts* (Edinburgh: Alembic Club Reprints, 1929), no. 16, 5–17, 18–24. See J. R. Partington, *A History of Chemistry*, 580, 672.

44. See Rocke, "The Quiet Revolution."

45. Williamson, quoted in Otto Theodor Benfey, "Concepts of Time in Chemistry," *JChem.Ed* 40 (1963): 574–577, on 574.

46. Schorlemmer, *The Rise and Development*, 183.

47. See Servos, *Physical Chemistry from Ostwald to Pauling*, 35.

chanics," "statics," and "dynamics" were for the most part used interchangeably.[48]

Thermodynamics applies to systems at equilibrium, focusing on initial and final states; it is a science that determines exact relations between energy and properties of systems without concerning itself with molecules or mechanisms. Thermodynamics says nothing about time, that is, about how long a reaction will take. In contrast, kinetics concerns itself with molecules and mechanisms; the methods and conclusions of kinetics, as Farrington Daniels and Robert A. Alberty put it, are inclusive, because they are "based on almost all of physical chemistry."[49]

Michael, interested in the hypothesis of the polymolecule and transition forms, was one of the first chemists to recognize clearly that some reactions are under kinetic control, yielding initially the thermodynamically less stable product. Modern kineticists concern themselves not only with the reactants and products that appear in the straightforward chemical equation but with intermediate steps in the reaction and with factors such as concentration, temperature, solvent, and catalyst which determine the rate.[50]

The roots of both chemical thermodynamics and contemporary kinetics both lie in the eighteenth-century ideas of chemical "affinity" and "force," transformed into nineteenth-century conceptions of "work" and "energy." Berthollet identified the fundamental difficulty for eighteenth-century theories of affinity in a critique that applied equally to early-nineteenth-century theories of electrochemical dualism. In "Recherches sur les lois de l'affinité" (1799), Berthollet wrote,

> The very term *elective affinity* must lead into error, as it supposes the union of the whole of one substance with another, in preference to a third; whereas there is only a partition of action, which is itself subordinate to other chemical circumstances, . . . care must be taken not to consider this affinity as a uniform force which produces compositions and decompositions. . . . Such a conclusion would lead us to neglect all the modification which it undergoes from the commencement of action to the term of equilibrium.[51]

Early investigations of the "forces of affinity" focused primarily on acid/base and metal/acid reactions, whereas Berthelot was to redefine affinity studies by focusing on organic systems and slow reactions. Berthelot, like the

48. See Maurice Lindauer, "The Evolution of the Concept of Chemical Equilibrium from 1775 to 1923," *JChem.Ed.* 39 (1962): 384–390; O. T. Benfey, "Concepts of Time in Chemistry," 574–577; Christine King, "Experiments with Time: Progress and Problems in the Development of Chemical Kinetics," *Ambix* 28 (1981): 70–82; and Keith J. Laidler, "Chemical Kinetics and the Origins of Physical Chemistry," *Arch.Ex.Sci.* 32 (1985): 43–75. Also, see M. H. Back and K. J. Laidler, eds., *Selected Readings in Chemical Kinetics* (Oxford: Pergamon Press, 1967).

49. Farrington Daniels and Robert A. Alberty, *Physical Chemistry*, 2d ed. (New York: John Wiley, 1963): 2.

50. D. Stanley Tarbell, "Organic Chemistry," 112.

51. C. L. Berthollet, *Researches into the Laws of Chemical Affinity*, 146, 154.

Danish chemist Julius Thomsen, regarded the heat evolved in a chemical reaction as a measure of chemical affinity, on the grounds that heat represented the *work* done by chemical forces. Thomsen and Berthelot similarly defined affinity as the force that unites the component parts of a chemical compound and that must be overcome by an equivalent force, the quantity of which can be measured by heat evolved.[52]

The independently published Thomsen/Berthelot studies ran afoul of organic chemists, physicists, and the new physical chemists. First, Thomsen's attempt to apply thermochemistry to structural chemistry foundered on several facts, not least among them his claim that the C-C bond and the C=C bond are identical in energy or heat content.[53] Second, Berthelot designated the chemical mass as simply the total quantity of substance in the system, whereas the new designation of "active mass" in terms of *molecular* concentration was to prove more fruitful. The Norwegians Cato Maximilian Guldberg and Peter Waage were to take this latter approach in the decades of the 1860s and 1870s.[54] Finally, the identification of "affinity" or "work" with heat alone was to prove erroneous. Criticized initially by the young Duhem in a thesis that Berthelot's colleagues rejected, Berthelot's principle of maximum work met more influential opposition in a paper by Helmholtz in 1882.[55]

Helmholtz, working not from chemical experimentation but from abstract physical reasoning about electromotive force, proved that affinity is measured not by the heat evolved in chemical reaction (in the galvanic battery) but by the maximum work produced when the reaction is carried out reversibly. One must distinguish, he argued, between that part of energy which appears only as heat and that part which can be freely converted into other kinds of work, that is, the "free energy" analogous to potential energy in mechanics. Conditions of chemical stability are not determined by heat production but by the production of a decrease in free energy, that is,

$$\Delta F \geqslant O.$$

In the next couple of years, Helmholtz applied his theory to electrochemistry and the calculation of heats of dilution, briefly returning to chemistry in 1887 in an experimental examination of the electrolysis of water to further support

52. Marcellin Berthelot, *Essai de mécanique chimique fondée sur la thermochimie*, and Julius Thomsen, "Ueber die Berthollet'sche Affinitätstheorie," *Ann.Phys.* 138 (1869): 65–102, discussed in Helge Kragh, "Julius Thomsen and Classical Thermochemistry," *BJHS* 17 (1984): 255–272. Also see M. G. Kim, "Practice and Representation," 136.

53. See Kragh, "Julius Thomsen," 264–266.

54. Kim, "Practice and Representation," 191; and Ihde, *The Development of Modern Chemistry*, 408.

55. On Duhem, see M. J. Nye, *Science in the Provinces*, 208–209.

the theory of 1882.[56] Helmholtz began but did not complete a manuscript entitled "Thermodynamische Betrachtungen über chemische Vorgänge," which was a project similar to Max Planck's *Grundriss der Thermochemie*, published in 1893 and enlarged in 1897 under the title *Vorlesungen über Thermodynamik*. Helmholtz praised Planck's work for avoiding special assumptions about atoms and ions, while expressing some skepticism about van't Hoff's theory of osmotic pressure and Arrhenius's theory of solutions.[57] The relevance of Helmholtz's work to chemistry was publicly acknowledged by Hofmann on the ceremonial occasion of Hofmann's seventieth birthday in 1891. The following year, the Deutsche Chemische Gesellschaft elected Helmholtz to honorary membership.[58]

In writing the *Etudes de dynamique chimique* (1884), van't Hoff drew on Helmholtz's 1882 paper but especially on the work of August Horstmann, a student of Bunsen, Clausius, and H. Landolt.[59] As has often been discussed, van't Hoff's was an ambitious and original synthesis of disconnected ideas and theories about opposing forces, equilibrium, active masses, work and affinity, electromotive force, and osmotic pressure. He demonstrated that the heat of reaction is not a direct measure of affinity but that the so-called work of affinity may be calculated from vapor pressures (the affinity of a salt for its water of crystallization), osmotic pressure (affinity of a solute for a solution), or electrical work in a reversible galvanic cell (which he showed to be proportional to the electromotive force).

Van't Hoff introduced the double arrow signs $\leftrightharpoons$ for denoting equilibrium, and he wrote the equation describing variation in the equilibrium constant as a function of temperature. His starting point was reaction velocities, rather than a balance of opposing forces. He used the symbol **A** for the work (*Arbeit*) that is done by the force of affinity (*Affinität*) that brings about the chemical reaction, and he recognized the need to quantify the role of the concentration of reactants in determining rates of reaction. Van't Hoff used the terms "monomolecular" and "bimolecular" for what now are classified as "first-order" and "second-order" reactions.[60] Thus, van't Hoff contributed to the foundations of modern chemical kinetics as well as chemical thermodynamics.

56. Hermann von Helmholtz, "Die Thermodynamik de chemischen Vorgänge," *Sitzungsberichte der Königlichen Akademie der Wissenschaften zu Berlin*, 2d ser., 1 (1882): 22–39; "Zur Thermodynamik chemischer Vorgänge (zweiter Beitrag)," ibid., 825–836; and "Zur Thermodynamik chemischer Vorgänge (dritte Beitrag), ibid. (1883): 647–665, discussed in Helge Kragh, "Between Physics and Chemistry: Helmholtz's Route to a Theory of Chemical Thermodynamics," ms. of May 21, 1990.

57. See Kragh, "Between Physics and Chemistry," 20–21, 26–27.

58. Ibid., 21.

59. August Horstmann, "Theorien der Dissociation" (1873), 26–41, and "Ueber ein Dissociationsproblem" (1877), 42–55, reprinted in *Abhandlungen zur Thermodynamik chemischer Vorgänge*, *Ostwald's Klassiker*, no. 137 (Leipzig, 1903). For an excellent short discussion of van't Hoff, see Servos, *Physical Chemistry from Ostwald to Pauling*, 24–34.

60. See Lindauer, "The Evolution of the Concept of Chemical Equilibrium," 388–389; and Laidler, "Chemical Kinetics," 48–52.

The term ''affinity'' gave way slowly to ''work'' and to ''free energy,'' with the adoption by chemists of the so-called Gibbs free energy function (**F**), rather than the expression (**A**) developed by Helmholtz (1882), van't Hoff (1884), and Nernst (1887).[61] Like many of the new physical chemists, Ostwald reveled in the new approach. ''As long as we sought to measure chemical 'forces,' '' he wrote, ''the theory of affinity made no progress.''[62] J. Willard Gibbs's method of making calculations at constant pressure and constant temperature was more useful than the free energy derivation by European scientists for constant volume and constant temperature. Thus, for chemists, **H** (heat of reaction), $\mathbf{c_p}$ (heat capacity at constant pressure), and **F** were more generally useful than **E**, $\mathbf{c_v}$, and **A**. With the standardization of free energies of formation of electrolytes in solution by Lewis and his American colleagues, thermodynamics became an ordinary tool for most chemists.[63] Indeed, Lewis said of this work that it had given him more personal satisfaction than anything else he had done in chemistry.[64]

Thus, in the decades after 1880, many physicists and some chemists especially interested in mathematical approaches to their subject matter turned their attention to the calculation of free energies in both physical and chemical processes, with renewed attention to heat capacities, electromotive force, and values for molecular weights and molecular concentrations. Studies of reaction rates became more and more common, following up on path-breaking work by the physicist L. Wilhelmy and the chemist Vernon Harcourt.[65]

Typically, when Noyes returned from Germany to teach at MIT, he reorganized the old course of ''theoretical chemistry,'' eliminating the subjects of structural theory and stereochemistry and now devoting fifteen of seventy-five class hours to chemical equilibrium.[66] However, a few years before Noyes left MIT for the California Institute of Technology, Walter Patrick, later professor of physical chemistry at Syracuse University and Johns Hopkins University, was unimpressed with the system of teaching undergraduates at MIT. As he wrote to his former mentor, F. G. Donnan, at Liverpool,

61. On Nernst, see Barkan, ''Walther Nernst,'' Chap. II.

62. W. Ostwald, ''Die Energie und ihre Wandlungen'' (1887), 185–206, reprinted in *Abhandlungen und Vorträge*, on 24; quoted in Kim, ''Practice and Representation,'' 380.

63. See Gilbert N. Lewis and Merle Randall, *Thermodynamics and the Free Energy of Chemical Substances* (New York: McGraw-Hill: 1923), 158–159, 171–174.

64. This statement appears in a letter to J. R. Partington, who intended to nominate Lewis for the Nobel Prize. Copy of Letter to Partington, 7 December 1928, G. N. Lewis Papers, BL.UCB.

65. L. Wilhelmy, a student of Victor Regnault, among others, worked out the first mathematical expression for the relationship between the speed of reaction and the quantity of reactants and products, measuring the hydrolysis of sucrose to glucose and fructose by change from dextro- to levo-rotatory forms. See F. L. Holmes, ''From Elective Affinities to Chemical Equilibrium: Berthollet's Law of Mass Action,'' *Chymia*, ed. Henry M. Leicester (Philadelphia: University of Pennsylvania Press, 1962), VIII: 105–146, on 129–130. On Harcourt, see Christine King, ''Time and Chemical Change: The Development of Temporal Concepts in Chemistry with Special Reference to the Work of Augustus Vernon Harcourt (Ph.D. dissertation, Open University, 1980).

66. See Servos, *Physical Chemistry from Ostwald to Pauling*, 61–62.

> I am in charge of the laboratory work of eighty men in theoretical chemistry. . . . Everything is reduced to a rigid system. . . . The students are required to solve all the problems in the text. . . . Of course they become adept in substituting into the stock formula, but it is a question if they really fix their knowledge by such a course of training. . . . They are required to solve many problems in relation to partial pressures, steam distillation, etc., and I think that if some of these processes were *explained*, their knowledge would be more useful to them. . . . I am going to give lectures on surface chemistry next semester. . . . I shall ground everything in the attraction between molecules.[67]

At Harvard, Theodore William Richards, like Noyes, inherited a course in theoretical chemistry. He renamed it physical chemistry. However, he cautioned students that the molecular kinetic hypotheses might prove ephemeral, and, to the young Lewis's consternation, Richards showed contempt for the notion of chemical bonds. ''Twaddle about bonds: A very crude method of representing certain known facts about chemical reactions. A mode of represent[ation] not an explanation.''[68] It was not so much that Richards sided with energeticists against kinetic and mechanical representations, but he did have a distrust of mathematical formulations too far removed from the laboratory. When J. Robert Oppenheimer enrolled in Richards's course in physical chemistry in 1925, he pronounced it ''a great disappointment, . . . a very meager hick course. . . . Richards was afraid of even rudimentary mathematics.''[69] Thus, physical chemistry by no means necessarily meant mathematical chemistry.

Richards's demand for an ''explanation,'' not a representation, was a valid concern among chemists concerned with the practical implications in laboratory work of mathematical equations and theoretical speculations. Could one predict and plan chemical syntheses on the basis of knowing the reaction pathway, step by step and molecule by molecule? And what triggered a chemical reaction? What made a stable substance transform itself and assume a new identity? Were there insights from experimental and theoretical physics which now could aid the chemistry of the late nineteenth century?

The Controversy over the Radiation Hypothesis

In a 1912 lecture in Paris, Arrhenius told his audience that the central problem of chemistry remained that of ''affinity'' or the ''cause of chemical

67. Letter from Walter A. Patrick to F. G. Donnan, dated October 16, 1915, Donnan Papers, UCL. Yet, as Servos has noted, Pauling found Noyes's method very much to *his* liking, and Noyes was already interesting his students in x-ray crystallography when Patrick wrote this letter. Private communication, 1 December 1991.

68. Quoted in Servos, *Physical Chemistry from Ostwald to Pauling*, 119.

69. Ibid., 81.

reaction."[70] At the time, one of Arrhenius's closest colleagues in Paris was the younger Sorbonne physical chemist Jean Perrin, recently returned from the first Solvay physics conference in Brussels. Perrin's laboratory was a meeting place for scientists interested in physical chemistry, with Einstein, Nernst, and Max Bodenstein among its frequent visitors.[71] Interrupted in his laboratory investigations by the demands of military service and acoustical research during the First World War, Perrin was to return in the 1920s to the problem of chemical activation, inspired by early radiation theory and quantum theory, as developed by Planck, Einstein, and Nernst.

Perhaps best known of Perrin's work is his spirited defense of kinetic theory and physical atomism entitled *Les atomes* (1913), in which he made use of his own work on Brownian motion, in combination with studies of cathode rays and x-rays, ionization, radioactivity, radiation, and quantum theory.[72] About the time of the 1911 Solvay physics conference, Perrin shifted from Brownian motion to work on thin films, fluorescence, and photochemistry, partly to test the early quantum theory and especially to study individual atom-based fluctuations.

Among French physical chemists, Perrin and his immediate circle of colleagues were unique in interesting themselves in kinetics and activation mechanisms. The other best-known practitioners of physical chemistry in France, Henry LeChatelier and Pierre Duhem, concentrated on studies of chemical equilibria and thermodynamic potential.

The direction of Perrin's research was adumbrated in *Les atomes* in a brief mention of Einstein's proposal that a molecule absorbs a quantum of energy when it undergoes photochemical change. Perrin suggested that absorption of radiation may be the key to the chemical destabilization of molecules. By the 1920s, his hypothesis had become radically bold: "The essential mechanism of all chemical reaction is therefore to be sought in the action of light upon atoms."[73]

Independently from Perrin, on the eve of the war, Bodenstein was studying thermal and photochemical reactions involving hydrogen and the halides (iodine, bromine, and chlorine), the decomposition of hydrogen halide, and the equilibrium between the two. In 1913, he announced that approximately one million molecules of hydrogen chloride gas are formed for every quantum of light absorbed in the hydrogen/chlorine system. Initially, Bodenstein suggested that the reaction proceeds through an ionized active intermediate. Later, fol-

70. Quoted in Charles Brunold, *Le problème de l'affinité chimique et l'atomistique: Etude du rapprochement actuel de la physique et de la chimie* (Paris: Masson, 1930): 3.

71. See Jules Guéron and Michel Magat, "A History of Physical Chemistry in France," *Ann.Rev.P.Chem.* 22 (1971): 1–25, 5, n. 6.

72. Jean Perrin, *Les atomes* (Paris: Alcan, 1913; 4th ed., 1914), and its analysis in Nye, *Molecular Reality*, 157–165.

73. Perrin, *Atoms*, trans. D. L. Hammick, 2d ed. (New York: Van Nostrand, 1923): 104.

lowing up on Nernst's hypothesis in 1918 of chain reactions involving the dissociation of molecules into atoms, Bodenstein offered a kinetic interpretation involving atomic collisions.[74]

Like so many physical chemists and physicists interested in kinetics, one of Perrin's starting points was Arrhenius's activity formula (1889) for the probability of transformation of a molecule at temperature T, where the probability is

$$k = Se^{-a/RT}$$

with a = quantity of energy; S = a constant for a given molecules; R = the gas law constant; and T = temperature.[75] Perrin and others took a to be the energy that moves a molecule from a stable to a transition ''critical'' state.

One of Perrin's students, the brilliant René Marcelin who perished in the First World War, set to work on the general problem, demonstrating that, in addition to the Arrhenius activation energy, the rate constant had to contain an activation entropy term.[76] In his thesis, defended in 1914, Marcelin developed a general theory of absolute reaction rates, describing activation-dependent phenomena by the movement of representative points in space. Eyring, who successfully incorporated a quantum mechanical treatment of particle interactions into a theory of absolute rates in the 1930s, later praised Marcelin's ''neglected'' work that had been accomplished well before physicists elaborated the second quantum theory.[77]

Work continued on the radiation hypothesis during the war, for example, in Perrin's laboratory by his student Nilratan R. Dhar and in the Muspratt laboratory at Liverpool, where William McCullagh Lewis succeeded Frederick George Donnan as director.[78] Following the war, Perrin found it easy to interest

74. See Hugh Taylor, ''Fifty Years of Chemical Kineticists,'' *Ann.Rev.P.Chem.* 13 (1962): 1–18, on 4. Bodenstein became director of the Institut für Physikalische Chemie at Berlin in 1923, succeeding Walther Nernst when Nerst became director of the physics institute. Also, see Laidler, ''Chemical Kinetics,'' 71.

75. Arrhenius's formulation is based on a prior idea and equation of van't Hoff in the *Etudes de dynamique chimique* (Amsterdam: Frederik Muller, 1884). See Arrhenius, ''Ueber die Reaktionsgeschwindigkeit bei der Inversion von Rohrzucker durch Säuren,'' *ZPC* 4 (1889): 226–248, partially translated in M. H. Back and K. J. Laidler, eds., *Selected Readings*, 31–35; also see Laidler, ''Chemical Kinetics,'' 42–75, on 55–57.

76. Job, ''Mécanismes chimiques,'' 145–147.

77. See René Marcelin, *Contribution à la cinétique physico-chimique* (Paris: Gauthier-Villars, 1914); Eyring, ''Physical Chemistry,'' 90. Also Guéron and Magat, ''A History,'' 6; and Karol T. Mysels, ''René Marcelin: Experimenter and Surface Scientist,'' *JChem.Ed.* 63 (1986): 740.

78. Nil Ratan Dhar, *Coefficient de température de réactions catalytiques* (Paris: Thèse d'Université, 1916); ''Catalysis, Pt. IV, Temperature Coefficients of Catalysed Reactions,'' *JCS.Trans.* 111 (1917): 707–762; and *The Chemical Action of Light* (London: Blackie and Son, 1931). And Alfred Lamble and William C. McCullagh Lewis, ''Studies in Catalysis, Pt. I, Hydrolysis of Methyl Acetate, with a Theory of Homogeneous Catalysis,'' *JCS.Trans.* 105 (1914): 2330–2342; and W. C. McCullagh Lewis, ''Studies in Catalysis, Pt. VII, Heat of Reaction, Equilibrium Constant, and Allied Quantities, from the Point of View of the Radiation Hypothesis,'' *JCS.Trans.* 111 (1917): 457–469.

his good friend and colleague André Job in working on the radiation hypothesis, since Job's primary interests were reaction mechanisms and hypothetical intermediates. Four papers, two by Perrin and two by Job, laid out the principal line of argument for the radiation hypothesis.[79]

As examples of molecular activation by radiation, Perrin cited molecular dissociation and tautomerism. Thus, an isolated molecule in vacuum may suddenly undergo a transformation, independent of any molecular collision. Marcelin had demonstrated in his thesis that when for some unspecified reason, the internal energy of molecules varies in conformity to the laws of statistical mechanics, a spontaneous reaction occurs and the reaction velocity obeys Arrhenius's law.[80] The ''unspecified reason'' now was specified as a quantity of radiation ''L'' which ''assumes a continuous series of values, corresponding to continually increasing kinetic energies W'.''[81] So, matter absorbs energy and then passes ''explosively'' from an intermediary critical state to a new stable state. Experimental verification was to be found by studies in photochemistry, where the velocity of reaction at low temperature depends only on the intensity of exciting light.[82]

From the beginning, there were objections to Perrin's line of argument as well as to the independent work of McC. Lewis and others. When Perrin requested him to read the original 1919 paper, Einstein wrote, ''Your opinion of the primary importance of radiation for all chemical reactions still appears doubtful to me,'' as it had when they first discussed it before the war. Objections to Perrin's presentation made at a 1921 Faraday symposium surfaced again at the first and second international Solvay chemistry conferences in 1922 and 1925.[83]

F. A. Lindemann, recently appointed director of the Clarendon Laboratory

79. Jean Perrin, ''Radiation and Chemistry,'' trans. H. Borns, *Trans.Far.Soc.* 17 (1921–22): 546–572, which is almost identical in its main arguments to Perrin's ''Matière et lumière: Essai de synthèse de la mécanique chimique,'' *Annales de Physique*, ser. 9, 11 (1919): 1–108. André Job, ''La mobilité chimique,'' 284–319, in Institut International de Chimie Solvay, *Rapports et discussions sur cinq questions d'actualité: Premier conseil de chimie, 21 au 27 avril 1922* (Paris, 1925; hereafter cited as *Solvay I*), on which ''Mécanismes chimiques'' also was based, 125–164, in *Formes chimiques de transition*, ed. J. Perrin and G. Urbain (Paris: Société d'Editions Scientifiques, 1931).

80. Perrin, ''Radiation and Chemistry,'' 547–548, citing Marcelin, 555 and n. 1, 555.

81. Ibid., 553.

82. Ibid., 551, 560–561.

83. Letter from Perrin to Einstein, 28 August 1919; Einstein to Perrin, 5 November 1919, Einstein Collection, Institute for Advanced Study, Princeton (now housed in Jerusalem), cited originally in Nye, *Molecular Reality*, n. 93, 177. See F. A. Lindemann, ''Note on the Significance of the Chemical Constant and Its Relation to the Behaviour of Gases at Low Temperatures,'' *Phil.Mag.* 39 (1920): 21–25, cited in M. Christine King and Keith T. Laidler, ''Chemical Kinetics and the Radiation Hypothesis,'' *Archive for History of Exact Sciences* 30 (1984): 45–86, on 55. And discussion, *Solvay I*, 320–334; and Institut International de Chimie Solvay, *Structure et activité chimique. Rapports et discussions, Deuxième conseil de chimie. 16 au 24 avril 1925* (Paris, 1926; hereafter cited as *Solvay II*), 399–416.

at Oxford, stuck to objections he had already given Perrin earlier: evidence was inadequate that the reaction velocity at low concentrations is independent of the number of molecular collisions; and, if reaction velocity is proportional to radiation density, then the inversion of sucrose studied by McC. Lewis should be greater in sunlight than in the dark. But this is not the case. In 1921, McC. Lewis suggested that sunlight is directed radiation, not temperature radiation, but Lindemann replied that it was inconceivable that a molecule could tell the difference.[84] Dhar reported success in showing that sucrose inversion *does* proceed more rapidly in tropical sunlight, but H. Austin Taylor could not duplicate Dhar's reported results.[85]

In dealing with the sunlight question, Perrin proposed the idea that only absorption bands at five or more critical frequencies trigger activation,[86] and he thought this hypothesis might also answer Einstein's objection that the density of light is too feeble at ordinary temperatures to activate molecules.[87] By the time of the 1925 Solvay conference, when Perrin delivered "Lumière et réactions chimiques," he was understood to be suggesting that the activating radiation belongs only to the infrared. But, in this case, objected A. Berthould, "I do not think Perrin's interpretation is general, and often the thermic acceleration of photochemical reactions is attributable to an increase of velocity in secondary reactions, not influenced directly by light."[88]

Questions about quantum relations, first broached by Langmuir at the 1921 Faraday symposium, were brought up by Arrhenius when Perrin read the absent Job's paper in 1922[89] and again by Eric Rideal in 1925. Like Langmuir, Rideal insisted on the need to integrate an explanation of Arrhenius's "active" molecules with explicit chemical reactions and with quantum energy relations.[90] On the one hand, Perrin claimed that a strength of the radiation hypothesis lay in its providing the means to deduce the laws of Planck and Bohr by equating the energy retained by a molecule as it passes from one state to another[91] to

$$L = h\nu.$$

On the other hand, Perrin made confused remarks from the standpoint of quantum physics: "We need not believe that the absorption of radiation is discon-

84. "Discussion on 'The Radiation Theory of Chemical Action,' " *Trans.Far.Soc.* 17 (1921–22): 598–606.

85. Ibid., written communication from Dhar, 603–604; and King and Laidler, "Chemical Kinetics," 56–59. H. Austin Taylor and Hugh S. Taylor were brothers who took Ph.D. degrees in McC. Lewis's department at Liverpool. Robert Robinson taught in that same department from 1915 to 1919.

86. *Trans.Far.Soc.* 17 (1921–1922): discussion, 598.

87. See Perrin, *Solvay II*, 336, n. 2.

88. Perrin, "Lumière et réactions chimiques," *Solvay II*, 322–398; Berthoud in discussion, 401.

89. *Solvay I*, 325.

90. *Solvay II*, 404.

91. Perrin, "Radiation and Chemistry," 556.

tinuous. . . . The molecule may always be likened to a complex resonator sensitive to diverse frequencies.''[92]

At the Faraday meeting, Langmuir had emphasized the necessity for chemists to explain phenomena in terms of quantum relationships, insisting ''A molecule cannot dissociate by a continuous process. . . . We ought to consider these phenomena frankly as quantum phenomena.'' For Langmuir, it made no sense to talk of a series of frequencies: ''It is very difficult to get a physical conception of a mechanism which involves the product of several intensities.''[93] To this, Perrin replied that he meant that the reaction takes place in ''steps'' in response to different radiation frequencies.[94] Langmuir remained unconvinced in 1929, when he wrote in an article, ''Modern Concepts in Physics and Their Relation to Chemistry,''

> At first it was proposed that the radiation is absorbed by the reacting gas to form excited molecules in accordance with Einstein's photochemical law. When this is found not to be in accord with experiment, the concept of radiation is altered repeatedly . . . to make the modified theory fit the facts. After this process has been carried on sufficiently it no longer has any meaning to ask whether the reaction is caused by radiation.[95]

Lewis wrote to Paul Ehrenfest that he thought he had ''hit upon something pretty fundamental.'' But Lewis had difficulty publishing his paper, ''On the Theory of Reaction Rate,'' which the editor and referees of the *Journal of the American Chemical Society* worried was too speculative and insufficiently oriented toward experimental verification.[96] Lewis ended by rejecting the radiation hypothesis, even issuing a press release to announce that a decision now had been made between the two rival theories of violent collision and radiation absorption in favor of the former.[97]

92. Ibid., 557.
93. Discussion, *Trans.Far.Soc.* 17 (1921–22): 601.
94. Ibid., 605.
95. Irving Langmuir, ''Modern Concepts in Physics,'' 385–396, on 391.
96. Letter from G. N. Lewis to Paul Ehrenfest, undated but probably 1925, G. N. Lewis Correspondence, BL.UCB. G. N. Lewis and D. F. Smith promised in their paper, ''The Theory of Reaction Rate,'' *JACS* 47 (1925): 1508–1520, to publish a demonstration that a range of frequencies of radiation affecting degrees of freedom in a molecule is responsible for chemical reaction. This paper was the subject of the letter, with anonymous referee's report, from Arthur B. Lamb to G. N. Lewis, 28 February 1925, G. N. Lewis Papers, BL.UCB. The referee said: ''No real unimolecular reaction has actually been observed; they have been shown to be merely catalytic; the idea that a unimolecular reaction is due to collision between a quantum and a molecule is not original with Lewis.''
97. Lewis's ''decision'' between rival theories was published in the paper written with Joseph E. Mayer, ''A Disproof of the Radiation Theory of Chemical Activation,'' *Proc.NAS* 13 (1927): 623–625. A copy of the news release is in the College of Chemistry Papers, 1923–1936, BL.UCB. Lewis wrote A. F. Joffe in fall 1927 of Mayer's failure to find a chemical reaction when a molecular stream is passed through a radiation field. Letter from G. N. Lewis to A. F. Joffe, 27 October 1927, Lewis Papers, BL.UCL.

Perrin's theory was flawed both in his failure to clearly express the radiation hypothesis in quantum terms and in his concrete examples of monomolecular reactions. Thomas Martin Lowry, recently appointed to a new chair of physical chemistry at Cambridge University, argued that Perrin's choices of chemical examples were unfortunate. On the one hand, dissociation of compounds like NH_4Cl or HgCl requires a catalyst; on the other hand, the two isomeric forms of acetoacetic ester are interconverted only when an alkaline catalyst is present, as demonstrated in Lowry's own work with isomeric forms of nitrocamphor.[98] Thus, according to Perrin,

$$A \rightarrow a \rightarrow A_1$$

as if there is a true chemical transformation, that is, as if an isomeric transformation could take place in an isolated molecule. This, Lowry continued, "is simply an imaginary phenomenon." The easiest isomeric transformations are those that include the migration of a proton, and these, said the expert on proton migration, are not spontaneous. The only clear case of a monomolecular reaction is in radioactivity.[99]

By 1925, as Christine King and Keith Laidler have described in their history of the radiation hypothesis, opinion was hardening in favor of the collision (kinetic) theory of chemical activation.[100] According to Taylor, this could be seen at the American Chemical Society meeting in Minneapolis in 1926, attended by Bodenstein, Michael Polanyi, Donnan, Langmuir, Cyril Hinshelwood, and Francis Perrin "representing his father."[101] By this time, Hinshelwood, a colleague of Lindemann at Oxford, experimentally demonstrated that what he took to be unimolecular reactions show the falling off in rate-coefficient at low pressures predicted by Lindemann's collision theory. In his *Kinetics of Chemical Change in Gaseous Systems* (1929), Hinshelwood ascribed reactions exclusively to collisions.[102]

Not only kinetics but also the role of polar catalysts like water increasingly was viewed as the cause of chemical activation, as we will see later. In his paper at the 1925 Solvay conference, Job brought into strong focus the role

98. Thomas Lowry, "Is a True Monomolecular Action Possible?" *Trans.Far.Soc.* 17 (1921–1922): 596–597.

99. Discussion, *Solvay II*, 414–415.

100. King and Laidler, "Chemical Kinetics," 77, 82. Also, see remarks by R. G. W. Norrish at the 1923 Faraday symposium on valency, in *Trans.Far.Soc.* 19 (1923–24): 520–521.

101. H. Taylor, "Fifty Years of Chemical Kineticists," 11.

102. See King and Laidler, "Chemical Kinetics," 69–73; Harold Hartley, "Schools of Chemistry in Great Britain and Ireland, XVI, The University of Oxford," *J. Royal Inst. Chem.* 79 (1955): 118–127, 176–184, on 180; Cyril N. Hinshelwood, *The Kinetics of Chemical Change in Gaseous Systems*, 2d ed. (Oxford: Clarendon Press, 1929). The Nobel Prize in chemistry for 1956 was awarded to Hinshelwood and N. N. Semenoff for their work in kinetics, including chain processes and collisions.

of electrical action in chemical transformation, citing Lowry's papers.[103] Some of Job's criticisms of the radiation hypothesis already had led to modification of Perrin's initial ideas,[104] and it appears that Job was losing confidence in radiation as an alternative to polarization in accounting for the dynamics of chemical reactions.[105]

Perhaps Perrin's continuing commitment is explained by the fact that his radiation hypothesis was an update of Marie and Pierre Curie's original explanation of radioactivity, that the apparently spontaneous emission of radiations and charged particles from molecules is activated by immanent radiations. The Curies supposed that radiations in the atmosphere or in the ether disturb the stability of the naturally radioactive elements. This hypothesis, like the radiation hypothesis of chemical activation, eventually succumbed to an explanation using electron configurations, adumbrated by the new quantum interpretation of matter and energy.

In addition, Perrin tended to interest himself in individual atomic events rather than in the molecule as a complex entity. The radiation theory of chemical activation attempted a mathematical representation in simple and general formulas, based on axiomatic principles. Perrin's work in radiation bears more resemblance to his reformulation of thermodynamics in the early 1900s than to his experimental work on x-rays and cathode rays in the 1890s or on colloids and Brownian motion in the early 1900s. What binds all his work together is interest in single events or the individual corpuscle.[106]

The radiation hypothesis fit within a long-standing tradition in France and in physical chemistry in general of empirical and mathematical studies of heat and light, including spectroscopy, magneto-optics, and the magnetic properties of bodies. The Sorbonne laboratory of Perrin's friend Aimé Cotton was the site for J. Lecomte's building an infrared spectrometer, with which he compiled lists of thousands of spectra in the 1920s. Jean Cabannes, who had worked with Perrin, quickly took up Raman spectroscopy, for which he developed a classical, nonquantum mechanical theory. Paul Langevin's laboratory at the Collège de France became a home for extensive spectroscopic investigations, many of them published in the *Journal de physique et le radium*, edited by Langevin and his son, with Jean Perrin among the advisory editors.[107]

103. Job, "Les réactions intermédiaires dans la catalyse," *Solvay II*, esp. 168–169, 172–173, 174, 193.

104. See Perrin, *Solvay II*, 336, n. 2.

105. Still, Perrin's commitment ran deep, and when he edited Job's lecture, "Mécanismes chimiques," for publication in a collection of essays after his death in 1928, Perrin added references to his own revisions of the radiation theory published in 1926 and 1928. He also appended a reprint of his 1928 article from the *Comptes Rendus* of the Academy of Sciences. See *Formes chimiques*, 154, n. 2; 156, n. 2; and 160–164.

106. Jean Perrin, *Traité de chimie physique: Les principes* (Paris: Gauthier-Villars, 1903), and "La fluorescence," *Annales de Physique*, ser. 9, 10 (1918): 133–159, on 133.

107. On infrared spectroscopy, see Yakov M. Rabkin, "Technological Innovation in Science:

A good deal of this work had no impact in the development of models of molecular structure and the elucidation of reaction mechanisms; one reason was Perrin's own coolness to quantum wave mechanics.[108] Another, according to Oxford's Harold Thompson, who studied with Nernst and Fritz Haber, was that researchers like Lecomte "did not know enough chemistry; he was a physicist."[109] Perrin, too, approached physical chemistry as a physicist, not as a chemist. He had little real interest or knowledge of organic chemistry. But what made his radiation hypothesis attractive to many chemists was his concern with transition states and the search for a scheme of pathways defining chemical kinetics.

By the late 1930s, a distinction began to be made between "radiation chemistry," on the one hand, and "photochemistry," on the other hand. Radiation chemistry came to mean the chemical effects of *ionizing* radiation, whereas photochemistry embraced the effects of light quanta or photons having energy levels associated with visible, ultraviolet, or x-ray parts of the spectrum. With the working out of Raman spectroscopy theory after 1928, it was realized that the absorption of infrared radiation leads to changes in vibrational and rotational molecular energy but not to chemical reactions.[110] It was the ionizing process associated with radiations that was a key to chemical activation.

Ions, Electrons, and Affinity

A third approach within the newly defined physical chemistry was to prove crucial to dealing with the old problems of affinity and reaction mechanisms. Like thermodynamics and radiation theory, it promised and eventually delivered a conceptual framework that constituted a truly theoretical chemistry. At the same time, this new ionic and electronic approach to chemical explanation served as an important testing ground for theoretical physicists primarily concerned with the physics tradition of ether- and electrodynamics. Helmholtz,

The Adoption of Infrared Spectroscopy by Chemists," *Isis* 78 (1987): 31–54; and S. Nunziante Cesaro and E. Torracca, "Early Applications of Infra-Red Spectroscopy to Chemistry," *Ambix* 35 (1988): 39–47. On LeComte, Cabannes, and Langevin, see Guéron and Magat, " A History," 7–8; also see Jean Cabannes, *Anisotropie des molécules: Effet Raman, Conférences faites au Conservatoire National des Arts et Métiers les 2 et 3 mai 1930* (Paris: Hermann, 1930); and Maurice Bourguel, "Applications de l'effet Raman à la chimie organique," *Bull.SCF* 53 (1933): 469–505.

108. On Perrin's reticence, see interviews with Edmond Bauer and Louis de Broglie, respectively, 8 January 1963 (T. S. Kuhn and T. Kahan) and 7 January 1963 (T. S. Kuhn, A. George, and T. Kahan), SHQP.

109. From Rabkin, "Technological Innovation in Science," 49.

110. For example, see Milton Burton, "Radiation Chemistry," *Ann.Rev.P.Chem.* 1 (1950): 113–132, on pp. 113, 116. Studies in radiation chemistry were renewed after the Second World War in Perrin's laboratory, which now was directed by Edmond Bauer, with Michel Magat one of the main researchers in the field. See Guéron and Magat, "A History," 20.

J. J. Thomson, and Niels Bohr were among the most notable physicists who contributed to this new theoretical chemistry. Nernst, Abegg, G. N. Lewis, and Langmuir were among the most influential physical chemists.

While the use of an electrical or electrostatic theory of chemical affinity declined decade by decade in the nineteenth century, the theory began to experience a small revival on two fronts in the 1880s. One source of revival was the lecture that Helmholtz gave to a large audience at the Chemical Society of London in 1881. Helmholtz firmly associated himself with a theory of electrical "particles," namely, that the ions produced in electrolysis carry discrete and indivisible "atoms of electricity" that are independent of the elementary substance with which they combine. He further identified these atoms of electricity with indivisible units of chemical affinity: "This is the modern chemical theory of quantivalence, comprising all the saturated compounds."[111]

Nernst took up this conception in the 1890s. He now used the word "electron" for Helmholtz's positive and negative "atom" of electricity, with the hypothesis that different elementary chemical atoms have varying affinities for combining with positive and negative electrons. Ions account for the formation of polar compounds, like the simple salts, as well as for ions in solution. Nernst assumed that a further, separate chemical or atomic affinity must be invoked in order to explain bonding in nonpolar molecules.[112] He declined to speculate further on the subject of affinity, although he nonetheless stated, "The ultimate goal of the theory of affinity [is] reducing the causes of material transformations to physical, well-reasoned causes."[113]

Nernst's point of entry into ionic and electronic theories in chemistry, then, was electrolysis and solution theory, in the mainstream of the "Ionist" dissociation theory. Indeed, van't Hoff similarly proposed an ionic theory of the polar molecule in 1895, speculating on the binding forces between O^+ and O^- ions in the O_2 molecule.[114]

Whether Helmholtz was sympathetic or not to this new mainstream in physical chemistry was important to its practitioners. He was perceived by contem-

111. Hermann von Helmholtz, "On the Modern Development of Faraday's Conception of Electricity," *JCS (London)*, 39 (1881): 277–304, on 303, reprinted in Nye, *The Question of the Atom*. See Kragh, "Between Physics and Chemistry," 12–14.

112. See Walther Nernst, "Die elektrolytische Zersetzung wässriger Lösungen," *Berichte* 30 (1897): 1563; also in *Theoretical Chemistry*, 390–392; discussed in G. V. Bykov, "Historical Sketch of the Electron Theories of Organic Chemistry," *Chymia* X (1965): 199–253, on 201; and Anthony N. Stranges, *Electrons and Valence. Development of the Theory, 1900–1925* (College Station: Texas A&M University Press, 1982): 77–80. Experiments on positive rays in the early 1900s showed the positive electricity never existed independently from particles of matter similar to ordinary atoms. Also see William Ramsay, "The Electron as an Element," *JCS* (1908): 774–788.

113. See Barkan, "Walther Nernst," 140 and 172, citing the *Theoretische Chemie*, 7th ed. (Stuttgart: Ferdinand Enke, 1913), quotation on 470.

114. See W. A. Noyes's résumé of the development of the electron theory of valence from the chemical point of view in discussion at the 1923 Faraday Society Symposium, *Trans.Far.Soc.* 19 (1923): 476–478. Also, Laidler, "Chemical Kinetics," 58.

poraries to be skeptical about the ionic theory of solutions and about the theory of osmotic pressure. However, Ostwald wrote Nernst that when they talked in Berlin in 1892, Helmholtz told him he had been convinced of the "absolute and total freedom of ions" from the beginning of his own electrolytic studies.[115]

Abegg, who became professor of physical chemistry at Breslau, was the first physical chemist to give a full-blown electron treatment of chemical affinity and valence, in a paper written with Bödlander in 1901. Their approach took seriously the family and periodic groupings in Mendeleev's periodic table of the chemical elements, devising a system of positive and negative valency numbers for each element to explain binding forces in polar, or "molecular," compounds. Abegg assigned to each element a positive valency, equal to the group number (e.g., Group I {sodium, potassium, etc}: +1; Group VII {fluorine, chlorine, etc.}: +7), and a negative valency, such that the sum of the positive and negative numbers, neglecting signs, was always eight.

Abegg designated each element's "normal" valence to be the smaller number, independently from sign. He allowed variable valency and a fundamentally electropolar bond. His reasoning was that all atoms have eight places for electrons. The positive valency (i.e., the periodic group number) is the number of such available places actually occupied in the neutral atom. The physicist Paul Drude theorized that a positive normal valency represents the number of easily detached electrons and a normal negative valency, the number of easily added electrons.[116] Abegg's 1901 paper was a starting point for Lapworth's 1901 paper on this subject (chapter 7).[117]

Electrolytic chemistry and solution theory continued to be a principal source of speculation about chemical bonding. As John Servos has noted, it was Noyes's attempt to visualize the difference between strong and weak electrolytes, to explain anomalies in the dilution law, that led him to make a distinction between "electrical molecules" and "chemical molecules" in the early 1900s.

Both detractors and proponents of ionist theory at the turn of the century

115. See Kragh, "Between Physics and Chemistry," 27; and Barkan, "Walther Nernst," 158–159, drawing on a letter from Ostwald to Nerst, 22 November 1892, Ostwald Papers, AAW, Berlin. The views at issue are found in J. H. van't Hoff, "Role of Osmotic Pressure in the Analogy between Solutions and Gases" (1887) and Svante Arrhenius, "On the Dissociation of Substances in Aqueous Solution" (1887), in *The Foundations of the Theory of Dilute Solution* (Edinburgh: Alembic Club, 1929), no. 19.

116. Richard Abegg and Guido Bödlander, *Zeitschrift für Elektrochemie*, 1901–1910; Abegg's best-known paper on this subject is "Die Valenz und das periodische System. Versuch einer Theorie der Molekular-Verbindungen," *Zeitschrift für anorganische Chemie* 39 (1904): 330–380; see Paul Drude, "Optische Eigenschaften und Elektronentheorie," *Ann.Phys.* 14 (1904): 677–725, 936–961, on 722. For discussion of this work, see Bykov, "Historical Sketch," 203; and Russell, *The History of Valency*, 272.

117. See Arthur Lapworth, "The Form of Change in Organic Compounds and the Function of the alpha-meta Orientating Groups," *JCS* 79 (1901): 1265–1284.

were concerned to explain why strong electrolytes do not behave as if they are completely dissociated. This anomaly in the theories of Arrhenius and van't Hoff was successfully resolved by Peter Debye and Erich Hückel in 1923, when they showed that ions in strong electrolytes have decreased mobility because individual ions accumulate groups of ions around them. In the meantime, Noyes focused the attention of his advanced students and associates at MIT on developing structural and molecular-kinetic models that would help "visualize" the difference between strong and weak electrolytes. He classified salts, inorganic acids, and bases as "electrical" molecules that are in a less intimate union of ionic parts than is the case in organic acids, the truly "chemical molecules."[118]

G. N. Lewis, W. C. Bray, and K. G. Falk developed theories of chemical valence addressing the problem of polar and nonpolar bonds. All were members of Noyes's MIT research group in the first decade of the 1900s, at a time when enthusiasm for the study of physics along with physical chemistry was at high tide in Noyes's laboratory. Noyes himself had been a student of Ostwald. Lewis had spent the academic year 1900–01 at the institutes of Ostwald and Nernst. Colloquium topics at MIT in the period from roughly 1905 to 1910 included both chemical and physical topics: Alfred Werner's coordination theory of valence, tautomerism, and the absolute size of atoms. A primary text for seminar discussion was J. J. Thomson's *Corpuscular Theory of Matter*.[119]

These speculations about the ionic, polar, or electronic nature of chemical bonding, which arose largely from solution theory, resulted mostly in static models of the chemical bond or atom structure. In contrast is another tradition, which is more closely identified with ether theory and electrodynamics. This tradition, too, may be associated with Helmholtz, especially by way of his contributions to nineteenth-century theories of a "vortex atom" that would explain chemical affinities as well as the origin of electromagnetism, radiation, and spectral lines.

Within this tradition, physical chemists took their cue from physicists, and among physicists, no one pursued the aim of unifying chemistry and physics more avidly than J. J. Thomson. As Helge Kragh has argued, just "as J. J. Thomson in 1882 had applied the aether vortex atom to chemistry, he and others now applied the electron atom to chemistry. It could even be argued that the electron was nothing but aether in disguise."[120] The physicist, once

118. See Servos, *Physical Chemistry from Ostwald to Pauling*, 122–129.

119. K. George Falk and John M. Nelson, "The Electron Conception of Valence," *JACS* 32 (1910): 1637–1654; W. C. Bray and Gerald E. K. Branch, "Valence and Tautomerism," *JACS* 35 (1913): 1440–1448; G. N. Lewis, "Valence and Tautomerism," *JACS* (1913): 1448–1455; and "The Atom and the Molecule," *JACS* 38 (1916): 762–785. See Servos, *Physical Chemistry from Ostwald to Pauling*, 131–133; and Robert E. Kohler, Jr., "The Origin of G. N. Lewis's Theory of the Shared Pair Bond," *HSPS* 3 (1971).

120. Kragh, "Aether," 61.

"thermochemist," Max Planck wrote Hendrik Antoon Lorentz in 1909 that in the matter of resolving the problem of energy transfer in matter, "there are better prospects of holding the electrons responsible . . . than the ether."[121]

Thomson's interest in a unified theory of matter for physics and chemistry was long-standing. In the preface to his book, *The Electron in Chemistry*, published in 1923 and based on lectures delivered at the Franklin Institute in Philadelphia, he was confident that "if the modern conception of the atom is correct the barrier which separated physics from chemistry has been removed."[122] Lectures that Thomson gave at Yale University in 1903 included his proposal of the "plum pudding" model of the atom. This strongly influenced physicists, physical chemists, and chemists interested in the problem of chemical affinity and valence bonds. In the plum-pudding model, Thomson associated the number of electrons in an outer ring of concentric rings of electrons with the chemical properties of the groups in Mendeleev's periodic table. Further, he postulated that Faraday tubes of force link atoms within the chemical molecule and that since the tubes are directional, rising out of positive charge and ending in negative charge, there must be polarities within the molecule, even in organic molecules. Thus, for the directional formulas

$$\begin{array}{ccccc} & H & & H & \\ & \nwarrow & & \nearrow & \\ H\leftarrow & C & \rightarrow & C & \rightarrow H \\ & \swarrow & & \searrow & \\ & H & & H & \end{array} \qquad \text{or} \qquad \begin{array}{ccccc} & H & & H & \\ & \nwarrow & & \nearrow & \\ H\leftarrow & C & \leftarrow & C & \rightarrow H \\ & \swarrow & & \searrow & \\ & H & & H & \end{array}$$

the central bond must be directional, because it represents a Faraday tube of force, with a source and a sink. The central carbon atoms in this case are not identical *electrically*, although they are identical *structurally*.[123]

At the University of Cincinnati, Harry S. Fry was among several American chemists who began using directional or arrow formulas. Another was Julius Stieglitz at the University of Chicago. Fry invented the word "electromers" for electrically different isomers and "electronic tautomerism" for nonisolable tautomers in dynamic equilibrium with each other. He applied the latter idea to benzene, arguing that benzene exists as six electromers in equilibrium with

121. Quoted in Barkan, "Walther Nernst," 258–259, from a letter from Planck to Lorentz, 10 July 1909, Roll III, Lorentz Papers, American Institute of Physics.

122. J. J. Thomson, *The Electron in Chemistry* (London: Chapham and Hall, 1923): preface. His earliest applications of physics to chemistry included *A Treatise on the Motion of Vortex Rings* (1883) and *Applications of Dynamics to Physics and Chemistry* (1888).

123. J. J. Thomson, *Electricity and Matter* (Westminster: Constable, 1904): 117–120, 133–135.

each other. The composite formula for the six forms is then a representation that shows similarities between the *ortho* and *para* positions, in contrast to the *meta* position, thus giving an explanation for the so-called Crum Brown-Gibson rule for orientation in disubstituted derivatives.[124]

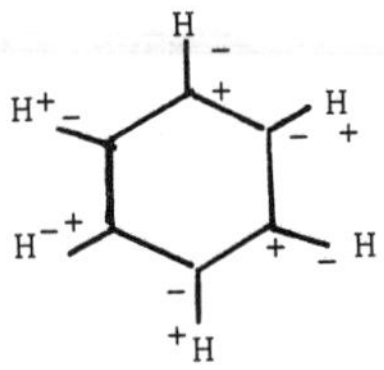

In further support of the notion of electronic tautomerism, Fry cited the 1905 paper by E. C. C. Baly and C. H. Desch in which they described keto-enol tautomerism as the result of the motions of a labile hydrogen atom that functions as a potential ion, "inasmuch as the bond of attraction or Faraday tube of force must be considered to be lengthened sufficiently to allow the interchange of the atom from the one position to the other within the molecule." Thus, the difference between "electrolytes" (electrical molecules) and nonelectrolytes (chemical molecules) has to do with conditions that determine the varying lengths of the Faraday tubes.[125] Baly and Desch's hypothesis was a result of their rumination on Wislicenus's earlier hypothesis (i.e., that the influence of "electrochemical polarization" explains the acidic properties of acetoacetic ester) in combination with thinking about J. J. Thomson's ideas.[126]

In the next few years after the Yale lectures, Thomson came up with several lines of evidence that the mass of an atom is due primarily to its positive charge and that the number of an atom's electrons is substantially smaller than in the plum-pudding model.[127] In addition, Thomson now argued against the existence of intramolecular ions after failing to find evidence that CO, HCl,

124. Harry Shipley Fry, *The Electronic Conception of Valence and the Constitution of Benzene* (London: Longmans, Green, and Co., 1921): 4–5, 48, 272–273, citing earlier papers from 1908–1911.

125. Fry, *The Electronic Conception*, 8–9.

126. On Johannes Wislicenus, see Lachman, *The Spirit of Organic Chemistry*, 67. E. C. C. Baly and C. H. Desch, "The Ultra-Violet Absorption Spectra of Certain Enol-Keto-Tautomerides, Pt. II," *JCS* 87 (1905): 766–784.

127. The evidence came from the scattering of x-rays by gases, the absorption of beta rays, and the dispersion of light by gases. Barkla demonstrated in 1911 that the number of electrons in lighter elements is approximately one-half the atomic weight, and Rutherford came to the same conclusion. See Stranges, *Electrons and Valence*, 51; and John L. Heilbron, "Rutherford-Bohr Atom," *American Journal of Physics* 49 (1981): 223–231, on 224–225.

H_2, Cl_2, and $Ni(CO_4)$ have internal charges.[128] Among chemists, William Noyes, Lauder Jones, Fry, and Stieglitz failed in efforts to isolate "electromers."[129] Léon Bloch's studies in Paris supported Thomson,[130] and G. N. Lewis arrived at the same view. Lewis also shared with Thomson a critical attitude toward Bohr's early theory of the atom. Sending Thomson reprints in 1919, Lewis wrote, "In looking over your recent paper in the *Philosophical Magazine*, I have been delighted to find what an extraordinary similarity there is between the ideas which were forced upon me chiefly from chemical considerations and those to which you have been led chiefly from physical considerations."[131]

In Germany, the physicists Johannes Stark and Hugo Kaufmann published papers in 1908 which portrayed electrons as dynamic binding agents between two atoms. Kaufmann took seriously Thiele's notion of partial valence and Werner's theory of divisible affinity, developing the model of a single electron dividing its lines of force among three atoms.[132] Stark proposed the model of an atom surface consisting of spheres or spherical zones of positive electricity; between or above these were small, pointlike negative valence electrons.

Stark classified chemical atoms into the categories of electropositive, electronegative, and electro-dual (*elektrozwiefach*), arguing that chemical reactions do not involve an interpenetration of the interiors of atoms but exchanges of electrons at their surfaces. A polar compound is one in which the valence electrons move away from the positive surface of one atom toward the positive surface of another; a nonpolar compound results from the valence electrons remaining more or less at an intermediate position between atoms.[133] In later discussions between Langmuir and Lewis about Lewis's claims for priority in what many were calling the Lewis–Langmuir theory of electron valency, Langmuir specifically reminded Lewis that already in 1908 Stark had identified a pair of electrons held in common between adjacent atoms with the valence bond.[134]

128. J. J. Thomson, "The Forces between Atoms and Chemical Affinity," *Phil.Mag.* 27 (1914): 758–789, and other articles that appeared during 1911–1914; see Stranges, *Electrons and Valence*, 174–175.

129. Stranges, *Electrons and Valence*, 163.

130. Léon Bloch, "Recherches sur les actions chimiques et l'ionisation par barbotage," *Ann.Chim.Phys.* 22 (1911): 370–417, 441–495.

131. Letter from G. N. Lewis to J. J. Thomson, dated 13 August 1919, Lewis Papers, BL.UCB. Lewis wrote a paper critical of Bohr in 1917, as did Thomson in 1919. See G. N. Lewis, "The Static Atom," *Science* 46 (1917): 297–302; and J. J. Thomson, "On the Origin of Spectra and Planck's Law," *Phil.Mag.* 37 (1919): 419–446.

132. Hugo Kaufmann, "Elektronentheorie und Valenzlehre," *PZ* 9 (1908): 311–314, discussed in Stranges, *Electrons and Valence*, 67.

133. Johannes Stark, "Die Valenzlehre auf atomistisch elektrischer Basis," *Jahrbuch der Radioaktivität und Elektronik* 5 (1908): 124–153; and *Die Elektrizität in chemischen Atom* (Leipzig: Hirzel, 1915), discussed in [Dr.] Ferdinand Henrich, *Theories of Organic Chemistry*, trans. from 4th rev. German ed. (1921) by Treat B. Johnson and Dorothy A. Hahn (New York: John Wiley, 1922): 92–106.

134. See Letter from Irving Langmuir to G. N. Lewis, 3 April 1920, Lewis Letters, BL.UCB.

Like J. J. Thomson, Kaufmann, and Stark, Bohr concerned himself with model building in an effort to unify the theories of the chemical and physical atom. He discarded altogether the ether model in favor of a corpuscular and quantum model. As John Heilbron and Kuhn argue on the basis of a memorandum that Bohr prepared for discussion with Rutherford, Bohr's starting point for the famous three-part paper, "On the Constitution of Atoms and Molecules," was a concern not with spectra (the subject of Pt. I) but with the relationship among electrons rings, the periodicity of the chemical elements, and the problem of chemical binding (Pts. II and III). Bohr employed the concept of affinity "work," calculating the quantity of work (energy) needed to remove an electron from diatomic hydrogen and, conversely, the heat released when two atoms of hydrogen form a molecule of hydrogen. He calculated that diatomic helium cannot exist and that diatomic oxygen does not dissociate into ions. He assumed that a pair of electrons is shared between two atoms in diatomic hydrogen and in other molecules, like methane. The electrons were modeled as a girdle of electrons rotating in a circle at a right angle to the axis connecting two atoms.

Bohr clearly distinguished chemical properties due to outer electrons from radioactivity due to the interior of the atom, substituting the notion of "shells" and "subshells" of electrons for the earlier idea of "orbits" or circles and relating the filling of these shells to properties of groups within the periodic table.[135] The full import of this paper was not appreciated until after the war, by which time Bohr had changed his mind about some aspects of the theory.

In the meantime, in 1916, Walther Kossel published a theory of the polar bond. Kossel's was an essentially electrostatic theory of the chemical bond, updated with the concept of the electron. Like some of his predecessors in applying electron theory to the chemical element, he observed the "rule of eight" by arranging electrons in concentric groups, each ring of electrons having a stable maximum of eight, except for the first case when the maximum is two. The loss or capture of electrons is a result of an atom's tendency to realize the electron configuration of a rare gas.[136]

Lewis, publishing in the same year, arranged eight electrons at the corners of cubes, rather than in spheres or circles, stressing the electron pair as the most fundamental of the electron groups. Thus, an octet consisted of four "duplets." Lewis stressed the sharing of a pair between two atoms as the fundamental bond of classical structural organic chemistry (the "chemical

135. Niels Bohr, "On the Constitution of Atoms and Molecules, Pt.I, Binding of Electrons by Positive Nuclei," *Phil.Mag.* 26 (1913): 1–25; Pt. II, "Systems Containing Only a Single Nucleus," *Phil.Mag.* 26 (1913): 476–502; "Pt. III, Systems Containing Several Nuclei," *Phil.Mag.* 26 (1913): 857–875. See John Heilbron and Thomas Kuhn, "The Genesis of the Bohr Atom," *HSPS* 1 (1969): 211–290.

136. Walther Kossel, "Ueber Molekülbildung als Frage des Atombaus," *Ann.Phys.* 49 (1916): 229–362, discussed in Russell, *A History of Valency*, 273; and Charles Brunold, *Le problème de l'affinité*, 59–80. Also, Stranges, *Electrons and Valence*, 227–229.

molecule''). He assumed that for some reason, Coulomb's law of electrostatic repulsion does not apply to electron particles at small distances, and he rejected the hypothesis of internally charged atoms, or ions, within molecules. Lewis insisted on *gradations* of electron-pair distribution between atoms, so that there is only one kind of bond, ranging from strongly polar to nonpolar.[137]

In 1919, Langmuir gave the names ''covalent'' and ''electrovalent'' to chemical bonds distinguished by the partition of the electron duplet between two atoms, consciously extending Lewis's theory. Langmuir emphasized the tendency of electrons to surround the nucleus in successive layers or ''sheaths''; the tendency of atoms to give up or take up electrons or to share electrons so as to form ''duplets''; and the tendency toward a minimum of residual charge on each atom and group of atoms. Electron rearrangement, said Langmuir, ''is the fundamental cause of chemical action.''[138]

But what triggers the process of electron rearrangement? What are the dynamics of the ''chemical'' electrons that account for the kinetic pathway of reaction? While a role for physical electrons, or electrical atoms, in the making (and breaking) of the chemical bond seemed more than a plausible hypothesis by 1915 or 1916, some very fundamental problems remained. Just as energy theory and radiation theory proposed by physicists and physical chemists needed to be applied in detail by chemists in everyday laboratory syntheses, so now the hypothesis of mobile electron pairs needed to be addressed in specific cases. Further, the question of intramolecular ions was a strong point of controversy among chemists. Do they exist or not? And under what circumstances? The American chemist S. J. Bates defined the problem clearly in 1914: ''On the whole the phenomena of physics are opposed to the view that in the molecule the atoms are charged with respect to one another. . . . Chemistry contributes the most satisfactory evidence in its favor.''[139]

There is considerable irony in the fact that it was to be practitioners of organic chemistry, more than those of inorganic or physical chemistry, who were to argue most forcefully in the first three decades of the twentieth century for a theory of reaction mechanisms that incorporated not just electrons but

137. Lewis had used the cubic octet in lecturing to undergraduates in 1902. See G. N. Lewis, *Valence and the Structure of Atoms and Molecules* (American Chemical Society, 1923): 29–30; ''The Atom and the Molecule,'' *JACS* 38 (1916): 762–785; ''Steric Hindrance and the Existence of Odd Molecules (Free Radicals),'' *Proc.NAS* 2 (1916): 586–592. Lewis conceded some polarity in covalent bonds, so that the C in CH_4 is slightly negative and the C in CCl_4 is slightly positive. Lewis's rejection of extreme, permanent polarity and polar valences was influenced by A. Werner's argument against permanently directed valences in *Neuere Anschauungen auf den Gebiete der anorganischen Chemie* (Braunschweig: F. Vieweg, 1905). See Stranges, *Electrons and Valence*, 173, 219.

138. Irving Langmuir, ''The Arrangement of Electrons in Atoms and Molecules,'' *JACS* 41 (1919): 868–934; and ''Isomorphism, Isoterism and Covalence,'' *JACS* 41 (1919): 1543–1559; also, ''The Structure of Molecules,'' *BAAS Rep.*, 1921, Edinburgh (London, 1922): 468–469. And see G. N. Lewis, ''The Static Atom,'' *Science* 46 (1917): 297–302.

139. S. J. Bates, ''The Electronic Conception of Valence,'' *JACS* 36 (1914): 789–793.

ions. The irony lies both in the fact that organic structural chemists so strongly opposed the early electrostatic theory of chemical reaction and in the fact that they often expressed skepticism about the whole theoretical project of a mechanics for chemistry. Although his attitude is not entirely representative, we note the statement by the organic chemist Baeyer at Planck's thesis defense in 1879 at Munich. Theoretical physics, pronounced Baeyer, is "a totally superfluous field, too remote from reality."[140]

In its disciplinary development at the end of the nineteenth century, physical chemistry served as a bridge, not a wedge, between the mathematical abstractions of theoretical physics and the metaphorical descriptions of organic chemistry. Not only through novel theories but also through control of new instrumentation, much of it electrical and optical in nature, physical chemistry was to revitalize and transform techniques in the chemical laboratory and theories of chemical explanation. Notably for our concerns, speculations about reaction mechanisms in hydrocarbon chemistry were to begin to proliferate in the early 1900s.

The historical, social, and intellectual elements of disciplinary identity for physical chemistry were well laid out in Europe, Great Britain, and the United States by the 1920s, and a new hybrid discipline of physical organic chemistry emerged in turn. However, more so than in some other scientific fields at this time, physical chemists often were to experience varieties of identity crisis. Were they physicists, or were they chemists? What *was* the difference between physics and chemistry? This is a question to which we return in the concluding chapters of this book.

We turn now to two schools of organic chemistry that specifically concerned themselves with "theory," one in Paris and one in London and Manchester, to analyze in detail different approaches of organic chemists to reaction mechanisms.

140. Quoted in Barkan, "Walther Nernst," 23, from Hans Hartmann, *Max Planck als Mensch und Denke* (Basel: Ott Verlag, 1953): 22.

6

The Paris School of Theoretical Organic Chemistry, 1880–1930

Theoretical Physics and Physical Chemistry at the End of the Nineteenth Century

The first chair of theoretical physics in France was the professorship established for Pierre Duhem in 1894 at the Bordeaux Faculty of Sciences.[1] Duhem was well known in French scientific circles not only as a physicist but as a physicist of exceptional mathematical skills who addressed himself early in his scientific studies to chemical problems. He wrote a controversial doctoral thesis (1886) in which he developed the concept of thermodynamic potential for chemistry and physics, and he later developed a treatment of equilibrium processes formally analogous to the mechanics of Lagrange. The goal was to make mechanics a branch of the more general science of thermodynamics, a science that embraces "every change of qualities, properties, physical state, chemical constitution."[2]

Like many of those involved in establishing the "new" physical chemistry in the 1880s, Duhem was primarily a physicist, and his conception of theo-

1. On chairs of theoretical physics in Europe at this time, see Christa Jungnickel and Russell McCormmach, *Intellectual Mastery of Nature: Theoretical Physics from Ohm to Einstein*, 2 vols., (Chicago: University of Chicago Press, 1986): II; and Paul Forman, John Heilbron, and Spencer Weart, eds., *Physics ca. 1900* V, *HSPS* (1975): 30–32.

2. From Pierre Duhem, *Traité élémentaire de mécanique chimique fondée sur la thermodynamique*, 2 vols. (Paris: Hermann, 1897), I: vi. The doctoral thesis, *Le potentiel thermodynamique et ses application à la mécanique chimique et à la théorie des phénomènes électriques* (1886), was rejected, partly because of its criticism of the "principle of maximum work" developed by the influential Collège de France chemist Marcellin Berthelot.

retical physics embraced chemistry through the program of thermodynamics. The fundamental principles of chemistry were to be found in the laws of thermodynamics; physical chemistry provided the framework for theoretical chemistry. It is significant that a starting point for Duhem's interests in theoretical physics and physical chemistry was Helmholtz's paper, "Die Thermodynamik chemischer Vorgänge," in which Helmholtz concerned himself with the distinction between chemical heat and voltaic heat in a battery.[3] It was precisely this problem that Ostwald regarded as the origin of modern physical chemistry, a scientific discipline to be organized by the principles of "energetics."[4] From energetics could be deduced the fundamental propositions of chemistry, not just laws of dynamics but even the laws of constant proportions, multiple proportions, and combining weights.[5]

As Christa Jungnickel and Russell McCormmach noted in their two-volume study of theoretical physics in Germany and Austria, many theoretical physicists in the 1880s—Helmholtz, Clausius, Gustaf Kirchhoff, Boltzmann, and Planck among them—worked on problems in physical chemistry; and the teaching of theoretical chemistry and physical chemistry was sometimes combined.[6] In the 1890s, there were still only a dozen or so university professorships defined as "physical chemistry." These included chairs held by Ostwald at Leipzig, Nernst at Göttingen (1905, Berlin), Wilder Bancroft at Cornell University, A. A. Noyes at the Massachusetts Institute of Technology, and Philippe A. Guye at Geneva. Arrhenius became director of the new Nobel Institute for Physical Chemistry in Stockholm; Donnan held the first professorship of physical chemistry in England, at Liverpool in 1904.

In France, the first university to establish a chair of physical chemistry was Nancy, in northeastern France. The Alsatian Paul-Thiébaud Mueller began lecturing on physical chemistry in Nancy's new Chemical Institute in 1890; a chair was established in 1899. The founding of a chair for physical chemistry at Nancy illustrates the competitive disciplinary politics of organic chemistry and physical chemistry. This first chair in physical chemistry in France was authorized as a replacement for the chair of organic chemistry, left vacant

3. See Donald G. Miller, "Duhem," in Charles C. Gillispie, ed., *Dictionary of Scientific Biography* IV: 225–233, on 228; and Emile Picard, *La vie et l'oeuvre de Pierre Duhem* (Paris: Gauthier-Villars, 1922): 3.

4. Wilhelm Ostwald, *Elektrochemie* (1896). See the discussion in Barkan, "Walther Nernst," 44–45. Ostwald's first chemical researches concerned chemical affinities; from these studies he went on to investigate electrolytic dissociation, electrical conductivity, mass action, reaction velocities, and catalysis. It was for work on catalysis that he was awarded the Nobel Prize in chemistry in 1909.

5. In Ostwald's Faraday Lecture of 1904, he demonstrated that these early-nineteenth-century laws could be deduced from the definition that "a substance or a chemical individual is a body which can form hylotropic phases within a finite range of temperature and pressure." In "Elements and Compounds," 185–201, in C. S. Gibson and A. J. Greenaway, eds., *Faraday Lectures 1869–1928* (London: The Chemical Society, 1928).

6. Jungnickel and McCormmach, *Intellectual Mastery of Nature*, II, 109.

when Albin Haller departed Nancy for Paris.[7] A hidden agenda for physical chemistry, Barkan has suggested, was the "revival of inorganic chemistry at the expense of the flourishing organic chemical establishment."[8]

The first formal course in physical chemistry at the University of Paris began in 1893 at the request of Salet, who already was teaching spectroscopy, photochemistry, and organic chemistry. When Salet died in 1894, opinion divided about the future of the course, namely, whether it should be taught by a physicist (physicist Edmond Bouty's view) or a chemist (chemist Charles Friedel's view).[9] In 1898, the position went to Jean Perrin, who had been trained as a physicist. Perrin's successes in the next decade made possible the establishment of a chair in 1910.[10]

From 1898 until the German invasion of 1940 (when he joined the Free French in New York City), Perrin, his close colleagues, and his students dominated the fields both of physics and of physical chemistry in France. Their base was the Ecole Normale Supérieure, the University of Paris, and the buildings of the Radium Institute, the Physical Chemistry Laboratory, and the Institute of Physicochemical Biology a block away from the Ecole Normale. In the early 1920s, the principal teachers of physics for University of Paris students preparing the *diplôme d'études supérieures* were Perrin, Gabriel Lippmann, Bouty, Marie Curie, and Henri Abraham; the principal teachers of chemistry were Haller, LeChatelier, Georges Bertrand, Urbain, and Lespieau.[11] Lespieau, who was director of the chemistry laboratory at the Ecole Normale Supérieure from 1904 to 1934, became the only chair holder in France in theoretical chemistry in 1921, as *professeur de théories chimiques.*

The aim of this chapter is to analyze aspects of late-nineteenth-century chemistry that led Lespieau to organize his own school of research in theoretical chemistry, to delineate the members and characteristics of this disciplinary school, and to assess its achievements over a period of some forty years. Particularly given the so-called positivist bias in French chemistry against the introduction of physical atomism and physical mechanisms into late-nineteenth-century chemical theory, the history of Lespieau's avowedly "theoretical" school of chemistry helps delineate styles and practices among specific, nationally distinct schools within the wider field of theoretical chemistry.

7. Discussed in Nye, *Science in the Provinces*, 43–44.

8. Barkan, "Walther Nernst," (1990), pp. 64–65.

9. Archives de l'Université de Paris, Faculté des Sciences, Procès-Verbaux du Conseil, 17 April 1894, 191–192.

10. See Nye, *Molecular Reality*.

11. A new *certificat de chimie physique et radioactivité* was available in 1920, with required courses by Perrin, Curie, and Urbain. Other areas of specialization for the certificates in physical sciences were mathematical physics, general physics, general electrotechnology, general chemistry, applied chemistry, biological chemistry, and mineralogy. From *Université de Paris: Livret de l'étudiant, 1920–1921* (Paris: Berger-Levrault, 1920).

French Chemistry and the Education of Robert Lespieau ca. 1890

Although Robert Lespieau became best known in France for a laboratory school that focused on reaction mechanisms in organic chemistry, he first gained international attention in another area. Van't Hoff spoke at the third Congress of Dutch Scientists and Doctors, in Utrecht in 1891, on recent accomplishments in physical chemistry, putting the young Lespieau in very good company.

> [In physical chemistry] we were, one must confess, quite in a hurry, while our more aged sister [physics] had already this stage behind it. And so it happened that, first gradually, after we had already enjoyed some results, mathematical-physical treatments followed: Max Planck, Duhem, Van der Waals, Lespieaux [*sic*], Lorentz, Boltzmann; but we can rest assured: osmotic pressure is well seen by all of them.[12]

Van't Hoff here referred to a lecture in which the twenty-six-year-old Lespieau had introduced the new physical chemistry of van't Hoff, Ostwald, and Arrhenius to colleagues in Friedel's Sorbonne laboratory. Among other things, Lespieau dealt in his paper with a problem posed by van't Hoff having to do with the validity of neglecting the difference in specific heats of water and ice, and Lespieau demonstrated that the term where this difference figures can be finally eliminated.[13] Who was Robert Lespieau?

Lespieau (1864–1947) was born into a family with strong military ties as well as an academic tradition. His mother, Clémence Lespieau-Theil, a forceful woman who did not hesitate to write the minister of education about appointments for her son, was the daughter of a Latin scholar. Lespieau's father was a veteran of the Crimean War and the siege of Paris; he spent time in Algeria; and in the 1880s, General Lespieau was head of the 27th Infantry Division of the 14th Army Corps, stationed at Grenoble. Lespieau's brother died in the Sudan in 1893; his sister was the wife of a general who became governor of Damascus. The family had considerable political connections, including friendship with Armand Faillières, at one time president of the French Republic.[14]

Lespieau brooked family tradition in 1886 by entering the Ecole Normale

12. Quoted from Ernst Cohen, *J. H. van't Hoff, Sein Leben und Wirken* (Leipzig: Akademische Verlagsgesellschaft, 1912), 298; in Barkan, "Walther Nernst," 60.

13. Robert Lespieau, "Sur la pression osmotique," Conférence au Laboratoire Friedel (Paris: Carré, 1892); discussed in Robert Lespieau, *Notice sur les travaux scientifiques* (Paris: Gauthier-Villars, 1910).

14. See Archives Nationales dossier F17 24392 (Robert Lespieau); C. Raveau, "Robert Lespieau," *Ann.AEENS* (1949): 25–26. Courtesy of ENS Bibliothèque des Lettres.

Supérieure rather than the Ecole Polytechnique.[15] His interests lay in the physical sciences, especially chemistry, but he found uninspiring the chemistry taught by the protégés of Henri Sainte-Claire Deville. (Deville, who had been in charge of the laboratory from 1851 to 1881, was succeeded, in turn, by his pupils Henri Jules Debray [1881–1888], Alphonse Alexandre Joly [1888–1897], and Désiré J. B. Gernez [1897–1904]).

This was the period during which Deville, Berthelot, Moissan, and other leading French chemists had persisted in the use of an outmoded chemical notation abandoned elsewhere.[16] By 1870 or so, the equivalent notation had disappeared in chemical journals outside France. French atomists sometimes used the tactics of the Sorbonne organic chemist Friedel, who wrote acetylene dichloride as $C_2H_2Cl_2$ for the *Berichte* of the Berlin Chemical Society but $C_4H_2Cl_2$ for the *Comptes rendus* of the Paris Academy of Sciences.[17]

This situation continued much too long. In 1900, Moissan published a fifteen-page memoir on ethyl fluoride in which no real formula even appears. In 1890, Berthelot was still writing water as HO but finally published a note in 1891 using the atomic notation [H_20]. However, he continued to write benzene as C_2H_2 (C_2H_2) (C_2H_2) as late as 1898; and there is not the least use of structural formulas in his discussions of isomerism in a 1901 text on hydrocarbons.[18]

In contrast, the chemists Lespieau admired were Wurtz, whose *La théorie atomique* (1879) "converted" him to atomistic ideas, and Grimaux, whose *Théories et notations* (1883) left Lespieau with the impression "that a sensible man could not help but adopt atoms."[19] Grimaux merited admiration for his modern approach to chemistry; after 1898, he also merited considerable sympathy when he was fired from the Ecole Polytechnique after signing the petition supporting Alfred Dreyfus.[20]

During the 1880s, while his father was stationed at Grenoble, Lespieau became personally acquainted with François Raoult, the Grenoble professor of chemistry whose experimental work on the measurement of freezing point depressions, vapor pressure depression, and boiling point elevation demonstrated the existence of submolecular particles or charged "radicals" (ions) in salt solutions. Raoult provided a novel, easy method of calculating molecular

15. Georges Dupont, "Robert Lespieau, 1864–1947," *Bull.SCF* 5, 16 (1949): 1–9, on 1.

16. See Nye, *Molecular Reality*; also Pierre Colmant, "Querelle à l'Institut entre équivalentistes et atomistes," *Revue des Questions Scientifiques* 143 (1972): 493–519.

17. See Albert Kirrmann, "La naissance des formules moléculaires en chimie organique," *L'Actualité Chimique*, no. 1 (January 1974): 45–49.

18. Ibid., citing Moissan, *Ann.Chim.Phys.* 19 (1890): 266; M. Berthelot, *Bull.SCF* 4 (1890): 809; M. Berthelot, *CR* 112 (1891): 829; M. Berthelot and E. Jungfleisch, *Traité élémentaire de chimie organique*, 4th ed. (Paris, 1898); M. Berthelot, *Les carbures d'hydrogène* III (Paris, 1901): 120.

19. Robert Lespieau, "Sur les notations chimique," *Revue du Mois* 16 (1913): 257–278.

20. See J. R. Partington, *A History of Chemistry* (New York: Macmillan, 1964), IV: 800, on Grimaux. And Robert Lespieau, "Poids moléculaires et formules dévelopées," 8-page extract from *Journal de Physique* (June 1901): 3–4.

weights of a soluble substance by measuring the depression in freezing point of the solvent caused by the soluble foreign substance or impurity. This work, Lespieau later remarked, was attacked by the chemists who were his teachers at the Ecole Normale, one of them, an adherent to the vapor density method of calculating atomic weights, stubbornly saying, "Bodies which are not volatile do not have molecular weight."[21]

Raoult's work received a rather different reaction outside France, and when young Lespieau met Raoult, while on his summer vacations in Grenoble in the late 1880s, Raoult had just become a collaborator with van't Hoff, Ostwald, and Arrhenius on the new *Zeitschrift für physikalische Chemie*. Raoult's name was one of only three French scientists listed among twenty-one on the title page of the first issue. Thus, Raoult, who had no advanced pupils in the Sciences Faculty at Grenoble, counted Lespieau as his student. Lespieau, in turn, introduced the methods of "cryoscopy" and "ebullioscopy" into Parisian laboratories. At Raoult's request, he also wrote an article on Raoult's apparatus for the Paris Chemical Society.[22] And through Raoult, Lespieau became thoroughly familiar with the latest developments in physical chemistry in the 1880s and 1890s.

In 1889, Lespieau successfully passed the *agrégation* exams in physical sciences and spent two years working in organic chemistry with Friedel, who succeeded Wurtz at the Sorbonne.[23] To his interest in the new physical chemistry, Lespieau added a passion for the stereochemistry of van't Hoff and Le Bel; and in 1892, he presented a lecture at Friedel's laboratory on the kinetics of tautomerism: "A body thought to be unique is in reality a mixture resulting from an equilibrium between two bodies which can be transformed back and forth in a reversible fashion." By representing these formulas in space, using the hypothesis of the tetrahedral carbon structure, Lespieau demonstrated how we can see that two forms should exist. This kind of thinking led him into problems for doctoral research, which would include the preparation of stereoisomers.[24]

From 1892 until 1904, when he became a lecturer at the Sorbonne, Lespieau taught chemistry at the Collège Chaptal and at the Ecole Primaire Supérieure Lavoisier. When he completed his dissertation in late 1896, his defense at-

21. Lespieau, *Notice* (1910), 10.

22. The other French scientists listed as members of the editorial board were Duhem and Berthelot. Robert Lespieau, *Notice sur les travaux scientifiques* (Paris: Gauthier-Villars, 1925): 12.

23. Friedel became internationally known for the synthetic method called the Friedel-Crafts Reaction using aluminum chloride as a catalyst in the the introduction of an alkyl or acyl group into benzene. James Mason Crafts was an American professor from MIT working with Friedel in 1877 at the Sorbonne. Crafts later became president of MIT.

24. Lespieau, *Notice* (1910), 10–12. The thesis concentrated on compounds of the propylene and propinol series; he prepared two dibromated propylenes and found stereochemical isomerism in one of them: $CH_2=CBr\text{-}CH_2Br$.

tracted a considerable audience because of his applications of Raoult's methods and use of the theory of stereoisomerism. The thesis was regarded as representative of "current chemistry." [25] In 1900, he was appointed assistant director of the Ecole Normale Chemistry Laboratory, with Raoult among his supporters for the position, and in 1904, he became the laboratory's director. He also received the 1904 Prix Jecker, one of the prestigious prizes of the Academy of Sciences.

The Practice of Organic Chemistry at the Ecole Normale, 1904–1934

There is extraordinary irony in the tradition of organic chemistry in France at the end of the nineteenth century. While Berthelot is known, among other things, for his contributions to organic synthesis, his control over the teaching of chemistry in Paris worked against the productive application of atomistic and structural theories that were widely used elsewhere. In the early 1900s, the best-known work in French organic chemistry was accomplished outside Paris, by chemists who had been hostile or indifferent to Berthelot's tutelage, most notably, by the 1912 Nobel Prize recipients Paul Sabatier (Toulouse) and Victor Grignard (Lyon and Nancy).[26] However, thermochemistry, then thermodynamics, flourished among French chemists, as did descriptive and analytic studies of mineral or inorganic chemistry.

The chemical laboratory at the Ecole Normale offered little excitement in the way of theory in the last decades of the nineteenth century. Under Deville, three fundamental themes in inorganic chemistry had been pursued: the preparation of aluminum; the exploration of metals in the platinum series; and the discovery of equilibrium in gaseous milieus, including the study of the dissociation of elementary vapors at high temperatures.[27] Neither Deville nor his pupils were interested in general theories. As noted in chapter 3, Lespieau later commented,

> My professors gave little thought to slowing down the lecture course a little in order to examine certain questions which should have been necessary. Four and a half hours was all the time devoted to generalities in the *classes spéciales*; if one had doubled this time it would not have been detrimental to the seventeenth property of chlorous anhydride that one might have had to omit.[28]

25. Report on Lespieau's thesis, dated 24 October 1896. AN F17/24392.

26. Grignard was recognized for the method of using alkyl magnesium halides in organic synthesis; and Sabatier, for the method of catalysis of hydrogenation and other reactions by finely divided nickel. See Nye, *Science in the Provinces*, for studies of the work and careers of Sabatier and Grignard.

27. Raymond Dulou and Albert Kirrmann, "Le Laboratoire de Chimie de l'Ecole Normale Supérieure, Notes historiques," *Bull.SAENS* 54 (September 1973), Lab.ENS.

28. Lespieau, "Sur les notations chimique," 259.

While Duhem and LeChatelier helped pioneer the application of thermodynamics to chemical processes, they concentrated on the study of chemical equilibria, not chemical mechanisms and kinetics.[29] In organic chemistry there was a need for the study of structure and mechanisms, and these kinds of studies were well under way in England, as we will see later.

One of Lespieau's first systematic ventures in designing a new teaching and research program for the Ecole Normale was the rewriting of a chemistry textbook, originally published by Joly, who had adhered to equivalent notation. The *Nouveau précis de chimie (notation atomique)* was revised in 1902 and appeared in three volumes. The third volume introduces the reader to fundamental principles of chemistry, following descriptions in the two earlier volumes of states of matter; experiments on air, water, and so on; descriptions of the properties of nonmetals and metals; reactions in organic chemistry; and industrial chemical processes.

Volume 3 explains the systems of molecular and atomic weight, valences, the atomic theory, the system of classification of the elements, and the laws of chemical equilibrium. Here we find Lespieau's view that the *goal* of chemistry is the *formule dévelopée*, not the *formule brute*, and that the atomic hypothesis gives us a striking interpretation and creates a *language* that is now adopted by all chemists, even those who reject the hypothesis of an indivisible primordial particle.[30]

Lespieau's view was that organic chemistry must make use of physical methods and of physical chemistry and that the experimental work of Graham, Williamson, Wurtz, and, more recently, Raoult had confirmed the superiority of the atomic hypothesis over chemical equivalents. As for the meaning of the constitutional (structural, developed) formula,

> We should use physical facts and the notion of valence in teaching as well as in research. It will remain in the mind of the student that the method of chemical science is not at all that of the mathematical sciences, that a formula is not demonstrated like a theorem, that one can simply find in it advantages and that

29. Jules Guéron and Michel Magat, "A History," 1–25.

30. A. Joly and R. Lespieau, *Nouveau précis de chimie (notation atomique)*, Redigé conformément aux programmes officiels du 31 mai 1902, 4th ed., 3 vols. (Paris: Hachette, 1904–05), quotations, III: 419, 437. The "brute" formula is of the type C_2H_6O; the developed formula

```
    H   H
    |   |
H - C - C - O - H
    |   |
    H   H
```

Renée Mestrallet Guerre also designates the "semi-developed formula" C_2H_5OH; in "Communication, Linguistique et Semiologie."

one day it will perhaps be abandoned for another formula more in harmony with the facts.[31]

This does not mean that Lespieau thought chemical formulas to be *only* conventions or heuristic devices that have no relation to ''reality.'' Two decades later, in a book entitled *La molécule chimique* (1920), he counseled that we are justified in seeing in the formulas currently in use ''an image of reality,'' although it is only a very crude image.'' For, he asked, if they had nothing to do with reality, wouldn't it be strange that such images result in so many successful predictions?[32]

Under Lespieau's directorship, the Ecole Normale laboratory was open to students in physics and mineralogy as well as chemistry.[33] The primary research program was the synthesis and study of nonsaturated organic compounds, along with the application to hydrocarbons of physical methods like Raoult's techniques. With his former student, Maurice Bourguel, Lespieau was one of the first people in France to apply Raman spectroscopy to organic analysis.[34]

A leading research idea in the laboratory was the hypothesis that atoms within a molecule are in constant motion, thus explaining the mechanisms of tautomerism. For example, the transformation of ammonium isocyanate to urea can be explained by motion of a hydrogen atom:[35]

$$O = C = N - NH_4 \quad \rightarrow \quad O = C \begin{matrix} \diagup NH_2 \\ \diagdown NH_2 \end{matrix}$$

Or, the conversion of the ketonic form of ethyl acetic ester to the enolic form can be explained:

$$CH_3COCH_2CO_2C_2H_2 \rightarrow CH_3C = \underset{\displaystyle\overset{|}{OH}}{C}HCO_2C_2H_5$$

Another leading research theme was the migration of the double bond, for example, in the saponification of an allyl nitrile to crotonic acid.[36]

31. Robert Lespieau, ''Poids moléculaires et formules développées,'' 6–8.
32. Robert Lespieau, *La molécule chimique* (Paris: Alcan, 1920): 250.
33. Lespieau took pride in the fact that not all his students became chemists. ''Il ne me gardent pas rancune de leur avoir dit que la chimie peut marcher seule, sans avoir besoin d'etre soutenue par sa soeur, la physique.'' He also commented that some of his students became naturalists and mineralogists, a field in which he also was interested. In Robert Lespieau, ''Remise de l'epée d'académicien à Robert Lespieau. Ecole Normale Supérieure 26–1-1935, Remerciements du recipiendaire.'' Two-page typescript. Courtesy of ENS Bibliothèque des Lettres.
34. Georges Dupont, ''Robert Lespieau,'' *Bull.SCF* (1949): 3.
35. Lespieau, *La molécule chimique*, 240.
36. Lespieau, *Notice* (1910), 14.

Most of Lespieau's students worked on research topics having to do with isomerism, unsaturated hydrocarbons, or catalysis, often focusing on reactions associated with the double and triple bond.[37] Of the six students who completed their doctoral research before the First World War, four survived the war: Charles Mauguin, Henri Pariselle, Georges Dupont, and Gustave Vavon. Pariselle later recalled with nostalgia their laboratory work together and their daily gathering for coffee and conversation at the café La Nature, with other students of "Normale Sup," around the influential Socialist thinker *citoyen* Lucien Herr.[38]

Theoretical Chemistry at the Ecole Normale, 1922–1934

After his promotion from lecturer to associate professor in 1912, Lespieau's laboratory work was halted during the war. He served at the Ecole de Pyrotechnie at Bourges. His students were sent to the front. Following the war, Lespieau was appointed professor of chemistry at the Ecole Centrale des Arts et Manufactures; and he was given the title at the Sorbonne of "professeur sans chaire." A decree of 1920 first permitted this new "chairless" category in higher education; it was a position that might be requested personally for a senior staff member who had been teaching courses for which there was no established chair or for whom an appropriate senior position could not be found otherwise. In 1922, Lespieau's title was changed to "professeur de théories chimiques."[39]

Lespieau's may have been the first professorship of "theoretical chemistry" anywhere. There was not a chair of theoretical chemistry in England, for example, until 1931.[40] That Lespieau wanted to have this title demonstrates his self-consciousness about the need for chemical "theories," not just chemical empiricism, to renew French chemistry. He had the sense that he was putting organic chemistry at the Sorbonne and the Ecole Normale on a new and different track.

Other professors of chemistry at the University of Paris and Ecole Normale

37. Charles Mauguin, *Amides bromosodés* (1910); Henri Pariselle, *Etude d'une glycérine en C4* (1911); Henri Duffour, *Derivées complexes de l'iridium* (1912); M. Viguier, *Aldehyde tétrolique* (1912); G. Dupont, *Gamma-Glycols acétyléniques* (1912); Gustave Vavon, *Reductions catalytique en présence noir de platine* (1913); Maurice Bourguel, *Carbures acétyléniques vrais* (1925); M. Faillebin, *Hydrogenation d'aldehydes et kétones en présence de noir de platine* (1925). Lespieau, *Notice* (1925); and Dulou and Kirrmann, "Le Laboratoire du Chimie," 12.

38. Henri Pariselle, "Georges Dupont," 29–30, in *Ann.AEENS* (1958). Courtesy of ENS Bibliothèque des Lettres. Also see Mary Jo Nye, "Science and Socialism," *French Historical Studies* 9 (1975): 141–169, on 142–145.

39. AN F17/24392 (Robert Lespieau) and AN AJ16/5738 (Georges Dupont).

40. This was the chair of theoretical chemistry at Cambridge University held by John Lennard-Jones, who pursued a research agenda very different from Lespieau. Christopher Longuet-Higgins became professor of theoretical chemistry at Cambridge, succeeding Lennard-Jones, in 1954.

in the 1920s were Haller in organic chemistry; Urbain in general chemistry, joined by Job who succeeded LeChatelier in inorganic chemistry in 1925; and Perrin in physical chemistry. Among these four and Lespieau there was a common purpose in overturning the old emphasis of the Berthelot generation on descriptive, empirical chemistry in a systematic tradition that avoided speculations about molecular structure and internal molecular mechanisms.

In preparing Ecole Normale students for the agrégation examination, Lespieau and his colleagues supervised preparation of a diploma, called the diplôme d'études supérieures, which was awarded following a research project by the student. In addition to the research project, the student also was required to discuss a topic proposed by the three-person faculty jury. During the 1920s, the questions for the student diplôme in Lespieau's laboratory focused on topics that partly or largely applied physical ideas or methods to chemical problems. These included catalysis, the measurement of magnetic fields, gas and vapor densities, stereochemistry of salt complexes, colloids, chemical equilibrium, diamagnetism, and, in one case, the origins of atomic notation.[41] Since the physicists Hélois Ollivier, Pierre Villard, and Georges Darmois regularly frequented the laboratory, students were expected to be familiar with ways in which physics works in concert with chemistry.[42]

Most of the research projects required synthesis of compounds and the duplication, or testing, of claims made by other researchers. Some projects required the students to verify the existence of isomers (Marius Faillebin, 1919; Rodolphe Garreau, 1921); many acquainted students with unsaturated systems (e.g., Daniel Chalonge, 1921; Albert Kirrmann, 1922; Charles Prévost, 1922; J. Vyon, 1925; Pierre Morelle, 1926); and some encouraged students to think about reaction mechanisms (Marcel Bouis, 1923; Jean Daujat, 1929).[43] Beginning about 1926, some students began acknowledging the help in their research of the two doctoral researchers, Kirrmann and Prévost.[44]

Pierre Piganiol, later a close friend of Prévost, recalled a description by another student in the chemistry laboratory in the early 1920s:

> [There was] a climate of familial sympathy and gaiety with several students preparing a *licence* certificate or a diploma of higher studies or a thesis with Lespieau—[he was] debonair, welcoming, amusing, hands in the pockets, muttering about with the two *préparateurs* who were always present, loved by all and altogether different: Bourguel, who died so prematurely, maintaining order,

41. These propositions are to be found in reprint copies of "mémoires" presented for the diplôme d'études supérieures. Twenty of these reprints were bound together and presented to Albert Kirrmann. Bound copy, Lab.ENS.

42. Dulou and Kirrmann, "Le Laboratoire de Chimie," 8–11.

43. Ibid.

44. Ibid. For example, Pierre Morelle, 1926; Jacques Ballet, 1927; Jean Grard, 1927; Jean Daujat, 1927; and Henri Journaud, 1929. Daujat and J. Wiemann, in 1929, also acknowledge help from Deluchat, Urion, Levaillant, and Grard. Joining Lespieau on juries were Haller, Péchard, Leduc, Urbain, Mme. Ramart-Lucas, and E. Blaise.

directing the work of the whole group; Prévost, whimsical, hardly conformist, full of humor, on reflection deep and penetrating.[45]

In the mid-1920s, then, the dominant figures in the laboratory were its director Lespieau, his chief assistant Bourguel, and the two advanced students Prévost and Kirrmann. The presence and influence of Dupont must also be noted. Throughout the 1920s, Dupont regularly spent time in Paris, a familiar figure in the laboratory of the Ecole Normale, where he had completed his doctorate on acetylenic glycols in 1913 and where he was to succeed Lespieau in 1934.

A victim of phosgene gas during the war, Dupont had been returned from the front and assigned to work in the factories of Tréfileries du Havre, where he created a laboratory of research and analysis for the production of steel and metal alloys. He was unusual among French academic scientists in his interest in industrial processes, and later, as an administrator in the CNRS, he pressed for better links between the universities and industry in France.[46]

In 1921, Dupont became a Sciences Faculty member at Bordeaux, and in 1925, he succeeded Maurice Vèzes in the chair of "mineral chemistry and physical chemistry." Bourguel at this time was named to Dupont's now-vacant lectureship at Bordeaux.[47] In addition to ongoing research on terpenes and resinous acids, pursued at the Institut du Pin at Bordeaux, Dupont worked on methods of column fractionation of hydrocarbons and on Raman spectroscopy, which became a major focus for research projects at the Ecole Normale laboratory. Bourguel was one of the first people in France to lecture on its usefulness for determining chemical identity and for demonstrating the individuality of particular functional groups.[48] The continued presence of both Dupont and Bourguel in the Ecole Normale laboratory during the 1920s demonstrates how some provincial scientists never in fact relinquished close ties to Paris.

As students who first entered the Ecole Normale immediately after the war, in the same class of 1919, Kirrmann and Prévost became fast friends. They learned much from Lespieau, Dupont, and Bourguel, they developed similar theoretical views and practical methods, and they defended their doctoral dissertations on the same day in June 1928. It is to their work that we now turn.

45. Pierre Piganiol, "Charles Prévost," *Ann.AEENS*(1985). Courtesy of ENS Bibliothèque des Lettres.

46. AN F17/25631 (Georges Dupont); and Micheline Charpentier-Morize, "La contribution des 'laboratoires propres' du CNRS à la recherche chimique en France de 1939 à 1973," ms., 12; published as No. 4, *Cahiers pour l'histoire du CNRS*, (Paris: Editions du CNRS, 1989).

47. Letter from Vigouroux and Richard to the Minister of Education, 9 January 1925. AN F17/25631.

48. Maurice Bourguel, "Applications de l'effet Raman à la chimie organique," 469–505.

The Theory of "Synionie" and French Theoretical Chemistry

Kirrmann entered the Ecole Normale in 1919 as part of a special group of five Alsatian students permitted to enter *hors concours*. Another student in the group was Alfred Kastler, whose later work was instrumental in the development of the laser and who received the Nobel Prize in physics in 1966 for his optical methods of studying Hertzian resonances in atoms.[49] Kastler and Kirrmann each had been conscripted into the German army during the First World War. They entered the Ecole Normale together, and they both taught at Bordeaux in the 1930s. During the Second World War, Kastler returned to work at the Ecole Normale. Kirrmann, who had been teaching at Strasbourg and who was Protestant, was deported to Buchenwald, where he was held from 1943 to 1945.[50]

As a young man, Kirrmann was somewhat unusual among French scientists in the 1920s in electing to spend a year in Germany, with Hans Fischer in Munich. He followed this with a second year of postdoctoral work at the Institut Rothschild (for physicochemical biology), which was co-directed by Perrin, Urbain, and André Mayer. In 1930, making use of the Dupont connection, Kirrmann was named to a Bordeaux lectureship and then to the Bordeaux professorship in mineral chemistry and physical chemistry, which Dupont relinquished to go to Paris. However, Kirrmann himself quickly moved to Strasbourg where he became professor of organic chemistry, a position he formally held until he moved to Paris, where he again succeeded Dupont, this time as professor, laboratory director, and associate director at the Ecole Normale.[51]

Charles Prévost (1899–1983), the son of a Parisian engineer, recalled the origins of his career in chemistry. As a lycée student, he had taken the baccalaureate examinations in both mathematics and philosophy. There followed the "special mathematics" class at the Lycée Louis le Grand which prepared students for the entry exams for the Ecole Normale and Ecole Polytechnique. "Like most of the students leaving the classes de spéciales, I thought myself to have a vocation as a mathematician."[52] But the lectures of the physicist

49. Alfred Kastler and Yves Noel, in *Hommage à Albert Kirrmann*, 18-page reprint, courtesy of ENS Bibiothèque des Lettres. Walter Sullivan, "Dr. Alfred Kastler, 81, Nobel Prize Winner, Dies," *New York Times*, January 6, 1984. Sullivan says there were five students admitted *hor concours*; Kastler says there were eight.

50. See Charles Prévost, "Notice nécrologique. Albert Kirrmann," *Bull.SCF* (1957), 1èr partie, 1451–1454. Charpentier-Morize has characterized Kirrmann as a Protestant of moral rigor and Prévost as a traditionalist Catholic. In private correspondence, 16 January 1991.

51. Ibid. Dupont was director of the Ecole Normale, but since professors of letters and sciences alternate in assuming the directorship, Kirrmann became vice-director, not director (1452).

52. Charles Prévost, *Notice sur les titres et travaux scientifiques* (Paris: Société d'Edition d'Enseignement Suprieure, 1967): 7.

Abraham and some practical experience in the physics laboratory first made Prévost take pause. Introduction to the chemistry laboratory completely reoriented Prévost, and when Robert Lespieau examined him for the certificate in chemistry, Prévost was offered a place in the laboratory for doing research for a diploma.

After completing his doctoral dissertation, Prévost, like Kirrmann, was given a post through the influence of Lespieau's research school. Vavon, one of Lespieau's former students who now taught organic chemistry at Nancy, recommended Prévost to the professorship of chemistry at the Nancy Pharmacy Faculty in 1929. In 1936, Prévost moved to Lille's chair of general and organic chemistry. The next year he returned to Paris as professeur sans chaire at the Sorbonne, where he finally received a chair of organic chemistry in 1953. During the Second World War, Prévost found himself under scrutiny after predicting in his lecture course that the Germans would lose the war because of the low octane content in their synthetic gasoline. He joined the Resistance forces near Champlits.[53]

As young men preparing their doctoral dissertations in the mid-1920s, Kirrmann and Prévost were influenced by physicists, especially Abraham and Eugène Bloch, as well as by chemists like Lespieau, Urbain, and Job.[54] Perrin presided over the jury for Kirrmann's thesis defense.[55] In characterizing his developing interests and aims, Kirrmann later said of his work,

> The dominant tendency of my studies has been not so much to obtain and describe organic compounds but . . . to penetrate their mechanisms. . . . For undertaking this kind of problem, the classic methods of organic chemistry are far from sufficient. Physicochemical procedures become more and more necessary. I have been led to use especially optical methods (the Raman effect and ultraviolet spectra) and electrochemical techniques (conductibility, electrode potentials, and especially polarography). . . . The notion of reaction mechanism led almost automatically to envisioning the electronic aspect of chemical phenomena. From 1927, and working in common with Charles Prévost, I have directed my attention on the electronic theory of reactions.''[56]

By the time Kirrmann and Prévost were defending their doctoral dissertations, they had reached the opinion that organic chemistry was in the process of evolving through three stages, one completed and one still in the future, that after a period in which organic synthesis had been the essential problem

53. See Pierre Piaganiol, ''Charles Prévost,'' 47.

54. See Kirrman's remarks, in *Hommage à Albert Kirrman.* And especially on Job, Albert Kirrmann, *Notice sur les titres et travaux scientifiques*(on 8), typescript of 20 pages, ENS, courtesy of ENS Bibliothèque des Lettres. Eugène Bloch taught physics at the Ecole Normale from 1913 on, and he became professor of theoretical physics and celestial mechanics at the Sorbonne in 1930. He succeeded Henri Abraham, one of Lespieau's old friends.

55. Prévost, ''Notice nécrologique,'' 1452.

56. Kirrmann, *Notice*, 8–9.

for the discipline, the time still had not arrived for a quantitative and mathematical stage of chemical explanation. "Our own epoch is that of qualitative explanation in which considerations of analogy are particularly fertile. The most perfect instrument is the study of chemical 'mobilities' " (Job's special interest). Their epoch, they thought, was a historical period in need of a general theory to coordinate the principles worked out in the last several years.[57]

How had they arrived at such a conclusion? In the mid-1920s, Kirrmann and Prévost both began working on problems posed by Lespieau on reactions of unsaturated hydrocarbons, in particular allylic and conjugated systems. Kirrmann's research topic became focused on bromated aldehydes and the "abnormal" reactions in which, for example, these aldehydes behave as if they were acid bromides, that is,

$$RCH_2-\underset{\displaystyle Br}{\underset{|}{C}}=O \text{ rather than } RCHBr-CH=O.$$[58]

Prévost took for his research project the preparation of derivatives of the tetratomic alcohol erythritol $C_4H_6(OH)_4$ and the study of the mechanism of bromine fixation. He, too, was struck by what appeared to be anomalies for his expectations. Thus, he obtained a 1,4 dibromide ($CH_2Br\text{-}CH=CH\text{-}CH_2Br$) instead of the 1,2 dibromide of erythrene (butadiene) expected on the basis of classical substitution theory. Similarly, instead of obtaining the expected 1,2 dibromide form of divinyl glycol, he obtained a 1,4 form.[59] Bouis, another research student working in the laboratory, found similar results for "anomalous" positioning of halogens.[60]

Students in Lespieau's laboratory were well prepared in the mid-1920s to think about intramolecular mobility and migrations not only of atoms or groups within the molecule but of electrons that were part of the molecular architecture. These ideas were among the principal topics at the Solvay Institute chemistry conferences in 1922 and 1925, and Job brought reports of the conferences immediately to students at the Ecole Normale and the Collège de France in lectures given during 1922–1925.

Job emphasized the reaction dynamics of inter- and intramolecular mobilities, the presence of unstable intermediary compounds (sometimes visible by a brief change in coloration), and the electronic or ionic character of transitory complexes. We are led to suppose, with the English chemist Lowry, he said,

57. Charles Prévost and Albert Kirrmann, "Essai d'une théorie ionique des réactions organiques" (2ème mémoire), *Bull.SCF* 4, 49 (1931): 1309–1368.

58. Albert Kirrmann, "Recherches sur les aldehydes a-bromées et quelques-uns de leurs dérivés" (dated 23-4-1928), *Ann.Chim.Phys.* 11 (1929): 223–286, e.g., 238.

59. Charles Prévost, "La transposition allylique et les dérivés d'addition des carbures érythréniques" (dated 29-1-1928), *Annales de Chimie* 10 (1928): 113–146, 147–181, 356–439.

60. Ibid., 116.

that the double bond is a polarized bond and that an acid is similarly polarized:[61]

$$O = O \text{ is } \underset{-}{O}-\underset{+}{O} \qquad \text{acid: } R'C\begin{matrix} \overset{-}{\diagup} O \\ \diagdown O \end{matrix} \quad \overset{+}{H}$$

Lowry, the holder of the new chair in physical chemistry at Cambridge University, spoke in Paris in March 1924 on aspects of the theory of valence, including the electronic theory of valence, and in December 1925 on optical methods of verifying structural chemistry and his own hypothesis of semipolar double bonds in organic compounds. The occasions were meetings of the Société de Chimique de France and the Société de Chimie Physique.[62]

Lowry praised the 1916 memoir of the American chemist Lewis as a "turning point in the history of chemistry" with its "plausible theory" of the electronic origin of the different types of chemical affinity and a clear differentiation between two kinds of valence, ionic and covalent. It is customary in mineral chemistry, he said, to consider reactions that occur between ions to be instantaneous, without attaching any importance to ionization in organic chemistry, except for the formation of salts from organic acids.

> My opinion is otherwise. I consider that 1) certain organic reactants which are the most active are already ionized. 2) Others owe their activity to a possibility of ionization, for example the influence of a catalyst. I conclude that in organic chemistry exactly as in mineral chemistry, reactions take place almost always between ions. Simply by counting electrons belonging to the atoms, one can establish the necessity of ionization.[63]

Lowry concluded with a discussion of "dynamic isomerism," which might be of three kinds: transfer of (1) a radical; (2) a hydrogen ion or proton; or (3) an electron. The latter two categories he called "prototropy" and "electrotropy."[64] Physical proofs for the ionic hypothesis were among Lowry's topics

61. André Job, "Mécanismes chimiques. Conférence faite au Collège de France (Laboratoire de M. Professeur Charles Moureu), le 10 mars 1923," 125–164, in Job, *Formes chimiques de transition*, ed. Jean Perrin and Georges Urbain (Paris: Société d'Editions Scientifiques, 1931); and "Les réactions intermédiaires dans la catalyse," rapport présenté au deuxième Conseil Solvay de Chimie tenu à Bruxelles du 16 au 24 avril 1925," 165–193, *Solvay II*, diagrams on 172–173, 174.

62. Thomas M. Lowry, "Nouveaux aspects de la théorie de la valence," *Bull.SCF* 35 (1924): 815–837, 905–921; Lowry, "Le dispersion rotatoire optique: Hommage à la memoire de Biot (1774–1862)," *JCP* 23 (1926): 565–585; and Lowry, "Preuves expérimentales de l'existence des doubles liaisons semi-polaires," *Bull.SCF* 39 (1926): 203–206. There is an account of the December 11, 1925, lecture, which notes discussion afterward. Dufraisse, Mauguin, Tiffeneau, and Job were among those present.

63. Lowry, "Nouveaux aspects," 817.

64. Lowry, "Nouveaux aspects," esp. 816, 827–828, 834, 836, and 908–909, 917.

for his lectures in 1925. Here he cited optical activity and measurements of ''parachor,'' thought to be a surface tension effect.[65]

There is no question that, indirectly or directly, Kirrmann and Prévost were influenced by Lowry's theories for explanation of reaction mechanisms. Another important influence was Dupont, with whom they talked at length in the laboratory and who published a paper in 1927 in which he attempted to combine the electron octet theory of valence and Bohr's hydrogen electron model with classical concepts of stereochemistry. Dupont also adopted without reservation Lowry's application of ionic radicals in hydrocarbon chemistry.[66]

At the time they were writing their doctoral dissertations, Kirrmann and Prévost conceived the idea of developing a general theory of reaction mechanisms in organic chemistry, using principles and notation in Prévost's dissertation. They were concerned to take into account the classical tradition of stereochemistry, in particular, the notion of the carbon tetrahedron, which commanded pride of place in French organic chemistry. They also wanted to develop ideas compatible with Perrin's radiation theory of chemical activation. As Prévost and Kirrmann put it, their goal was to construct a conceptual edifice *plus commode et plus élégant* than the array of theories presently available in organic chemistry.[67]

Their joint effort was published in the form of three memoirs in the *Bulletin* of the Société Chimique de France in 1931 and 1933. The joint publication would have appeared earlier, before their departures from Paris to Munich and Nancy, respectively, had it not been for the fact that the manuscript was in press at the time that a fire destroyed the printing house.[68]

In their introduction to the memoirs, Kirrmann and Prévost were at pains to state that theirs was not simply a ''compilation of numerous memoirs published, especially abroad, on the question.'' They stressed that this was a ''purely theoretical work'' and that it was new in character, while building on previous hypotheses. Echoing a sentiment expressed by several chemists at the Solvay chemistry meetings, they also suggested that chemists' work now provided challenges and guidance to problems that must be tackled by physicists.[69]

Their starting point was the fundamental hypothesis that the reactions of organic chemistry are reactions of ions, that is, that at the moment of reaction,

65. Lowry, ''Preuves expérimentales.''

66. Georges Dupont, ''Sur la théorie électronique de la valence: Essai de représentation stéréochimique des éléments,'' *Bull.SCF* 4, 41 (1927): 1101–1137, quotation on 1107.

67. Prévost and Kirrmann, ''Essai,'' 1368.

68. The three memoirs are Prévost and Kirrmann, ''Essai d'une théorie ionique des réactions organiques: Premier mémoire,'' *Bull.SCF* 49 (1931): 194–243; ''Essai . . . Deuxième mémoire,'' *Bull.SCF* 49 (1931): 1309–1368; and ''La tautomérie anneau-chaine, et la notion de synionie. Troisième communication sur la théorie ionique des réactions organiques,'' *Bull.SCF* 53 (1933): 253–260; quotation from 1931, 194.

69. Prévost and Kirrmann, ''Essai,'' 196.

the sections (*tronçons*) of the hydrocarbon molecule possess localized electric charges.[70] Physical proof exists for this hypothesis: for example, that with the exception of saturated carbides, molecules with pronounced chemical reactivity have a permanent electric moment. This was a subject on which Kirrmann had just published a review paper in the *Revue générale des sciences*.[71]

They further argued that the ionized molecule was the ''active molecule'' referred to in Perrin's recent theoretical work on the energy of activation and that the energy required to cause the ionization must come from circumstances exterior to the molecule, for example, the presence of another molecule or catalyst or radiation.[72]

The theory they offered rested on a new and more general law than Lowry's notion of prototropy, which designated hydrogen or proton migration. In his dissertation, Prévost introduced this innovation on the grounds that Lowry's term was too limiting, whereas a new term, *synionie*, would not imply the sign of the mobile radical. In particular, they had in mind a principle (or law) and a theory that would cover the case of hydroxyl (OH^-) and halogen (especially Br^-) migration, which they had studied in their research on alcohols and brominated hydrocarbons.

For the molecule of form X—A—B = C, the departure of negative X creates the ions

$$\overset{-}{\mathrm{X}} \quad \text{and} \quad \overset{+}{\mathrm{A}}-\mathrm{B}=\mathrm{C};$$

but this form becomes

$$\overset{+}{\mathrm{A}}-\overset{-}{\mathrm{B}}-\overset{+}{\mathrm{C}}$$

by induction, which is also the activated form of

$$\mathrm{A}=\mathrm{B}-\mathrm{C}-\mathrm{X};$$

so we say that $\mathrm{X}-\mathrm{A}-\mathrm{B}=\mathrm{C}$ and $\mathrm{A}=\mathrm{B}-\mathrm{C}-\mathrm{X}$ are two ''synionic isomers.''[73] An example from experimental work is the existence of the synionic isomers

70. They distinguished the molecule ''in repose'' from the molecule in action. They also suggested use of circled + and - symbols to indicate ''tendency'' to charge rather than actual charge. As we will see later, the ''curly arrow'' and the $\delta+/\delta$-notation, not their symbols, were to accomplish this notation. Prévost and Kirrmann, ''Essai,'' 201, 208–210.

71. Albert Kirrmann, ''Le moment électrique des molécules,'' *Revue Générale des Sciences* 39 (1928): 598–603. He cites work including Paul Langevin's on the theory of the Kerr Phenomenon. He also calculates that the average calculated dipole length is on the order of tenths of an angstrom, compatible with the estimated distance of an angstrom between atoms, e.g., in hydrogen chloride (602).

72. Prévost and Kirrmann, ''Essai,'' 197, 200, 1324, 1339.

73. Prévost, ''La transposition allylique,'' 199–120.

$$CH_3 - CHCl - CH = CH_2 \text{ and } CH_3 - CH = CH - CH_2Cl.$$

They also defined a state of *métaionie* in which no structural formula for the so-called tautomers corresponds to the "real" structure of the reacting molecule. The Kekulé forms of benzene are examples of métaionie.[74]

The general theory of ionization and the new terminology were meant to cover and coordinate three different types of ionic reactive states, the allylic (involving carbon, hydrogen, oxygen, and sulfur), cyanhydric (involving nitrogen), and ring-chain or cyclization,[75] as well as migration of differing atoms and functional groups. In addition, Prévost and Kirrmann stressed the problematic character of the electron doublet representation, an "image, certainement fausse" because the doublet does not really have a fixed position but corresponds "to an average position in the oscillations of which the molecule is the seat." It was through spectroscopy focused on the infrared spectrum and the Raman effect that they thought the true movement would be made precise in an "electrogeometric synthesis."[76]

The "French" and "Anglo-Saxon" Schools of Theoretical Organic Chemistry

The joint memoirs of Prévost and Kirrmann self-consciously presented a general theory of organic chemistry that constituted an application of physical methods and principles to the problem of organic reaction mechanisms. However, the language system devised by Prévost and Kirrmann was not adopted by chemists in general, and that part of their notation which was new was not used outside France. Indeed, there was very little interest in their work *inside* France.

The year following the publication of the last installment of the Prévost-Kirrmann theory, there appeared in *Chemical Reviews* the paper that was to set up a system of explanation language for organic reaction mechanisms that became dominant in the next many decades. This was Ingold's "Principles of an Electronic Theory of Organic Reactions," which included a "nucleophilic"/"electrophilic" classification and notation, including the δ^+/δ^- notation for fleeting charge.[77]

Curiously, the Ecole Normale school of chemistry now repeated the history of the past generation, in which many French chemists persisted in the use of

74. Prévost and Kirrmann, "Essai," 224, 240.
75. Prévost and Kirrmann, "La tautomérie," 253.
76. Ibid., 1324–1325. Prévost and Kirrmann, "Essai," 1357; also Prévost and Kirrmann, "La tautomírie," 259.
77. Christopher K. Ingold, "Principles of an Electronic Theory of Organic Reactions," *Chemical Reviews* 15 (1934): 225–274.

a language and theory (''equivalents'') not in use elsewhere. Their ''ionic'' theory was everywhere else the ''heterolytic'' theory; their ''synionie'' was the S_N1 or E1 group of reactions. The elsewhere ubiquitous curly arrow and electron octet hardly appeared in French papers and textbooks. Many organic chemists in France followed the influential lead of Grignard who, in 1935, in the introduction to the multivolume *Traité de chimie organique*, wrote,

> As for the new electronic theories, they are not sufficiently developed for serving as the basis for speculations in organic chemistry, despite all the promises they offer to chemists. In this text, they will not be laid aside systematically, but for the moment will remain discreetly in the background; and it is still the very fruitful conception of Le Bel and van't Hoff which will constitute the surest guide for us.[78]

Prévost and Kirrmann did little to develop further the theory of synionie in the 1930s and 1940s, although Prévost returned to these questions after an international meeting at Montpellier in 1948 in which Ingold participated. Prévost remained convinced that the S_N1/S_N2 categories were too rigid and did not adequately describe the nuanced spectrum of electronic mechanisms.[79] His criticism ignored the fact that the classification was not meant to be a rigorous set of inferential principles, that the explanatory descriptions S_N1 and S_N2 ''are no more than classifications in terms of extreme types.''[80]

Ingold's *Structure and Mechanism in Organic Chemistry* appeared in 1953. It became one of the fundamental handbooks of organic chemists—a piece of classic literature for the discipline of physical organic chemistry. It included both in its text and in its network of citations a history of developments in physical organic chemistry. As in Ingold's 1934 paper, this history made no reference to Prévost and Kirrmann or to any French theories of reaction mechanisms.[81] Prévost and some of his protégés were bitter about this, and they often referred to an ''Anglo-Saxon'' ''school'' of chemistry that ignored their contributions to the theory of reaction mechanisms.[82] They nonetheless professed admiration for the work of Ingold, whom Prévost and Kirrmann cited

78. Victor Grignard, ed., *Traité de chimie organique* I (Paris: Masson, 1935): x–xi.

79. Conversation with Professor Georgoulis, Paris, 11 June 1987. Also, Prévost, *Notice*, 13.

80. See D. W. Theobald, ''Some Considerations on the Philosophy of Chemistry,'' *Chemical Society Reviews* 5 (1976): 203–213, on p. 213.

81. Christopher Ingold, *Structure and Mechanism in Organic Chemistry* (Ithaca: Cornell University Press, 1953).

82. Charles Prévost refers to the ''Anglo-Saxon school'' often, for example, in ''La valence et l'enseignement,'' *L'Information Scientifique* 6 (1951): 14–18, on 15–16: ''We must keep those teaching methods which are good . . . in guarding against our falling into the exaggerations of the Anglo-Saxon school which from the start sacrifices descriptive chemistry, the solid part, to 'up-to-date' [original in English] theoretical chemistry which is less definitively established.'' Pierre Piganiol wrote in his obituary notice for Prévost, ''After the war . . . the foreign schools [of chemistry] took the lead, and British 'fair-play' [original in English] proved more a myth than a reality in this regard.'' (Piganiol, ''Charles Prévost,'' 47.)

as early as 1931 in their joint memoirs.[83] Prévost and Ingold exchanged some correspondence, and in 1954 Prévost proposed Ingold's name to the Nobel Chemistry Committee in Stockholm.[84]

During the 1930s, it was mostly French work in the fields of infrared and Raman spectroscopy, paramagnetism and diamagnetism, and electrical measurements that brought recognition to the chemical laboratory of the Ecole Normale and, for that matter, to the laboratories of physical chemistry and physics in Paris. Kirrmann's laboratory at Strasbourg also pursued work in this field, as did his and Dupont's former colleagues at Bordeaux. In a 1938 paper on Raman spectroscopy, Kirrmann linked their recent accomplishments specifically to the earlier studies of Bourguel.[85] For physical organic chemistry in France, as for physical mineral chemistry, the principal problems lay in kinetics and in establishing, as Kirrmann put it, "the scale" for applying chemical schemas: "It has been necessary to appeal to physics to establish molecular dimensions in organic as well as in mineral chemistry."[86]

Students of French chemistry after the Second World War were to feel that they had lost step with the progress in theoretical chemistry elsewhere, especially in the understanding of organic reaction mechanisms and in the application of quantum mechanics to chemical problems. In a historical essay on the development of their field in France, the physical chemists Jules Guéron and Michel Magat wrote that apart from the laboratory of Perrin's successor, Edmond Bauer, no one in France during the late 1940s engaged in work on the theoretical aspects of physical chemistry, with the exception of a quantum chemistry group started by Raymond Daudel.[87] Most of those interested in "theory" were placed in the "Applied Mathematics" division of the CNRS, or they worked in the Commissariat à l'Energie Atomique after the war. "It is something of a paradox," note Guéron and Magat, "that the creation of the

83. Prévost and Kirrmann cite Ingold's experiments providing evidence of the inadequacy of Lapworth and Robinson's theory of alternate induced polarity in their 1931 memoir (1320). Their reference is to Ingold, "The Mechanism of, and Constitutional Factors Controlling, the Hydrolysis of Carboxylic Esters. Pt. I. The Constitutional Significance of Hydrolytic Stability Maxima," *JCS* (1930): 1032–1057, on 1037.

84. Letter from Christopher Ingold to Prévost, dated 29 July 1946; and a letter from the Secretary of the Nobel Committees of the Swedish Royal Academy of Sciences to Prévost, dated 16 January 16 1954. Copies of these letters were given to me by Constantin Georgoulis, who completed a doctoral dissertation at Paris in 1960 on the kinetic study of reaction schemas for allylic transpositions, under the direction of Prévost. Another of Georgoulis's teachers was Paul Job, cousin of André Job.

85. Albert Kirrmann, "Structure chimique et effet Raman," *Journal de Physique et le Radium* ser. 7, 9 (1938): 48S. Bourguel became *professeur sans chaire* at the Sorbonne in 1932, shortly before his untimely death the following year. See Robert Lespieau, "Notice sur les travaux de Maurice Bourguel," *Bull.SCF Memoires* 53 (1933): 1145–1153.

86. Albert Kirrmann, *Chimie organique,* 2 vols. (Paris: Colin, 1947), I: 71. Kirrmann uses both Ingold's terminology and his and Prévost's terminology in this text. In Charles Prévost's *Leçons de chimie organique*, 4 vols. (Paris: Société d'Enseignement Supérieure, 1949–1953), he also uses the term "desmotropie" and "pseudoforme" from their joint memoir.

87. Guéron and Magat, "A History," 12.

Atomic Energy Commission in 1946 added to the difficulty of restarting research in fundamental physical chemistry.''[88]

It was only in 1959 that Guy Ourisson renovated the study of organic chemistry by organizing the Groupe d'Etudes de Chimie Organique (GECO), to facilitate contact with chemists abroad and to rethink the theoretical foundations of modern organic chemistry.[89] That this was a project sorely needed is indicated by Micheline Charpentier-Morize's reflections on her doctoral thesis defense in 1958. Prévost presided over the jury. When she referred to the possible existence of a ''π-complex'' to explain reactivity, Prévost exclaimed, ''Madame, if I have one reproach for you, it is that you know the modern theories too well.''[90]

Lespieau's Research School and the Discipline of Theoretical Chemistry

Robert Lespieau's aim to establish a disciplinary specialization of ''chemical theories'' in France was partially realized in the work of some of his students, especially Dupont, Prévost, and Kirrmann. For the first time, a clearly defined research school in France practiced the art of ''theoretical chemistry'' in their study of organic structure and reaction mechanisms. They self-consciously employed physical methods and apparatus, and they stayed in contact with a small network of physicists who were teachers, friends of Lespieau, or immediate colleagues. They had a laboratory terrain that was the home meeting place, no matter what their current affiliation. They had a common history that could be traced back generation by generation in the Ecole Normale laboratory to Berthollet, the ''father'' of chemical mechanics.

This research school also had a common language, which included not just the ''argot'' of the ''archicubes'' of the ''Normale Sup,'' but the language of ''atoms,'' ''electrons,'' and ''synionie.'' There were adversaries who had to be overcome: the bugbears of their youth (Deville, Berthelot, Jungfleisch) and the competitors of their maturity (Ingold and the ''Anglo-Saxon school''). They shared physical methods and instruments and learned techniques and ideas from each other; they collaborated in their research work and supported one another in their academic appointments and professional careers; and they controlled access to their school and to its rewards through maintaining its direction in their own hands.

But the two generations that pursued theoretical chemistry in Lespieau's laboratory from roughly 1900 to 1935 were isolated. The First World War and

88. Ibid.

89. Charpentier-Morize, ''La contribution,'' 46, n. 66.

90. Personal correspondence, letter of 16 January 1991.

its legacy of Germanophobia cut them off from ordinary discourse with many of their German and Austrian colleagues for a decade. Many fewer English and American scientists, as well as northern European scientists, studied in France during these years. The dropoff is remarkable in comparison with the nineteenth century.

Hardly any French scientists studied abroad. Kirrmann was unusual in his decision to spend a year in Munich in 1930, but he had, after all, been born in German Alsace. John C. Smith notes in his history of Oxford's Dyson Perrins Laboratory, directed by Robert Robinson in the 1920s and 1930s, that there was a great mixture there of ages and nationalities among the twenty or so research students each year but never, until 1947, a French person.[91] This insularity contributed to the closure of the boundaries of the research school associated with Lespieau's laboratory at the Ecole Normale Supérieure and to its exclusion from the wider disciplinary history of physical organic chemistry and theoretical chemistry.

Finally, this was a school that defined theoretical chemistry as a kind of chemistry (and a kind of science) that aimed at abstract knowledge in the purest sense. This science could not yet achieve the level of abstraction, or the level of prestige, of mathematics, but it was more free than other kinds of chemistry from practical ends and applications.

This does not mean that Lespieau's students agreed about collaboration between university and industry. Dupont, who fostered industrial research both at Havre and Bordeaux, set up liaisons between the university and industry when he was an administrator at the CNRS. However, according to his student Constantin Georgoulis, Prévost abhorred any collaboration or funding from industry.[92]

We conclude with Kirrmann's views on the agenda for organic chemistry in the early 1950s. Fundamental problems of organic chemistry lay, he said, in three areas: the accomplishment of organic synthesis, the elucidation of molecular structure, and the inquiry into reaction mechanisms. The study of mechanisms, Kirrmann wrote,

> is, more than that of synthesis, deeply impregnated with the spirit of pure science, of which the unique goal is to know and not to create. . . . It is still the case that the achievements of chemistry, in the face of problems of extreme complexity, inevitably are of a qualitative order on the whole, and that it appeals exclusively to the *esprit de finesse*. Physics, in contrast, being able to have the habit of

91. J. C. Smith, xx, in "The Development of Organic Chemistry at Oxford," 2 pts., Robert Robinson Papers, Manuscripts A.6 and A.7, RSL.

92. Conversation with Professor Georgoulis, Paris, 11 June 1987.

> limiting its parameters, has more easily attained the quantitative stage, and the geometric spirit suffices for it on the whole for it to progress on its way.[93]

Kirrmann and his colleagues in theoretical chemistry at the Ecole Normale Supérieure still continued to look forward to the day when theoretical chemistry and theoretical physics would share laurels for *l'esprit de géometrie* in understanding the natural world.

93. Albert Kirrmann, "Aspects actuels de la chimie organique," *Revue Philosophique* 156 (1966): 53–58.

1. Antoine Laurent Lavoisier and his wife Marie-Anne Pierrette, née Paulze, the father figure (joined by a mother figure) of modern chemistry. Painted in 1788 by Jacques-Louis David. *Courtesy of the Metropolitan Museum of Art, New York City.*

2. Commemorative medal in honor of Marcellin Berthelot. In this design, the chemist is seated in front of apparatus from his laboratory while, behind him, *la Verité* unveils herself and *la Patrie,* coiffed in the phrygian bonnet of the Revolution, presents Berthelot with a crown of laurel leaves. *Courtesy of Jean Jacques and the Collège de France, Paris.*

3. Participants in the second conference (April 1925) of the Institut International de Chimie Solvay in Brussels. The topic was "Structure and Activity," and four papers were devoted to activation or mechanism in chemical reactions. Henry Armstrong and Jean Perrin are seated at the center section of the adjoining tables. André Job and Thomas Martin Lowry are to Perrin's left. *Courtesy of the Instituts Internationaux de Physique et Chimie (Solvay), Brussels.*

4. Jean Perrin in his laboratory at the rue Cujas in 1924. *Courtesy of the late Francis Perrin.*

5. Albert Kirrmann and Charles Prévost at center of group of colleagues in 1950. *Courtesy of Mme. Noémi Guern-Prévost.*

6. Charles Prévost during 1936–1937, shortly before his return to the University of Paris from the Faculty of Sciences at Lille. *Courtesy of Mme. Noémi Guern-Prévost.*

7. Arthur Lapworth, chairholder in both physical chemistry and organic chemistry at the University of Manchester. *Courtesy of the Royal Society of Chemistry, and courtesy of the History of Science Collections, University of Oklahoma. From Alexander Findlay and William Hobson Mills, eds.* British Chemists *(London: The Chemical Society, 1947).*

8. Thomas Martin Lowry, the first professor of physical chemistry at the University of Cambridge. *Courtesy of the Royal Society, London.*

9. Gertrude Walsh Robinson and Robert Robinson met as students in Chaim Weizmann's chemical laboratory at Manchester. Like Edith Hilda Usherwood Ingold and Christopher K. Ingold, the Robinsons co-authored some of their research. *Courtesy of AP/Wide World Photos.*

10. Christopher K. Ingold, during 1921–1922, when he was a lecturer at Imperial College, London. *Courtesy of Keith U. Ingold.*

11. Sir Christopher K. Ingold, following his investiture at Buckingham Palace in 1958. *Courtesy of the Chemistry Department of University College London.*

12. Robert S. Mulliken, a founder of the theory of molecular orbitals. *Photograph taken by Samuel A. Goudsmit. Courtesy of American Institute of Physics Emilio Segrè Visual Archives.*

13. John Clarke Slater, theoretician of quantum chemistry and solid-state physics. *From Massachusetts Institute of Technology. Courtesy of American Institute of Physics Emilio Segrè Visual Archives.*

14. Charles Coulson in the 1930s, when he was a student of mathematics, physics, and chemistry in Bristol and Cambridge. *Courtesy of Mrs. Eileen Coulson.*

7

The London-Manchester School of Theoretical Organic Chemistry, 1880–1930

In his inaugural lecture for the chair of theoretical chemistry at Oxford in 1973, Charles Coulson recalled a student asking him, ''Why are you not located in physical chemistry?'' Nernst and other physical chemists at the turn of the century had thought of physical chemistry as theoretical chemistry. In contrast, by 1970, as Coulson put it, ''the theoretical chemist is concerned not just with physical chemistry but with all branches of chemistry.''[1].

Coulson's generation of theoretical chemists was equally at home in physics, mathematics, and chemistry. What clearly distinguished his generation from the preceding generation of theoretically minded chemists was the use of quantum mechanics. In turn, what distinguished this early-twentieth-century generation from *its* predecessors and from practitioners of nineteenth-century physical chemistry was the incorporation into theoretical chemistry of organic chemistry.

We turn now to an analysis of English chemists who provided the first systematic interpretations of chemical reaction mechanisms in which the molecule was modeled as a dynamic system of positive nuclei and negative electrons. While their approach was informed by physical ideas and theories, it was unarguably a *chemical* approach, consistent with classical nineteenth-century chemistry, from which it developed, and with quantum chemistry, which it helped to construct.

Our focus is on a relatively small network of chemists associated with two

1. C.A. Coulson, *Theoretical Chemistry: Past and Future*, ed. S. L. Altmann (Oxford: Clarendon Press, 1974). In his inaugural lecture for the chair of theoretical physics at King's College, Coulson noted that the first bill he received at the college was for a calculating machine, and it was made out to the department of ''Theatrical Physics.'' In C. A. Coulson Papers, Bod.Oxford.

strong traditions of organic and physical chemistry, one at the Central Technical College of London, later Imperial College, and the other at Owens College, later the University of Manchester. We will trace the principal characteristics of the emerging disciplines of physical organic and theoretical chemistry in England during the late nineteenth century and early twentieth century through an account of the individuals, educational experiences, and research agendas associated with this larger network that we will call the London-Manchester school. The boundary points for this group are the careers of Edward Frankland (1825–1899), on the one hand, and Christopher Ingold (1893–1970), on the other, with Thomas Martin Lowry, Arthur Lapworth, and Robert Robinson at the center. Ingold's work will be analyzed in more detail in chapter 8.

The London-Manchester Network (and the German Connection)

In the first half of the nineteenth century, institutions in metropolitan London dominated British teaching and research in chemistry. These institutions included the hospitals, the colleges of the University of London, the Royal Institution, the Royal College of Chemistry, and the Chemical Society of London, founded in 1841.[2] The most prestigious English institutions, Oxford University and Cambridge University, did not encourage specialized scientific studies, especially not chemistry.

The creation of Owens College at Manchester in 1851 proved a boon to chemical education and research in England. The college was instituted under John Owens's trust fund to instruct "young persons of the male sex" in "such branches of learning and science as are or may be taught in the English Universities" and, like University College in London, it freed university education from religious tests.[3] Professorships were established in classics, mathematics and natural philosophy, mental and moral philosophy, English language and literature, and chemistry.[4]

Frankland provides an early link between London and Manchester chemistry and between British and Continental chemistry. He was the first professor of chemistry at Owens College, in 1851. He had studied with Lyon Playfair at the Royal School of Mines in London, where he became fast friends with the young German student Hermann Kolbe. Frankland went with Kolbe to

2. Robert F. Bud and Gerrylynn K. Roberts, *Science vs. Practice: Chemistry in Victorian Britain* (Manchester: Manchester University Press, 1984): 48. Also Gerrylynn K. Roberts, "The Establishment of the Royal College of Chemistry: An Investigation of the Social Context of Early-Victorian Chemistry," *HSPS* 7 (1976): 437–485.

3. See G. Norman Burkhardt, "The University of Manchester," *Chemistry and Industry* (1949): 427–429, on 427.

4. Ibid., and Robert Kargon, *Science in Victorian Manchester: Enterprise and Expertise* (Baltimore: Johns Hopkins University Press, 1977): 155–156.

Marburg, where he worked for several months in the laboratory of Robert Bunsen, a friend of Playfair's. After a brief return to England, Frankland became the first Englishman to receive a doctoral degree at Marburg, in 1850.[5] His Continental education also included study with Liebig at Giessen in the autumn of 1849 and, more dramatically, a few brief weeks in Paris when revolution broke out during the summer of 1848. Despite distractions, he managed to hear lectures by Edmond Fremy and Dumas.[6]

Frankland acquired a considerable reputation from his very first collaborations with Kolbe on the preparation of acetic acid from alkali and methyl cyanide. His interest in organic radicals and his attempts to isolate the free ethyl radical led to the discovery of the organometallic compounds and the general theory of combining capacities or valences.[7] His interests were strongly oriented toward chemical theories.

When he took over the teaching of chemistry at Owens College in Manchester, the college building was the large house in which Richard Cobden had lived. Frankland's lecture room could hold 150 students; the 51-foot by 21-foot laboratory had room for 42 students and was possibly the best in Great Britain. A study of the chemistry examination set between 1851 and 1857, when Frankland resigned, shows that much time was spent on physics, including questions on atomic theory, latent and specific heat, and the gas laws. A course for 1856–57 included the relations to chemical phenomena of heat, light, electricity, cohesion, and gravity. Unfortunately, the number of students at Owens College in all courses dropped in the mid-1850s. In 1858, the chief local newspaper pronounced the college a "mortifying failure." Frankland grew discouraged, writing Bunsen, "It must be in the highest degree satisfactory to you to be surrounded by such a number of students so many of whom are engaged in original research."[8]

Frankland left Owens College in 1857 to teach at St. Bartholomew's Hospital in London, and in 1865, he succeeded Hofmann at the Royal School of Mines, when Hofmann left London to return to Germany. Frankland's successor at Owens was Henry Roscoe, who had just returned from studying with Bunsen at Heidelberg and who was the son of a distinguished Lancashire family (and the uncle of Beatrix Potter).

5. See Herbert McLeod, "Edward Frankland," *JCS* 87 (1905): 574–590, 574–575; and Colin A. Russell, *Lancastrian Chemist* (Milton Keynes: Open University Press, 1986).

6. Frankland's diary includes an account of his arrival in Paris on June 23, 1848, and being taken by surprise by revolutionary events. I read the transcribed typescript, for 1848–49, in the Royal Society Library, MS 221 XII b.9.

7. See Nye, *Science in the Provinces*, 170–171; Edward Frankland, "On a New Series of Organic Compounds Containing Metals," *Phil.Trans.* 142 (1852): 417. Also, G. N. Burkhardt, "Schools of Chemistry in Great Britain and Ireland—XIII. The University of Manchester (Faculty of Science)," *J. Royal Inst. Chem.* 78 (1954): 448–460, on 448.

8. See Peter John Davies, "Sir Arthur Schuster, 1851–1934," 2 vols. (Ph.D. dissertation, University of Manchester Institute of Science and Technology, 1983), Chap. 5, pp. 6, 14. Quotations, from Burkhardt, "Schools of Chemistry," 427; and Kargon, p. 161.

Most of Roscoe's students aimed at careers in manufacturing, industry, or business; a few were premedical students, and eight of sixty students in 1870–71 intended to be professional or academic scientists.[9] Roscoe emphasized careful, precise training and required his pupils to enter all analyses in a general logbook as well as in a private notebook. He aimed to create at Manchester a true chemical school, and he thought carefully about how to do it.

> The personal and individual attention of the professor is the true secret of success; it is absolutely essential that he should know and take an interest in the work of every man in his laboratory. . . . The professor who merely condescends to walk through his laboratory once a day . . . is unfit for his office, and will assuredly not build up a school.[10]

We will return to Roscoe and Manchester shortly.

After Frankland's move to the Royal College of Chemistry, which had become part of the Royal School of Mines, the school moved to South Kensington (1872), where it became known in 1885 as the Normal School of Science, then, in 1890, the Royal College of Science. The City and Guilds Central Institution, later the Central Technical College, opened across the street from the Royal School of Mines in 1884. Frankland's former pupil, Henry Armstrong, became the Technical College's professor of chemistry.

Armstrong had enrolled in the Royal School of Mines in 1865, which was the first year Frankland taught there. After two years, Frankland strongly urged Armstrong to study in Leipzig with Kolbe. He thought Leipzig the best chemistry laboratory in Germany.[11] From Germany, Armstrong wrote his parents in February 1870 that he wanted to remain in Germany as long as possible to prepare himself well in chemistry and that he wanted to pursue studies in

> the physical direction, a thorough knowledge of Physics being now as necessary as Chemistry to a chemist. Frankland's advice was "work Physics practically if possible." The great reputation which he justly enjoys is due to his having combined the two; curiously enough the same is visible in almost all of those who have been pupils of Bunsen.[12]

Regarding the relations between physics and chemistry, Bunsen is supposed

9. Kargon, *Science in Victorian Manchester*, 168, 179–181.

10. Quoted in D. Bettridge, "The Teaching of Chemistry in Victorian and Edwardian Times," *RIC Reviews* 3 (1970): 161–176, on 170. Also see H. E. Roscoe, *The Life and Experiences of Sir Henry Enfield Roscoe* (London: Macmillan, 1906).

11. A letter from Edward Frankland to Henry Armstrong, dated 12 January 1869, reports that the principal of Owens College and Roscoe found Leipzig, on the whole, to be the best of the Continental laboratories they visited. RSL, MM.10.93.

12. Quoted in J. Vargas Eyre, *Henry Edward Armstrong, 1848–1937: The Doyen of British Chemists and Pioneer of Technical Education* (London: Butterworths Scientific Publications, 1958): 52.

to have said, "Ein Chemiker der kein Physiker ist, ist gar nichts" (A chemist who is not a physicist is nothing at all).[13]

By the 1890s, Armstrong's laboratory in South Kensington housed a lively chemical group that regularly published papers and reports and that enjoyed more prestige than any other chemistry department in London. Armstrong's son, aptly named Edward Frankland Armstrong, wrote with some pride from Germany that his father's laboratory seemed to be dominating the pages of the *Proceedings* of the Royal Society.[14] *The Central*, an alumni magazine begun in 1903, touted the work of the chemistry school: "Even at the risk of being accused of undue elation, we must express our satisfaction at the . . . list as being one of which even the best German university Laboratory might be proud[!]"[15]

Armstrong's pupils included Ida Smedley (later Maclean), William J. Pope, Lapworth, Lowry, and F. P. Worley (later professor of chemistry at Auckland University in New Zealand). Smedley and Lapworth each were to teach at Manchester, where one of Lapworth's students was Robinson, the later 1947 Nobel Prize winner. Lowry became the first professor of physical chemistry at Cambridge University (1920), where Pope was Jacksonian Professor of Chemistry after teaching at the Manchester Municipal School of Technology in the early 1900s.

Across the street from Armstrong's laboratory at the Royal College of Science in the 1890s was the student Jocelyn F. Thorpe, who was to teach briefly at Manchester and then return to South Kensington in 1913 as professor of chemistry at Imperial College, which by then had incorporated the Central Technical College and the Royal College of Science (in 1907). Thorpe's most illustrious pupil was Ingold,[16], later professor of organic chemistry at Leeds University and then at University College, London. These are among the main figures in the network we call the London-Manchester school of theoretical chemistry.

Lest the presence of women students in Armstrong's laboratory suggest that he was a supporter of women's professional careers, his views were clearly expressed in his BAAS presidential address in Winnipeg in 1909.

> The subject [of women chemists] has been brought before the chemical world in England recently by the application of a number of women to be made Fellows of the Chemical Society. Many of us have resisted the application because we were unwilling to give any encouragement to the movement which is inevitably leading women to neglect their womanhood. . . . If there be any truth in the doctrine of hereditary genius [he thinks there is], the very women who have

13. J. R. Partington, *A History of Chemistry*, IV: 282.
14. Letter from Edward F. Armstrong to H. E. Armstrong, [summer] 1900, ICL.
15. *The Central* 1, no. 1 (1903): 5–6, ICL.
16. See W. H. Brock, ed., *H. E. Armstrong and the Teaching of Science, 1880–1938* (Cambridge: Cambridge University Press, 1973): 15.

shown ability as chemists should be withdrawn from the temptation to become absorbed in the work, for fear of sacrificing their womanhood; they are those who should be regarded as chosen people, as destined to be the mothers of future chemists of ability.[17]

Nonetheless, women students continued to make some inroads in the chemical laboratory and in chemical careers. Some collaborations led to marriage. Robinson and his fellow Manchester student, Gertrude M. Walsh, are one example. Ingold and Edith Hilda Usherwood, a postgraduate student at Imperial, are another. Both Gertrude Robinson and Hilda Ingold maintained active professional careers, sometimes working with their husbands and sometimes working independently.

There were other family connections in the London-Manchester network, including two father-son teams, William Henry Perkin, Sr./William Henry Perkin, Jr., and Henry Edward Armstrong/Edward Frankland Armstrong. Perkin, Jr., became professor of organic chemistry at Manchester University in 1892, following studies under Bunsen.

Thus, this history of London-Manchester chemistry illustrates lines of genealogy both in the biological sense and in the disciplinary sense, with overlapping traditions carried by disciplinary practitioners moving from London to Manchester and back again. We also can see that there was a strong German connection.

From 1840 until the First World War, nearly eight hundred Britishers and Americans earned doctoral degrees in chemistry at one of the twenty active German universities, more than half of them at one of four German universities: Göttingen, Leipzig, Heidelberg, and Berlin.[18] At least seventy-four British chemists were trained on the continent during 1828–1939. Of these, thirty-nine came under the influence of Liebig, Bunsen, and/or W. Wislicenus. Hardly any British students studied in Paris after Wurtz's death.[19]

In addition, just as members of Hofmann's "inner ring" of advanced German students often directed work in Hofmann's London laboratory from 1845

17. H. E. Armstrong, "Presidential Address," Chemistry Section, *BAAS Rep. Winnipeg (1909)* (1910): 420–454, on 451. It was not until 1950 that the Alembic Club at Oxford admitted women. See J. C. Smith, "The Development of Organic Chemistry at Oxford, Part II," typescript, Robinson Papers, RSL, on 54. Kathleen Lonsdale, an x-ray crystallographer, was the first woman elected to the Royal Society of London; Dorothy Crowfoot Hodgkin was the first woman chemist elected.

18. Paul R. Jones, "The Training in Germany of English-Speaking Chemists in the Nineteenth Century and Its Profound Influence in Britain and America," paper presented at First International Summer Institute in the GDR: Philosophy and History of Science, Leipzig, June 29, 1988. Jones examined copies of these dissertations, which are listed in the "Bibliographie der Dissertationen amerikanischer und britischer Chemiker an deutschen Universitäten, 1840–1914," Deutsches Museum, München, 1984.

19. B. N. Clark, "The Influence of the Continent upon the Development of Higher Education and Research in Chemistry in Great Britain during the Latter Half of the Nineteenth Century" (Ph.D. dissertation, University of Manchester, 1979), list, chap. 2, 2–9.

to 1865,[20] so Englishmen sometimes directed research in German laboratories. For example, S. F. Kipping's student work in the Munich laboratory of Baeyer was directed by a more senior Englishman, "Privatdozent" William Henry Perkin, Jr.[21]

It is interesting to note, however, that by 1900, Edward Frankland Armstrong was less keen on the caliber of German scientific work than his father. The younger Armstrong complained that van't Hoff rarely talked with the students in his Berlin laboratory and that van't Hoff's personal research on salts was of little real interest. The Germans were not keeping up with foreign publications, and there was a strong anti-English bias due to the Boer War, Edward Amstrong wrote his family. Most of the good students in the Berlin laboratories, particularly in Emil Fischer's, were foreigners or men from fields other than chemistry. The younger Armstrong found Germans unenterprising in their doctoral research, simply pursuing their degree by following orders for the preparation "of ethyl, butyl, propyl, hexyl, etc., derivatives of a known methyl acid."[22]

Nor were Edward Armstrong's unenthusiastic reactions to German training unique in this period. In the next three decades, there was to be much less exchange and interchange in Anglo-German chemical education, partly for reasons that helped bring on the First World War or flowed from that war. There also was the perception that German chemistry was no longer on the cutting edge of physical chemistry and especially of chemical *theory*. Harry Clary Jones of Johns Hopkins University wrote his colleague, Edgar Fahs Smith, at the University of Pennsylvania from Germany in the summer of 1904:

> It has been just ten years since I left Germany as a student [with Ostwald in Leipzig], and I can see very marked changes during that time. Ostwald has lost all interest in the experimental sciences, and is devoting all of his time and energy to philosophy and painting. . . . Van't Hoff is working along quietly in Berlin in a few small rooms, has no students and wants none, and thus it goes with the physical chemists. The Germans seem to be keeping up their former interest and activity in organic chemistry. . . . But, all in all, Germany has lost much of her prestige in chemistry. . . . I am fully convinced that there is more good work

20. See Michael N. Keas, "Herbert McLeod and the Inner Ring at the Royal College of Chemistry," paper delivered at Midwest Junto of the History of Science, Kansas City, April 1990.

21. See [Sir] Robert Robinson, *Memoirs of a Minor Prophet: Seventy Years of Organic Chemistry* (London: Elsevier, 1976): 22–23. On Robinson, see Trevor I. Williams, *Robert Robinson: Chemist Extraordinary* (Oxford: Oxford University Press, 1990). This biography does not contain a detailed account of Robinson's laboratory work and published scientific papers.

22. Letters from Edward Frankland Armstrong to his father, H. E. Armstrong, dated 11 February 1900 and [summer] 1900. ICL. The younger Frankland also wrote his father of increasing anti-Semitism among students in the Berlin laboratories, ascribing it to the view that "nearly all the Berlin chemists are Jews." Letter of 23 December 1900.

being done in America in physics, physical chemistry, and inorganic chemistry than in Germany.[23]

English chemists beginning their studies or their careers in the early 1900s most frequently received all their training in English laboratories from men who themselves had studied chemistry in German laboratories. This was the case for Robinson, Lapworth, and Lowry and for the next generation that included Ingold.[24]

Problems of Molecular Structure and Dynamics ca. 1900

Of his teaching methods, Henry Armstrong said in 1933, "I taught very little chemistry but a great deal of chemical method."[25] Well known for a sharp tongue and his critical attitude toward ionist theories in physical chemistry, Armstrong imported into English chemistry the legacy and views of Kolbe, the archcritic of Kekulé, van't Hoff, and Wislicenus.[26] He followed the precedents of Liebig and Kolbe for ridiculing opponents by writing "A Dream of Fair Hydrone" and "The Thirst of Salted Water."[27] Brodie, the conservative and eccentric Oxford chemist, found in Armstrong a kindred soul, writing him in 1874 that he was glad to see Armstrong avoiding "pictorical chemistry." And Lodge, who supported Armstrong in initial doubts about Arrhenius's theory of ionization, wrote Armstrong twenty years later to protest that he was going too far, that it now was understood that atoms are charged and how they are charged.[28] Nor did Armstrong like the convention of electron dots:

> When, following Odling, we represent valency by dashes written after the elementary symbol, we give clear expression by means of a simple convention to certain ideas that are well understood by all among us who are versed in the fact; to speak of electrons and use dots instead of dashes may serve to mislead the unwary . . . into a belief that we have arrived at an explanation of the phenomena.[29]

23. Quoted in Erwin N. Hiebert, "Nernst and Electrochemistry," 180–200, in George Dubpernell and J. H. Westbrook, eds., *Selected Topics in the History of Electrochemistry*.
24. For a discussion of patterns of study in Germany by American predoctoral and postdoctoral chemists, see John Servos, *Physical Chemistry from Ostwald to Pauling*.
25. Brock, *H. E. Armstrong*, 72–73.
26. Ibid., 8.
27. Henry E. Armstrong, "A Dream of Fair Hydrone" and "The Thirst of Salted Water," in his *The Art and Principles of Chemistry* (New York: Macmillan, 1927).
28. Letter from Benjamin Brodie to H. E. Armstrong, February 25 1874, ICL; and letters from Oliver Lodge to H. E. Armstrong, 11 May 1887 and 21 September 1908, ICL.
29. Armstrong, "Presidential Address," 432.

Just as he had been encouraged to study physics by Frankland and by Bunsen, so Armstrong encouraged his students to study physics before embarking on a chemical career. That the two disciplines are "inextricably . . . mixed up" was clear, he wrote Joseph Larmor.[30] But chemistry is not *reducible* to physics: the molecular architecture of space formulas and tetrahedral models of organic chemistry have "more than analogical significance," and "[physicists'] dynamical views must so far as possible be adapted to [chemistry], not the other way around."[31] Armstrong observed that if Helmholtz had written "that organic chemistry progresses steadily but in a manner which, from the physical standpoint, appears not to be quite rational," then this remark "must be regarded as little more than a confession that he was out of his depth."[32] Like Kolbe, Armstrong emphasized the " 'chemical feeling' aspect of theory."[33]

In 1895, Armstrong published some twenty-five papers jointly with members of his chemical laboratory. A report to the alumni in 1903 mentions work in five subjects undertaken by Armstrong and some of his protégés: radioactivity (H. Armstrong and Lowry), dynamic isomerism (Lowry), stereochemistry of noncarbon elements (Pope), the chemistry of camphor (M. O. Forster), and tautomerism (Lapworth).[34]

It was in the areas of "dynamic isomerism" and tautomerism, including the study of camphor derivatives, that some of the most original contributions of Armstrong's laboratory were to be made and that the careers in organic chemistry of Lapworth and Lowry were to become firmly oriented toward physical chemistry and chemical *theories*. The problems these two addressed had to do with puzzling behavior of compounds containing the carbonyl or carboxyl group (C = O; COOH), on the one hand, and, on the other, the so-called conjugated systems of alternating single and double bonds (−CH = CH−CH = CH−). Among the compounds most closely studied were ones related to acetoacetic ester ($C_6H_{10}O_3$) and benzene (C_6H_6), along with crotonic acid (CH_3-CH=CH-COOH) and vinyl-acetic acid (CH_2=CH-CH_2-COOH). What was fascinating about these substances was the contrast between appearance and reality. In the case of benzene, two apparently different structures have the same properties (the mono-substituted derivatives that should be different are identical [the 1,6 and 1,2 mono-substituted derivatives]). In the case

30. Letter from Henry Armstrong to Joseph Larmor, 21 October 1905, RSL, Lm17.

31. Henry Armstrong, "Presidential Address," 423–424.

32. Ibid., 423. Helmholtz's remark can be found quoted in Leo Königsberger's life of Helmholtz, *Hermann von Helmholtz*, trans. Frances A. Welby (Oxford: Clarendon Press, 1906): 340: "Chemists must be allowed to form hypotheses after their fashion, since the whole extraordinarily comprehensive system of organic chemistry has developed in the most irrational manner, always linked with sensory images, which could not possibly be legitimate in the form in which they are represented."

33. Henry Armstrong, "Presidential Address," 423.

34. Eyre, *Armstrong*, 125. *The Central* 1 (1903): 5–6. ICL.

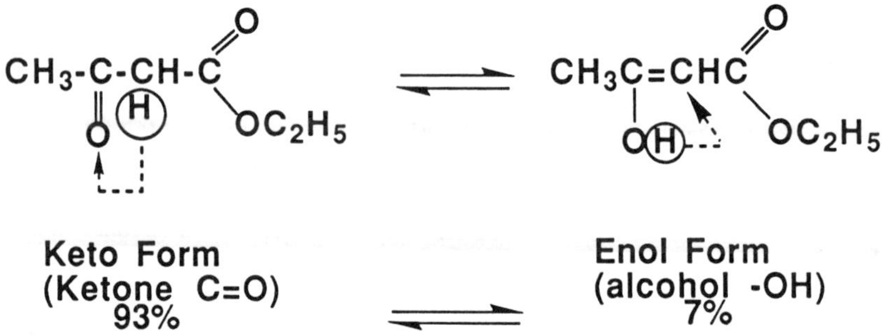

Figure 6. Tautomerism: Acetoacetic Ester or Ethylacetoacetate. Drawings courtesy of Glenn Dryhurst.

of acetoacetic ester, one substance acts as if it has one or the other of two totally different structures.

The problem of benzene is the more familiar of the two problems. As we have seen in chapter 5, to explain the unexpected behavior of benzene, which appears to contain alternating single and double bonds, Kekulé introduced the hypothesis that the benzene ring behaves as if it consists of successive "phases" in which what are described as valence bonds oscillate around the ring as the result of the number of mechanical contacts carbon and hydrogen atoms have with one another in unit time, so that on average, all the bonds are identical.[35] Around 1900, on the basis of experimental investigations, not on the basis of Kekulé's mechanical hypothesis,[36] Kekulé's two hexagon formulas, Ladenburg's prism formula, and Armstrong's centric formula were all in common use among chemists.

For acetoacetic ester, the mystery lay in the two rather different forms the "ester" took: the so-called *enol* (hydroxy) form $CH_3C(OH)=CH\text{-}COOC_2H_5$ and the *keto* (ketone) form $CH_3COCH_2COOC_2H_5$.[37] (See fig. 6.) As mentioned briefly in chapter 5, in 1885, Laar at Bonn coined the word "tautomerism" for the behavior of acetoacetic ester, (*tautos*, the same; *meros*, part), hypothesizing that the phenomenon of two structures in one is due to a continual oscillation of a hydrogen atom between two positions in the molecule, with an accompanying change in the bonds between the atoms. Of course, this hypothesis appears to have been influenced by Kekulé's earlier mechanical hypothesis.

35. Kekulé, *Liebig's Ann.* 162 (1872): 77; abstracted by Henry Armstrong in *JCS* 10 (1872): 612.

36. See Alan J. Rocke, "Kekulé's Benzene Theory," 145–161.

37. As discussed previously, A. Geuther described the former, Edward Frankland and B. F. Duppa the latter. The names *enol* and *keto* were invented by Julius W. Brühl in 1894. See J. R. Partington, *History of Chemistry*, IV: 814.

By 1900, it was beginning to be understood that the tautomeric forms of acetoacetic ester are in something like reversible equilibrium with each other.[38] The two forms can be separated from each other, the unsaturated hydroxy compound as an oil and the ketonic ester as a crystalline solid at −78 degrees Centigrade. However, on allowing either form to stand at room temperature for some time, the liquid acetoacetic ester identical to the usual compound is produced.[39] Following the kinetics of this process and understanding its mechanics was a substantial challenge. The chemist Arthur Lachman, for example, wrote in 1899, "This new theory is to be placed upon the same plane as Kekulé's benzene theory in importance."[40]

Lapworth and Lowry in London: Theories for Organic Chemistry

It was precisely a dynamic view of the molecule, replacing the static concept of the carbon tetrahedron, that was taken by Lapworth and Lowry, as they worked in Armstrong's laboratory in the 1890s and early 1900s.

Lapworth had first studied at St. Andrews and Birmingham, where his father was professor of geology, before obtaining his D.Sc. at the University of London in 1895. After teaching at the School of Pharmacy in Bloomsbury and then becoming head of the Chemistry Department at Goldsmiths' College (an institution for students in art and textile manufacture), he became a senior lecturer in inorganic and physical chemistry at the University of Manchester in 1909 and succeeded William Henry Perkin, Jr., in the Manchester chair of organic chemistry in 1913. He relinquished this chair in favor of assuming the chair of physical and inorganic chemistry in 1922, in order that his former pupil, Robinson, might return to Manchester.[41]

Beginning in the late 1880s, under Armstrong's direction, Lapworth studied isomeric changes in the naphthalene series and in camphor derivatives. While he naturally was influenced by the views of his teacher, Armstrong, he by no means adopted all these views. Smedley, his fellow student, later recounted to friends that Lapworth refused to show Armstrong his Ph.D. dissertation, remarking, "It is my work and I have no intention of changing it."[42]

In an 1898 paper, Lapworth sought to come up with a rationale to "demonstrate that it is possible to refer the majority of reactions and changes in

38. Peter Conrad Laar, *Berichte* 18 (1885): 648; 19 (1886): 730. Alexander Findlay, *A Hundred Years of Chemistry* (London: Duckworth, 1937; 2d ed., 1948): 119; and J. W. Baker, *Tautomerism* (London: Routledge, 1934).

39. See, e.g., James Bryant Conant, *Organic Chemistry: A Brief Introductory Course*, rev. with Max Tischler (New York: Macmillan, 1936): 175.

40. Arthur Lachman, *The Spirit of Organic Chemistry*, 60–88.

41. See Robinson, *Memoirs of a Minor Prophet*, 69; Martin Saltzmann, "Arthur Lapworth: The Genesis of Reaction Mechanism," *JChem.Ed.* 49 (1972): 750–752.

42. Robinson, *Memoirs of a Minor Prophet*, 69.

organic chemistry to necessary variations of *one or at most two simple laws*'' [my emphasis].[43] To explain the substitution mechanism, Lapworth boldly attempted to develop a general theory that combined mechanical folding and flipping motions of tetrahedral hydrocarbon valence structures with that of individual, labile atoms or groups.

But in a paper published in 1901, Lapworth dropped the stereochemical model of transference of the mobile group and focused on the principle of atom or group migration, adding the notion of instantaneous and fleeting electric charge in organic molecules dissolved in weak electrolytes. The paper contained the surprising claim that to this cause ''the majority of changes in organic compounds may be most probably assigned.''[44]

In his teaching and in his private notes, Lapworth began using a system of [+] and [-] labeling signs to express the ''relative'' or ''latent'' polar characteristics displayed by atoms at the instant of chemical transformation, especially a ''scheme'' expressing an enhanced positive polar character of hydrogen atoms in relation to a ''key'' carbonyl group. Lapworth developed the idea that this key atom or key group had an influence extending over a fairly long range in a hydrocarbon molecule, especially where there are double bonds in conjugating, or alternating, positions. He applied the same reasoning to the conjugated straight chain and to benzene, claiming that there is an exact parallel between the ready substitution of the *alpha*-hydrogen atom in ethyl crotonate and the hydrogen atoms in the methyl group of *ortho* and *para* nitrotoluene.[45] Now one rule—Lapworth's rule—would work for both aliphatic and aromatic chemistry.[46]

$$\overset{+}{H} - \overset{-}{\underset{\gamma}{CH2}} - \overset{+}{\underset{\beta}{CH}} = \overset{-}{\underset{\alpha}{CH}} - \overset{+\,-}{COOEt}$$

Lowry eventually followed similar thinking about ions and charges in organic molecules. The son of a Methodist minister, Lowry entered the Central Technical College in 1893 on a Clothworkers' Scholarship, and he carried out

43. Cited in Saltzmann, ''Lapworth,'' 751.

44. Arthur Lapworth, ''The Form of Change in Organic Compounds, and the Function of the *a-meta-* Orientating Groups,'' *Trans.CS* 79 (Pt. II) (1901): 1265–1284, esp. 1265–1267, 1276–1277.

45. Recalled in Arthur Lapworth, ''Latent Polarities of Atoms and Mechanism of Reaction, with Special Reference to Carbonyl Compounds,'' *Mem.Manchester LPS* 64, no. 3 (1919–20): 1–5.

46. Ibid., 7. Another rule that had to be subsumed under a general theory, as well as its exceptions, was the Brown-Gibson Rule (*JCS* 61 [1892]: 366): When X is naturally regarded as a derivative of HX, then C_6H_5X produces *ortho* and *para* derivatives; when X is naturally regarded as a derivative of HOX, then C_6H_5X gives *meta* derivatives.

research in Armstrong's laboratory from 1896 to 1913. He became a lecturer at a Methodist school, the Westminster Training College, in 1904 and a lecturer in chemistry at St. Guy's Hospital in 1913. He was the first teacher of chemistry in a London medical school to be named professor of the University of London,[47] and in 1920, he was awarded the first chair in physical chemistry at Cambridge University, established with aid from the British Oil Companies. At this time, Imperial College (J. C. Philip, 1913) and the University of Liverpool (Donnan, 1904) were among the few institutions with chairs in the discipline of physical chemistry.[48]

Lowry is best known to chemistry students through the tradition of eponymony, since the proton theory of acidity is known as the "Brönsted/Lowry theory" of proton donors. His most important experimental investigation likely was a long series of studies on optical rotatory dispersion.[49] For our purposes, there is special interest in his discovery of mutarotation in camphor derivatives and his theory of dynamic tautomerism, which led him to an ionic theory of organic reaction mechanisms.

In 1899, Lowry discovered the change in the rotatory power over time of a solution of nitrocamphor in benzene, an effect previously encountered only with aqueous solution of sugars. He named this effect "mutarotation," and its discovery was taken as a prominent achievement for Armstrong's laboratory research group.[50] Lowry ascribed the phenomenon to tautomeric conversion (from a CH-NO_2 form to a C = NO-OH form), that is, the shift of a hydrogen atom and the shift of a double bond. In 1909, he and Desch concluded that this reversible transformation occurs very quickly because they could not find an ultraviolet absorption spectral band characteristic of either isomer.[51] But what *triggered* this reversible transformation?

The surprising answer evolved for Lowry as he discovered that mutarotation ceases when traces of nitrogenous bases such as ammonia are removed from

47. See C. B. Allsop and W. A. Waters, "Thomas Martin Lowry," 402–418, in Alexander Findlay and William Hobson Mills, eds., *British Chemists* (London: The Chemical Society, 1947): 402; and W. J. Pope, "Thomas Martin Lowry," *Obituary Notices of Fellows of the Royal Society*, 2 (1936–1938): 287–293.

48. Donnan studied in Germany under Wislicenus, Ostwald, and van't Hoff during the period 1893–1897. He succeeded William Ramsay in the chair of general chemistry at University College, London, in 1913. Arthur Lapworth wrote to congratulate Donnan for his appointment to Ramsay's chair, noting that Manchester might benefit from Donnan's good fortune, since the Muspratt laboratory had attracted potential researchers away from Manchester. Letter from Arthur Lapworth to F. G. Donnan, 24 March 1913, Donnan Papers, UCL.

49. See Partington, *A History of Chemistry*, IV: 856. Lowry published *Optical Rotatory Power* (London: Longmans, 1935). See also Lowry, "La dispersion rotatoire optique: Hommage à la mémoire de Biot (1714–1862)," Conférence faite le 9 décembre 1925 devant la Société de Chimie Physique, *JCP* 23 (1926): 565–585.

50. Allsop and Waters, "Thomas Martin Lowry," 409. Eyre: "Armstrong christened this mutarotation" (223).

51. See J. W. Baker, *Tautomerism*, 5–6.

the solvent and when the reaction is carried out in silica vessels. Thus, he concluded, it is catalytic activity of amine bases that triggers the mutarotation phenomenon.[52]

In 1905, a BAAS subcommittee on "dynamic isomerism" was established which included Armstrong (chairman), Lowry (secretary), and Lapworth. In a report of 1909, Lowry summarized two different kinds of isomeric changes, one involving the "oscillatory transference" of an atom of hydrogen from carbon to oxygen, as in ethyl acetoacetate (acetoacetic ester), or from nitrogen to oxygen, as in isatin, or from oxygen to another oxygen, as in *para*-nitrosophenol. A second kind of isomerism was ascribed to the rearrangement of bonds in a molecule without any substantial alteration in the relative positions of atoms, as in Kekulé's well-known hypothesis for benzene.[53]

As Lowry continued to work in London, many of his London colleagues were leaving, several of them for Manchester. Lapworth moved to Manchester in 1904, joining Pope at the Manchester School of Technology (later UMIST). Frederic Kipping, who had collaborated with Pope and Lapworth in London, became professor of chemistry at University College, Nottingham, in 1897. Kipping, Perkin, Jr., and Lapworth acquired long-standing family ties by marrying three sisters, the Holland sisters of Liverpool.[54] And Smedley, who had been studying the relationship between optical properties and the constitution of unsaturated compounds in Armstrong's laboratory in the late 1890s, became an associate at Newham College, Cambridge, in 1899 and then assistant lecturer in chemistry at Manchester from 1906 to 1910.[55] There she found a

52. See Lowry's summary of this work (from *JCS* 75 [1899]: 218), "On Chemical Change," *The Central* 13 (1916): 25–41, ICL.

53. Thomas Lowry, "Dynamic Isomerism." Report of the Committee, consisting of Professor H. E. Armstrong (Chairman), Dr. T. M. Lowry (Secretary), Professor Sydney Young, Dr. C. H. Desch, Dr. J. J. Dobbie, Dr. M. O. Forster, and Dr. A. Lapworth, *BAAS Rep. Winnipeg (1909)* (1910): 135–143, on 136.

54. Kipping, a Manchester native, studied with Roscoe and Schorlemmer, then spent a year (1886) in Munich at Adolf von Baeyer's laboratory where Perkin, Jr., was Privatdozent and von Baeyer's assistant. Kipping completed his London doctoral degree in 1887 and was Armstrong's assistant from 1890 to 1897. See Partington, *A History of Chemistry*, IV: 851. And Robert Robinson, *Memoirs*, 22–23.

55. Ida Smedley entered the Central Technical College in 1899–1900. She was an associate at Newham College, Cambridge (1899–1903), a researcher at the Davy-Faraday Research Laboratory (1905), an assistant lecturer in chemistry at Manchester (1906), a Beit Memorial Medical Research Fellow (1910), and a member of the research staff in the Biochemical Department at the Lister Institute. She wrote with her husband, Hugh MacLean, *The Lecithins and Allied Substances* at roughly the same time she completed her doctoral thesis at University College, London. In 1915, she received the $1,000 Ellen Richards Research Prize, awarded by the Naples Table Association for the "best thesis written by a woman (of any nationality) on a scientific subject embodying new observations and new conclusions." The prize was awarded six times during the period 1902–1924: to Florence Sabin (Johns Hopkins University Medical School, 1903); Nettie Stevens (Bryn Mawr, 1905); Florence Buchanan (University College, London, 1910); Ida Smedley MacLean (University College, London, 1915); Eleanor Carothers (University of Pennsylvania, 1921); and Mary Laing (University of Bristol, 1924). See Margaret Rossiter, *Women Scientists in America: Struggles and Strategies to 1940* (Baltimore: Johns Hopkins University Press, 1982): 48. And *The Central* 13 (1916): p. 117, ICL.

thriving scientific center, staffed by familiar faces from London. Let us return now to Manchester.

Manchester Chemistry and Physics in the Early Twentieth Century

When Pope, Lapworth, Smedley, Robinson, Walsh, and other chemists and chemistry students arrived in Manchester in the early 1900s they found "the Manchester of the *Guardian* and the Hallé Concerts and of Miss Horniman's Repertory Company, [and] the Manchester of cotton, liberalism, Free Trade, and the Manchester School."[56] A principal benefactor of the Hallé Orchestra was physics professor Arthur Schuster,[57] Longworthy Professor of Physics and Director of the Physics Laboratory. Another musical enthusiast was William Henry Perkin, Jr., Professor of Organic Chemistry and Director of the Schorlemmer Chemistry Laboratory. It was said that Perkin had given up the violin when he could no longer practice on his uncle's Guarnerius. So he turned to the piano, and at Manchester, he and Lapworth, a violinist, organized musical evenings, playing chamber music with professional friends from the Hallé Orchestra.[58]

On the scientific side, in the early 1900s, a new physics laboratory had just been completed, the fourth largest in the world after the Johns Hopkins University, Darmstadt, and Strassburg physics laboratories. A new electrochemical laboratory was opening for work complementing the physics department's course in chemical physics and the chemistry department's course in metallurgy.[59]

A former student of Roscoe's, Schuster had been appointed to a new chair in applied mathematics at Manchester in 1881, following his studies at Heidelberg, Berlin, and the Cavendish Laboratory. One of his competitors for the position was J. J. Thomson, who had briefly been a student in engineering at Manchester under Osborne Reynolds. In his inaugural address, Schuster freely

56. A. S. Russell, "Lord Rutherford: Manchester, 1907–1919: A Partial Portrait," 87–101, in J. B. Birks, ed., *Rutherford at Manchester* (London: Heywood and Co., 1962), on 88. The Gaiety Repertory Theater specialized in Shaw and Galsworthy, according to J. L. Heilbron, *H. G. J. Moseley: The Life and Letters of an English Physicist, 1887–1915* (Berkeley, Los Angeles, and London: University of California Press, 1974): 57.

57. Heilbron, *H. G. J. Moseley*, 57.

58. See John Greenaway, "Memorial Notice. William Henry Perkin," 7–38, in *The Life and Work of Professor William Henry Perkin* (London: The Chemical Society, 1932): 12–13; and [Sir] Robert Robinson, *Memoirs*, 27. As a talented pianist, Robinson later joined Lapworth, a violinist, in a shared love of Mozart. See Williams, *Robert Robinson*, 22–23.

59. See Kargon, *Science in Victorian Manchester*, 228. And P. J. Davies, "Schuster," chap. 7, 96–97.

interchanged the terms "applied mathematics" and "mathematical physics." He gave up his chair to H. Lamb in the mathematics department to become the Longworthy professor of physics in 1888.

Schuster had an informed interest in chemical problems. He taught a course in chemical physics that was attended by chemistry honors students. Topics included spectrum analysis, saccharimetry, and electrolysis. In 1894, he organized a fortnightly "physical colloquium" for second- and third-year physics students joined by some chemistry honors students. He closely followed the chemical research of his colleagues, Roscoe and Harold Baily Dixon. In interpreting results in his special field of spectroscopy and in electrochemistry, Schuster visualized an "ionic" mechanism of electrical conduction in liquids and gases.[60]

It was Schuster who wooed Rutherford to Manchester, after Rutherford had turned down offers at King's College and Yale University because of their inadequate research facilities.[61] Rutherford quickly built up his own staff and investigators in the field of radioactivity at Manchester, with some fifteen to twenty people engaged in full-time research in radioactivity during the period 1908–1919. This work included the study of alpha particles, the scattering of these particles by metallic screens; the nuclear theory of the atom (1913); and the work culminating in the disintegration of the nitrogen nucleus in 1917. Among those working in the laboratory were Hans Geiger, R. B. Boltwood, Kasimir Fajans, George Hevesy, E. N. da C. Andrade, James Chadwick, George C. Darwin, Edward Marsden, Harry Moseley, and Bohr.[62]

Rutherford's attitude toward chemistry was stereotyped by his jokes and barbs occasionally directed at his chemical colleagues. The later Manchester physicist P. M. S. Blackett recounted the famous crack, "All science is either physics or stamp collecting,"[63] and it was said that Rutherford chafed at receiving the 1908 Nobel Prize in chemistry, rather than in Physics. In a lecture in which he described his theory of the nuclear atom, he joked that the "nucleus is a round, hard object—just like Professor Perkins' head."[64] However, Rutherford expressed great respect for his chemist collaborator Frederick Soddy and for other chemists, as well.

There were students and researchers in Rutherford's laboratory who knew a good bit of chemistry. Bohr remarked that he was particularly helped at Manchester in his thinking about the physical and chemical properties of the elements by Hevesy, who "distinguished himself among the Manchester group

60. See Davies, "Schuster," chap. 6, 12–15, 26; chap. 7, 52, 61–63.

61. Kargon, *Science in Victorian Manchester*, 233.

62. See essays by H. R. Robinson (53–86) and A. S. Russell (87–101) in Birks, ed., *Rutherford*.

63. Essay by Blackett (102–113) in Birks, ed., *Rutherford*, 108.

64. Quoted in J. C. Smith, "Development of Organic Chemistry," Pt. II, 57 (A.6-A.7 in Robinson Papers, RSL).

by his uncommonly broad chemical knowledge.''[65] Bohr's three-part paper, ''On the Constitution of Atoms and Molecules,'' appeared in 1913 following a sojourn at Manchester, in which he began to relate Thomson's ring model for the atom, the periodic properties of the chemical elements, and chemical binding in molecules. Bohr made trips back and forth between Denmark and Manchester in 1913, and he succeeded Darwin during 1914–1916 as Schuster reader.[66]

Rutherford continued Schuster's practice, begun in 1905, of organizing short lectures and experimental demonstrations to keep people up to date on radioactivity. Robinson greatly admired Rutherford, whom he saw socially during these years.[67] The Friday colloquia series continued to be frequented by chemists as well as by mathematicians. These were preceded by ''an enormous tea-party, generally presided over by Lady Rutherford,'' recalled the physicist Harold R. Robinson. ''I think we all felt that we were living very near the center of the scientific universe.''[68]

Chemists shared the euphoria. When Roscoe succeeded Frankland in 1857, he established a chemistry department that became one of the leading departments in the country, with new buildings (1872) that included his own private laboratory, two teaching laboraories, and more than twenty small research laboratories and staff rooms. By 1887, the number of students in the department had reached 120, and the department had an honors program for the best of them. One of Roscoe's assistants, Schorlemmer, a student of Heinrich Will and Kopp, was the first chair holder in organic chemistry in Great Britain, at Manchester, in 1874. Roscoe and Schorlemmer's *A Treatise on Chemistry* (1877) became a classic textbook.[69]

By 1900, Dixon had succeeded Roscoe, and Perkin, Jr., had succeeded Schorlemmer. The Schorlemmer laboratory and the Perkin laboratory (named for Perkin, Jr.,'s father) provided facilities for organic teaching and research; the Frankland and Dalton laboratories (originally built in 1872) were for undergraduates. The private library and laboratory of E. Schunck were bequeathed to the university and moved there from the moors of Kersal. The John Morley laboratory for organic chemistry was completed in 1909.[70]

This was a school that had close ties to local industries. Roscoe made

65. Bohr's essay (114–167) in Birks, ed., *Rutherford*, 117.

66. Ibid., 122, 130, 138; Bohr, ''On the Constitution of Atoms and Molecules.'' And John Heilbron and Thomas Kuhn, ''The Genesis of the Bohr Atom,'' 245–250.

67. Robinson, *Memoirs*, 69. ''I wish to say that at the top of my list of scientific heroes I find Ernest Rutherford and Louis Pasteur.''

68. H. R. Robinson, in Birks, ed., *Rutherford*, 72–73.

69. G. N. Burkhardt, ''The School of Chemistry in the University of Manchester,'' *Journal of the Royal Institute of Chemistry* (September 1954): 448–460. This is part of the Schools of Chemistry in Great Britain and Ireland Series (XIII). On 450.

70. Ibid., 451; and J. C. Smith, ''The Development of Organic Chemistry at Oxford,'' Pt. I, 18, RSL.

himself available to industrialists and the local government, for example, giving advice to Charles Macintosh on his rubber solution for waterproofing cloths.[71] Among other projects, Thorpe and Perkin, Jr., worked on a method of fireproofing ''flannelette,'' the cotton material substituted for wool in inexpensive underwear.[72] Thorpe and Perkin, like their successors at Manchester, served on the Dyestuffs Research Committee for Imperial Chemicals, headquartered in Manchester.[73]

Robinson entered Manchester as a chemistry undergraduate in 1902 because his father thought chemistry might be useful in the family business of manufacturing surgical dressings.[74] One of his teachers was Dixon, expert in physical chemistry and the chemistry of combustibles, who had studied with Vernon Harcourt at Oxford. Another teacher was Thorpe, about whom it was said that he made two new substances each day. Chaim Weizmann, later the president of Israel, directed the Schorlemmer laboratory.[75] But it was Perkin's lectures (''a miracle of clear exposition'') that persuaded Robinson to follow organic chemistry. ''The only fault that could be urged against the lectures,'' a student said, ''was that they made organic chemistry appear too easy.''[76]

In Weizmann's laboratory, Robinson met Gertrude Walsh, who was a research assistant to Weizmann and recently had been teaching chemistry at the Manchester High School for Girls. Her first two papers were written with Weizmann in 1910 and 1911. She and Robinson married in 1912.[77] Another comrade was the later Nobel Prize winner Walter N. Haworth. Robinson became a friend, too, with Walsh's close friend, Ida Smedley, from London. And he formed a close friendship with a senior mentor, Lapworth, who then was teaching inorganic and physical chemistry and had introduced chemical thermodynamics into the undergraduate course.[78]

While Robinson was completing his thesis work under Perkin's direction, on the chemistry of brazilin and the structure of alkaloids, he discovered that Lapworth had a highly personal and successful method of deciding whether a reaction would ''go'' or not; it turned out to be the scheme of alternating

71. Kargon, *Science in Victorian Manchester*, 175; Davies, ''Schuster,'' chap. 7, 16.

72. Greenaway, on Perkin, 26.

73. Thorpe, like some other English students in Germany, had combined university studies with industrial work experience, in his case working at the Badische Chemical Works at Ludwigshafen. See Greenaway, on Perkin, 26.

74. Martin D. Saltzman, ''Sir Robert Robinson—a Centennial Tribute,'' *Chemistry in Britain* (June 1986): 543–548, on 543; and Williams, *Robert Robinson*, 5–9.

75. Robinson, *Memoirs*, 14–17, 47–50, 66. Weizmann studied chemical change induced by bacteria, which he demonstrated to be a means of manufacturing acetone during the First World War (47).

76. Greenaway, on Perkin, 23.

77. See Williams, *Robert Robinson*, 19–20. Also, W. Baker, ''Obituaries. Lady Robinson,'' *Nature* 173 (1954): 566–567.

78. Lapworth's work in this period included a series of papers with J. R. Partington, W. J. Jones, and E. Newbery on hydrogen ions, with particular reference to solvation and to ester hydrolysis. Burkhardt, ''Schools of Chemistry,'' 454.

polarities in a chain of atoms.[79] Lapworth's influence is first seen clearly in Robinson's 1911 paper suggesting a reaction mechanism for cotarnine synthesis: there is the fleeting formation of a positive hydrocarbon ion that reacts with a negative halide ion, followed by migration of the halide from nitrogen to carbon in the product.[80]

After several years at the University of Sydney, Robinson became professor of organic chemistry at Liverpool from 1915 to 1920 and, then, through Lapworth's good offices, professor of organic chemistry at Manchester in 1922 after brief appointments at the British Dyestuffs Corporation and St. Andrews University near Edinburgh. He moved to University College, London, in 1928, and two years later he succeeded Perkin, Jr., at Oxford University, where Perkin had moved in 1912.[81] By 1930, three schools of organic chemistry were recognized as the premier schools in Great Britain: Oxford, Imperial College, and Manchester. Linus Pauling and Robert Millikan despaired, rightly as it turned out, of enticing young Alexander Todd to Caltech in 1938 when they heard of an opening at Manchester which they knew to be "much more in the chemical swim than Caltech."[82]

During Robinson's five years as professor of organic chemistry at Manchester, he wrote, or co-wrote, 105 papers.[83] According to Smith, who was with him both at Manchester and at Oxford, Robinson was "a great leader who inspired his students."[84] According to Todd, who studied with him at Oxford, Robinson was

> the most inspiring director of research with whom I have ever come in contact. Certainly one had to be reasonably tough and independent to appreciate him fully. . . . He was interested in many topics and his interest would flit from one to the other with great frequency. . . . His collaborators had to get accustomed to being alternately badgered about their progress several times a day, and being almost totally ignored for weeks on end.[85]

He "was very emotional and impatient with those holding different views,"

79. Ibid., 453; Martin D. Saltzman, "Sir Robert Robinson" 543–548, on 543; from Robinson's obituary notice of Lapworth.

80. See Jennifer Seddon, "The Development of Electronic Theory in Organic Chemistry," (Honours Thesis, St. Hugh's College, Oxford University), 13.

81. After Robinson's departure and contemplating his own retirement, Lapworth concentrated on maintaining the prestige and contributions of the Manchester school. In 1933, Ian Heilbron was appointed professor of organic chemistry, and in the same year, Michael Polanyi became professor of physical chemistry.

82. See Smith, "Development of Organic Chemistry," Pt. II, 3; and Alexander Todd, *A Time to Remember: The Autobiography of a Chemist* (Cambridge: Cambridge University Press, 1983): 44–45.

83. See figures in Burkhardt, "Schools of Chemistry," 455; and Greenaway, on Perkin, 32. It is not clear whether the 1908–1913 figures are for organic chemistry only or for all labs.

84. Smith, "Development of Organic Chemistry," Pt. II, 8.

85. Todd, *A Time to Remember*, 27.

Todd recollected. "[A] genius unrivalled in his own field of endeavor, Robert Robinson could suddenly become shy, gauche, jealous, selfish, cruelly unfair, and unable to brook the mildest criticism," was Smith's candid assessment.[86]

The 1947 Nobel Prize in chemistry was to be awarded to Robinson to honor his work in the synthesis of natural products, investigations that he pursued all through his career, and a field in which his wife both collaborated and worked independently.[87] Yet, in his memoirs, Robert Robinson wrote that he considered the development of an electronic theory of reaction mechanisms "my most important contribution to knowledge."[88] This suggests the seriousness with which he viewed scientific theories and his belief that scientific glory and reputation rest on theories, not discoveries. Let us turn now to these theories.

A Theory of Chemical Dynamics

After Robinson's appointment at Liverpool, he and Lapworth engaged in a lively correspondence on the means of applying polar and electron ideas in chemistry, especially as begun by Thomson and by American chemists like Fry at Cincinnati (discussed in chap. 5). Robinson claimed later that it was at Liverpool, during 1916–17, that the idea came to him that that partial valencies were not possessed by atoms in addition to their normal valency but as the result of subdividing or splitting the normal valency.[89] "A second stage of my theoretical ideas was the recognition that when a bond is divided, it will be *ipso facto* polarized. At first, two dotted lines marked [+] and [-] were used, but later the dotted line marked [+] was not included because it represented a quantity in defect." These two ideas, he claimed, were published together only in 1922 but were used in lectures at Liverpool as early as 1916.[90]

In a 1917 paper written with Gertrude Robinson, they stated that

> partial valency . . . shall be considered to be derived from the normal valencies and that the dissociation necessarily weakens the bond between the carbon and iodine atoms so that the complete symbol is
>
> . . . CH3 . . . I . . . ;

86. Ibid.; Smith, "Development of Organic Chemistry," Pt. II, 27.

87. Gertrude Walsh Robinson developed a new method for the synthesis of high-molecular-weight fatty acids, and she was active in penicillin research during the Second World War. Baker writes (1954): "It is perhaps not generally known that she was the first chemist to prepare synthetical material with genuine antibiotic character of the penicillin type" (567).

88. Robinson, *Memoirs*, 184.

89. Ibid.

90. Ibid.

> partial dissociation may be a stage in a complete process of electrolytic dissociation. . . . Reactions between ions are not excluded, but regarded as the limiting case, and further it is recognized that the symbols which are used to express the activated condition of molecules can represent only a first approximation to the actual distribution of affinity.[91]

Robinson referred in very general terms to electrons, saying that ''cohesion via partial valencies'' is followed by ''movement of electrons along a potential gradient.''[92]

In the correspondence with Lapworth, Robinson wondered how he arrived at specification of the key atom or reactive center. Lapworth expressed doubt about the usefulness of the notion of ''latent valency,'' particularly as Robinson used the concept for nitrogen and oxygen. Robinson preferred using both the notions of polarity and conjugation, rather than limiting himself to one or the other. Robinson insisted on the value of the idea of hydrocarbon ring formation via partial valency because it enables the chemist ''to envisage the transference of a group from one position to another without that horrid idea—the jump.''[93]

Lapworth liked the use of polarization symbols, so that if vinyl bromide yields different forms of dibromide depending on conditions, this can be understood simply as the result of its constitution as

$$\underset{+}{CH_2} = \underset{-}{CHBr} <-> \underset{-}{CH_2} = \underset{+}{CHBr}.$$

Still Lapworth recognized that he could not consider ''the whole thing as purely electrical (à la Fry)'' because of cases where electrical predictions result in wrong predictions. And Lapworth used conjugation symbols, for example, in what he called his ''postmanteau expressions'' containing ''all the possible (or most) of the rationally arranged closed circuits having signs in due order.'' For the camphor series, in which the halide group is ''torn away,'' ''I picture this happening thus'':

91. R. Robinson and G. M. Robinson, ''Researches on Pseudo-Bases,'' *JCS* 111 (1917): 958 ff., quoted in Robinson, *Memoirs*, 187–189.

92. I am grateful in some of the following to the analysis by Seddon [now Curtis], ''The Development of Electronic Theory in Organic Chemistry.'' Cited in Seddon, 17.

93. Letter from Robert Robinson to Arthur Lapworth, 15 February 1919, Robinson Papers, D. 38, RSL.

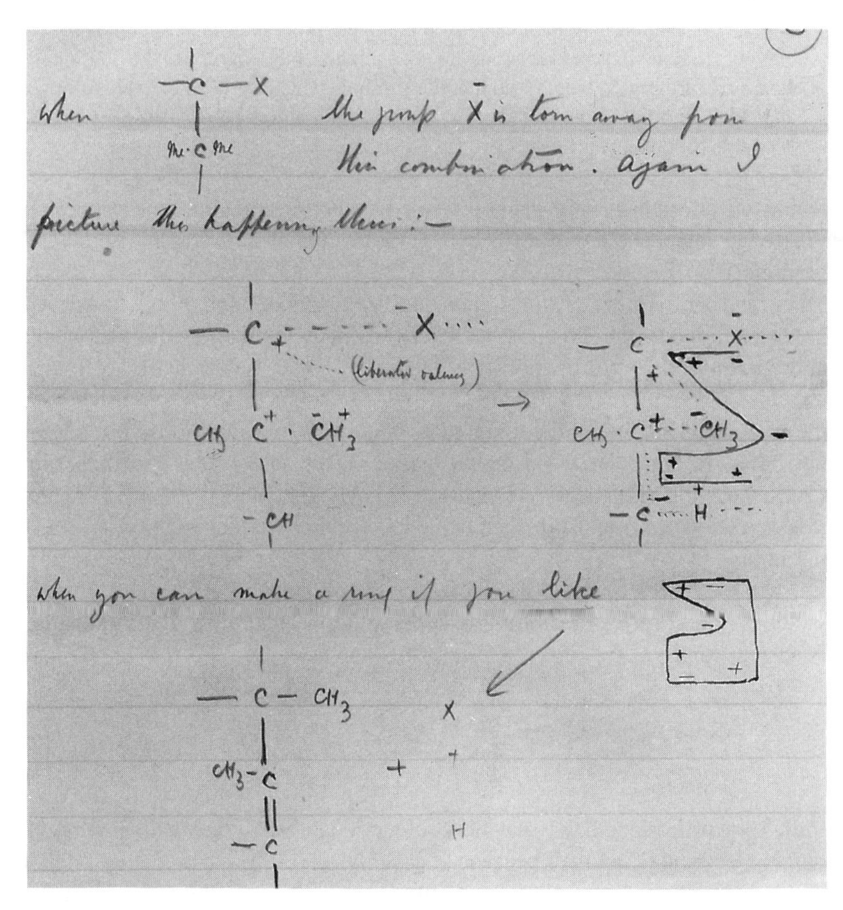

Figure 7. Diagram from Arthur Lapworth's letter of 26 February 1919 to Robert Robinson, in correspondence exchanging ideas and visual representations about reaction mechanisms. From Robert Robinson Papers. Permission of the Royal Society, London.

"Where do we differ?" Lapworth wrote Robinson.

I think largely here. You invoke the latent valences of nitrogen and oxygen for example, in phenol and aniline, in many cases where I do not think it necessary. Now after twenty-five years worrying over the theories of organic chemistry I have become convinced that it is not always desirable to try and make one explanation do everything.[94]

Lapworth continued to caution Robinson that it was desirable to use reaction signs as generalizations of facts rather than as explanations based on hypotheses.[95]

On March 16, 1920, the year that Robinson left Liverpool for what turned out to be one year with the British Dyestuffs Corporation in Manchester, the two friends both presented papers at a meeting of the Manchester Literary and Philosophical Society. Lapworth gave "alternating" polarity and the influence of a "key-atom" primary roles in initiating and determining the course of reaction. Robinson identified the activation of molecules with the rearrangement of valences "most probably synonymous with changes in position of the electrons," so that the active molecules are polarized and contain partially dissociated valences.[96]

For representing the splitting of valence, Robinson used two *or more* dotted lines not necessarily of equal value but together quantitatively equivalent to the normal unit of valency. Looped dotted lines were used in which the loop could be opened up to produce partial valencies of opposite sign.[97]

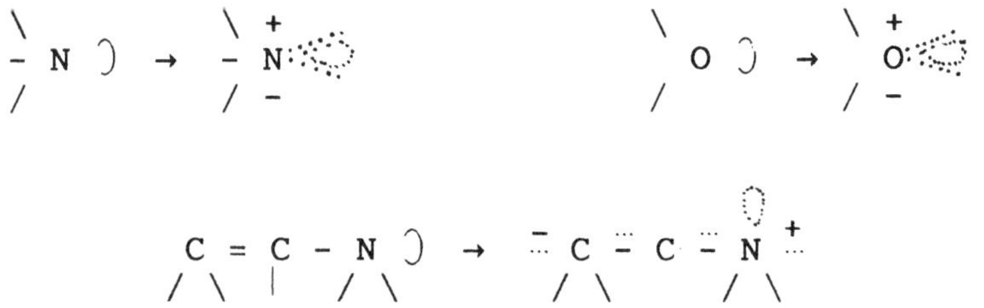

```
\             \ +                 \           \ +
- N )   →     - N:⁛               O )   →     O:⁛
/             / -                 /           / -

                                            ⁛
                                                  +
     C = C - N )   →    ⁻ C ⁻ C ⁻ N  ⋯
    / \  |  / \            / \       / \
```

Robinson defined conjugation as the transfer of free partial valency to an adjacent atom or to the end of a chain of atoms: "It is the explanation of action at a distance in a molecule."[98] By assuming the possibility of subdivision of valency, reactions could be represented as "being almost continuous and with

94. Letter from Lapworth to Robinson, 26 February 1919; same series as above.
95. Letter from Lapworth to Robinson, 3 March 1919; Same series as above.
96. Arthur Lapworth, "Latent Polarities of Atoms," 1–16; Robert Robinson, "The Conjugation of Partial Valencies," ibid., no. 4 (1920): 1–14. This paper is reproduced in Robinson's *Memoirs*.
97. Robinson, "Conjugation," 1–2.
98. Ibid., 3.

many intermediate phases.''[99] Key-atoms, as in Lapworth's schema, were designated by overhead dots. Specific reactions were treated with the aim of bringing them under one system of explanation.[100]

In 1921, Robinson and Lapworth discovered the electron valence-bond theories of Lewis and Langmuir when Langmuir lectured before a joint session of the chemistry and physics sections of the BAAS. Langmuir laid out his theory in this Edinburgh lecture, including the tendency of atoms to give up or take up electrons or to share electrons so as to form ''duplets,'' that is, two-electron bonds. Electron rearrangement, said Langmuir, ''is the fundamental cause of chemical action.''[101]

By the end of October 1921, Lapworth worked out an explanation of his induced alternate polarities based on the Lewis-Langmuir theory. He now divided one valency into two, not three, partial valencies, each equal to one shared electron. He wrote Robinson that he had mentioned his theory of alternating polarities to Langmuir and that Langmuir ''is sure to be 'on' it although he hadn't a ghost of an idea at the time how to apply his theory to explain the 'principles.' ''[102] Lapworth encouraged Robinson to be innovative and to challenge Langmuir's ''obsession'' with the ''duplet idea.''[103]

In a 1922 paper written with W. O. Kermack, Robinson used the Lewis-Langmuir approach to give physical meaning to partial valency (so that the double bond [=] was equivalent to four-electron sharing; $\dot{-}\dot{-}\dot{-}$ to three-electron sharing, – – – to two and · · · to one electron). The activated or extreme polar stage of a molecular form was described as the result of electron rearrangement, the reagent ''taking advantage of a momentary manifestation of polarity.''[104] Benzene was written with six three-electron bonds, an ''aromatic sextet.''[105]

In Jennifer Seddon's analysis of Robinson's work, she notes that he now

99. Ibid., 4.

100. Ibid., 6–10 ff.

101. Irving Langmuir, ''The Structure of Molecules,'' *BAAS Rep. Edinburgh. 1921* (1922): 468–469. G. N. Lewis, ''The Atom and the Molecule,'' 762–785; Irving Langmuir, ''The Structure of Atoms and the Octet Theory of Valence,'' *Proc. NAS* 5 (1919): 252–259, and *JACS* ''The Arrangement of Electrons in Atoms and Molecules,'' 41 (1919): 868–934, and ''Isomorphism, Isoterism, and Covalence,'' *JACS*, 41 (1919): 1543–1559.

102. Letter, handwritten, from Lapworth to Robinson, 30 December 1921, Robinson Papers, D.40, RSL. (This letter includes a P.S.: ''I was extremely obliged for your very useful letter about Ingold, etc.'') And, see G. Norman Burkhardt, ''Arthur Lapworth and Others: Structure, Properties, and Mechanism of Reactions of Carbon Compounds; Some Developments 1898 to 1939, with Particular Reference to Arthur Lapworth and his Work,'' typescript, 150 pp., with addenda, written 1973–1978, bound and deposited at Royal Society of London, 1980; 88–89.

103. Letter from Lapworth to Robinson, 17 June 1922, D.41, Robinson Papers, RSL.

104. W. O. Kermack and Robert Robinson, ''An Explanation of the Property of Induced Polarity of Atoms and an Interpretation of the Theory of Partial Valencies on an Electronic Basis, *JCS* 121 (1922): 427ff.; quoted in Seddon, ''Development of Electronic Theory,'' 25. W. O. Kermack was a member of the Royal College of Medicine at Edinburgh.

105. In a letter from Lapworth to Robinson, 17 June 1922, Lapworth recommends J. J. Thomson's recent paper in the *Philosophical Magazine* which supports this idea.

began using partial valencies with more confidence as a guide to his investigations of natural products, for example, in planning reactions in alkaloid synthesis and the synthesis of tertiary amines as well as in explaining the final stage of the formation of benzoyl benzoin from mandelonitrile and sodium ethoxide.[106]

Manchester-London Controversies, 1922–1928

While Robinson was at Manchester, from 1922 to 1928, he and Lapworth frequently were seen intently discussing reaction mechanisms. As G. Norman Burkhardt recalled, they were to be found ''in the corner of the staff common room covering old envelopes with [+] and [-] signs, partial valencies, equations, diagrams, and arrows of many shapes indicating electron drifts and availabilities.''[107] By the mid-1920s, they also were thoroughly embroiled in a controversy with Ingold and Bernard Flürscheim, an independent chemist working in his private laboratory. The controversy, a vicious one concluded by charges of intellectual theft and deceit, was carried out in meetings of the Chemical Society of London and in the pages of the *Journal of Chemistry and Industry*.[108] Robinson's students found themselves listening to stimulating but confusing and scathing accounts of theories of reaction mechanisms. One student later recalled an occasion when Robinson said something like, ''Dr. X should really be associated with this work, but the only thing that Dr. X should be associated with is the effect of boot leather.''[109] One wonders what Robinson *really* said.

Members of the Chemical Society who witnessed many exchanges among Flürscheim, Ingold, Lowry, Lapworth, and Robinson became impatient to end it all. A motion was introduced in late 1925:

> That henceforth the absurd game of chemical noughts and crosses be tabu within the Society's precincts, and that . . . no further contributions to the mystics of Polarity will be received, considered, or printed by the Society.[110]

The editor of the *Journal of Chemistry and Industry* ended the published controversy, by fiat, in 1926.

106. Seddon, ''Development of Electronic Theory,'' 27–28.

107. Burkhardt, ''Arthur Lapworth and Others,'' 31. Also quoted in Martin D. Saltzmann, ''Sir Robert Robinson,'' 545.

108. Arthur Lapworth, letter to the editor, *Chemistry and Industry* 44 (1925): 83–84; Or see Saltzmann, ''Sir Robert Robinson,'' 564, also by Lapworth.

109. W. Cocker, ''Chemistry in Manchester in the Twenties, and some Personal Recollections,'' *Natural Product Reports. Royal Society of Chemistry* 4 (1987): 67–72, on 71.

110. Quoted in J. Shorter, ''Electronic Theories of Organic Chemistry: Robinson and Ingold,'' *Natural Products Reports. Royal Society of Chemistry* 4 (1987): 61–66 on p. 63.

We will not look at the reaction-mechanism controversy in detail here. It has been treated thoroughly elsewhere by Martin Saltzman and John Shorter, among others.[111] However, it is important to make a few comments about the debates and their outcome.

When the controversy began, Ingold had recently moved from London to the chair of organic chemistry at Leeds University. He was a former student and protégé of Thorpe, who had left Manchester in 1913 to teach at Imperial College. Except for two years in chemical industry, following his Ph.D. degree in 1918, Ingold worked largely with Thorpe at Imperial College from 1913 until 1924, serving as the first lecturer in the newly established Whiffen Laboratory.[112] Ingold collaborated with his wife, Hilda Usherwood Ingold, in some of the controversial papers of the 1920s.

Thorpe's laboratory was always a busy one, with links to chemical industry and many foreign students, especially Indians, in residence. During the First World War, the Imperial College laboratory had produced drugs, including novocaine, and gas explosives. Thorpe himself was a member of the Anglo-Iranian Oil Company and a member of the Dyestuffs Committee for Imperial Chemical Industries, Ltd. Usually kindly and genial, he could also be tough. He frequently assisted Soddy in examining D.Phil. candidates at Oxford. One candidate

> emerged pale from his exam to say, I wish I could have had the two profs in for a few drinks in college, to mellow them before the fray, little knowing that the two professors had before the exam rounded off an ample meal with a bottle of vintage port.[113]

Ingold's first papers in Thorpe's laboratory were quite classical, combining stereochemical ideas with traditional notions of affinity forces. Like Thorpe, Ingold was skeptical about simple electrical explanations, for example, in ring formation.[114] He initially viewed valence and reaction theory through the theoretical lenses of Thiele and Werner, with whom his informal mentor, Flürscheim, had studied before taking a degree at Heidelberg in 1901 and setting up a private laboratory in Hampshire in 1905.

As early as 1902, Flürscheim offered a "theory," not just a rule, for ori-

111. Saltzman, "Sir Robert Robinson," 543–548; Saltzman, "The Robinson-Ingold Controversy: Precedence in the Electronic Theory of Organic Reactions," *JChem.Ed.* 57 (1980): 484–488; and J. Shorter, "Electronic Theories of Organic Chemistry," 61–66.

112. Saltzman, "The Robinson-Ingold Controversy," 485.

113. Smith, "Development of Organic Chemistry," Pt. I, 43; also G. A. R. Kon and R. P. Linstead, "Jocelyn Field Thorpe," in *British Chemists*, ed. Alexander Findlay (London: The Chemical Society, 1947): 369–401, on 371–373.

114. Seddon, "Development of Electronic Theory," 32–33, citing Ingold, "The Conditions Underlying the Formation of Unsaturated and of Cyclic Compounds from Halogenated Open-Chain Derivatives. Pt. I. Products Derived from alpha-Halogenated Glutaric Acids," *JCS*, 119 (1921): 305–329, "Pt. II. Products Derived from alpha-Halogenated Adipic Acids," 951–970.

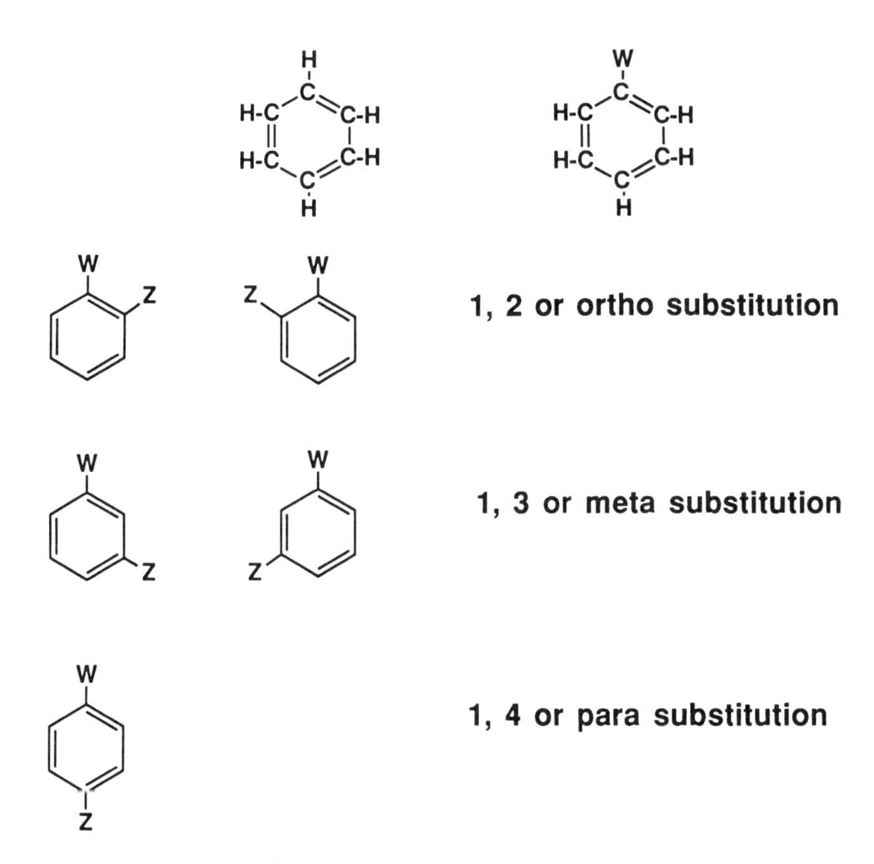

Figure 8. Aromatic Substitution Patterns. Drawings courtesy of Glenn Dryhurst.

entation in aromatic substitution (fig. 8) on the basis of the principle that chemical affinity is continuously divisible, distributed partly in chemical "bonds," with the rest free. Flürscheim inferred that there is an excess of free affinity at the *otho, para* positions when an oxygen atom sets up a disturbance of the ordinary affinity distribution at one carbon site in the benzene ring. Longer than Ingold, Flürscheim rejected the Lewis-Langmuir theory of the shared-electron bond on the grounds that the bond must be electrostatic in nature, "since physics [has] provided no other strong forces."[115]

In 1923, the Faraday Society arranged a conference on the electronic theory of valence, hosted by Lowry at Cambridge University. The temperature was 86 degrees in the shade, and the lecture theater was even hotter. Thomson, who chaired the first session, invited his colleagues to take off their coats but

115. See Christopher K. Ingold, "Bernard Jacques Flürscheim," *JCS* (1956): 1087–1089.

left his own on.[116] Lewis presented a paper on valence and the electron as the opening address.[117]

The meeting at Cambridge was a heated affair in more ways than one. Lowry noted that Thorpe and his colleagues at South Kensington opposed his views but that here at Cambridge, ''we already have a programme that is almost too big for us, in following up other lines of research suggested by the theory.''[118] Lapworth and Robinson joined Thorpe (and Lewis) in criticizing Lowry's ideas. The Manchester chemists argued that Lowry's principles were not sufficiently general, in contrast to their own.[119]

Challenged by Thorpe (''The polarity theory explains everything and predicts nothing''),[120] Robinson made two specific predictions for substitution in nitrobenzene and chlorobenzene derivatives, which he said would be verified in his laboratory by Mr. Oxford.[121]

Although the proceedings of the conference make no reference to Ingold, a later reviewer of the proceedings volume included Ingold with the Thorpe camp at the conference:

> The impression left on our minds—probably an inaccurate one—is that Thorpe and Ingold believe that in organic chemistry we do not know much more of the fundamentals than that the carbon atom is tetrahedral in its grasp, and that it is no good talking gaily about the different kinds of valency and pretending we do.[122]

However, as discussed in chapter 8, this critique is clearly a misreading of Ingold and Thorpe. While Ingold did not speculate about electron or electrical mechanisms, his papers of 1922–1925 combined ideas about intramolecular polarity, affinity, tautomerism, and atom or group mobility. Showing clear influences from Thorpe, on the one hand, and from Bernard Flürscheim, on the other, Ingold's ideas follow, too, from those of the London doyen of chemistry, Henry Armstrong, who had suggested that *meta* substitution was due to the inhibition of reactivity in the *ortho*, *para* positions.[123]

116. Burkhardt, ''Arthur Lapworth and Others,'' 89.

117. Ibid., 89. See *Trans.Far.Soc.* 19 (1923): ''The Electronic Theory of Valency,'' July 13–14, 1923, Department of Physical Chemistry, University of Cambridge, 450–543, on 477.

118. Thomas Lowry, ''Introductory Address to Part II: Applications in Organic Chemistry of the Electronic Theory of Valency,'' 485–487, *Trans.Far.Soc.* 19 (1923): 487. ''Intramolecular Ionisation in Organic Compounds,'' ibid., 487–496; and ''The Transmission of Chemical Affinity by Single Bonds,'' ibid., 497–502.

119. Lapworth and Robinson, ''Remarks on some Recent Contributions to the Theory of Induced Alternate Polarities in a Chain of Atoms,'' *Trans.Far.Soc.* 19 (1923): 503–505, on 504–505.

120. Ibid., 527.

121. That the *para* chlorobenzyloxy group would have a greater orienting effect than *meta* and that the *para*nitrobenzyloxy group would have a weaker orienting power than *meta*. Ibid., 529.

122. Quoted in Shorter, ''Electronic Theories of Organic Chemistry,'' 63.

123. Robinson mentions Armstrong, in *Memoirs*, 224.

In the ensuing duel of laboratories following the 1923 Cambridge meeting, Lapworth, Robinson, and their collaborators mostly bested Thorpe, Ingold, Flürscheim, and their collaborators in predicting the outcome of the same reactions on the basis of rival theories. Ingold and his Leeds group made several mistakes in analyzing the identity of reaction products, which caused them some embarrassment. Robinson was forced by constant challenges to improve the capabilities of his predictions.

But Ingold's triumph came in finally seeing the advantages of Robinson's explanation system, revising it, and substituting a new and clearer language and classification of types of reaction mechanisms. Lapworth, Robinson, and their collaborators referred to Ingold's ''conversion'' experience, a conversion in which Paul eventually helped create the myth of his role not as saint but as savior.

The conversion experience is found in Ingold's response to a paper presented by Robinson at the Chemical Society in the summer of 1925 and sent to Ingold before its publication in 1926. Robinson's paper, written with J. Allen, A. E. Oxford, and John C. Smith, classified conjugated systems into nine categories of reactants, two of them ''anionoid'' and the rest ''cationoid.'' ''Crotonoid'' and ''crotenoid'' were two of the nine types. This was a detailed and cumbersome classification, based on studies of crotonic acid, amino acids, and their salts, in which crotenoid was an instance of anionoid (electron donor) reaction and crotonoid of cationoid (or electron acceptor) reaction.

Crotonic acid	CH3−CH = CH−COOH		
C = C−O−	Crotenoid (anionoid)	[−] /C = C−O−/[+]	(later, "nucleophile")
C = C−C = O	Crotonoid (cationoid)	[+] /C = C−C = O/[−]	(later, "electrophile")

In 1925, Robinson first published his schematic ''curly arrows,'' indicating electron drift or displacement,[124], defining nitrosobenzene as a crotenoid reactant.

Robinson's theory was one of electron displacements propagated by electrostatic induction along the length of a carbon chain. As a hydrogen atom at one end of a carbon chain becomes a weakly held proton, electron charge drifts

124. Robinson, ''Polarisation of Nitrosobenzene,'' *Journal of the Society of Chemistry and Industry [Chemistry and Industry]* 44 (1925): 456–458.

to the other end of the chain through the conjugated structure. These changes result in an active polarized form of the molecule. One process initiates another in a process of mutual adjustment. This is polarization on demand, not a permanent polarized state of the molecule.[125] In the 1926 paper, Robinson and his co-authors used their classification to explain experimental results reported in three earlier papers and to explain substitution patterns in the benzene ring.

Shortly after Ingold received the preprint of the 1926 paper, he sent a paper to the Chemical Society which was published later in 1926. Here he dealt with the nitration of *ortho* aminophenol derivatives, adopting the kind of explanation that appeared in Robinson's paper and rationalizing the meaning of the so-called free and bound affinity, which he and Flürscheim previously had been employing, by the new electronic interpretation. As discussed in chapter 8, the paper included Robinson's curved arrows but introduced a new delta (δ) notation for polarity as well as other new terms.[126] The Manchester group referred to this paper as ''Ingold's conversion.''[127]

As is well known, Robinson became convinced that Ingold had not given him proper credit for his, and Lapworth's, role in the development of an electronic theory of organic reaction mechanisms. In reply to a furious letter from Robinson, Thorpe tried to reassure Robinson about Ingold.

> When I read the *electronic* part I regarded it as based on your views and, in fact, to offer support to them on the experimental side. It seemed to me that your views were so well known that no one would fail to see the source of inspiration.[128]

However, within the next few years, Ingold's dominance in the field of organic reaction mechanisms theory became clearly established, following a 1933 paper on tautomerism in which he introduced the terms ''nucleophilic'' and ''electrophilic'' and a 1934 article in *Chemical Reviews* systematizing ''Principles of an Electronic Theory of Organic Reactions.'' Burkhardt, one of Lapworth's collaborators, said, ''It was a complete takeover of terminology at the right time.''[129]

125. James Allen, Allen E. Oxford, Robert Robinson, and John C. Smith, ''The Relative Directive Powers of Groups of the Forms RO and RR'N in Aromatic Substitution. Part IV. A Discussion of the Observations Recorded in Parts I, II, and III,'' *JCS* 129 (1926): 401–411, reprinted in *Memoirs*, 212–214.

126. E. H. Ingold and C. K. Ingold, ''The Nature of the Alternating Effect in Carbon Chains. Part V. A Discussion of Aromatic Substitution with Special Reference to the Respective Roles of Polar and Non-Polar Dissociation; and a Further Study of the Relative Directive Efficiencies of Oxygen and Nitrogen,'' *JCS* 129 (1926): 1310–1328.

127. Burkhardt, ''Arthur Lapworth and Others,'' 140.

128. Letters from Jocelyn Thorpe to Robert Robinson, 4 July 1926; from Thorpe to Robinson, 11 July 1926; from Thorpe to Ingold, 11 July 1926; and from Ingold to Thorpe, 12 July 1926. Robinson Papers, D.33, RSL.

129. Burkhardt, ''Arthur Lapworth and Others,'' 143. C. K. Ingold, ''The Significance of Tautomerism and of the Reactions of Aromatic Compounds in the Electronic Theory of Organic Reactions,'' *JCS* (1933): 1120; and C. K. Ingold, ''Principles of an Electronic Theory of Organic Reactions,'' *Chemical Reviews* 14 (1934): 225–274.

Lapworth, then Robinson, well before Ingold, were leaders in the quest for a theoretical chemistry that would set out guiding principles for organic chemists. Lowry, too, played an important role in introducing into organic chemistry the hypothesis of ions and then electron-induced "polarization." All three introduced ideas diverging from the geometrically based classical affinity concepts of Werner and Thiele. Lapworth and Robinson, more than Lowry, envisioned a dynamic hydrocarbon molecule in which continuous and gradual motions produce discrete and sharp effects.

The systematic linking of physical chemistry and qualitative theoretical organic chemistry was a next step, one taken by Ingold but not by the older protagonists of the London-Manchester school. Lapworth retired in the early 1930s; Lowry died in 1934. Robinson's laboratory research programs at Manchester, University College, and Oxford were now to stress classical methods of degradative and synthetic organic chemistry. He was little interested in polarimetry, spectrometers, or pressure hydrogenators; and he was slow to adopt methods of ultraviolet and infrared spectroscopy as aids in structural work.[130]

Robinson continued to think pictorially and visually in his work, having little interest in mathematical representations of chemical problems. He specifically eschewed interest in the applications of quantum theory to chemical problems after an exchange with Walter Hückel. In 1931, Hückel published a theoretical paper on the directive influence of substituent groups attached to a benzene molecule. Robinson and Lapworth argued on chemical grounds that Hückel's conclusions were wrong, but Hückel proved intransigent until his assumptions eventually were shown to be invalid. Alexander Todd and John Cornforth ascribed to this Robinson's subsequent "estrangement from quantum theory."[131]

Some Comparisons and Preliminary Conclusions

An influential textbook in physical organic chemistry by Lapworth's student, William Waters, appeared in 1935 with an introduction by Lowry giving an account of recent developments as Lapworth saw them.[132] But, as Robinson knew, the history of a theory and the history of a discipline in large measure is created by historical introductions and the systems of citations in the most-cited articles and books in a field.

130. See Todd, *A Time to Remember*, 26–27, commenting on the Dyson Perrins laboratory at Oxford in 1931.

131. See the account in Williams, *Robert Robinson*, 78–79; and [Lord] Alexander Todd and [Sir] John Cornforth, "Robert Robinson," in *Biographical Memoirs of Fellows of the Royal Society* 22 (1976): 465–478.

132. William A. Waters, *Physical Aspects of Organic Chemistry* (London: Routledge, 1935).

In contrast to the complaints of Prévost and Kirrmann, which we saw in chapter 6, Robinson did not claim that Ingold failed to cite his and Lapworth's work but that Ingold failed to cite it fairly and sufficiently. Ingold's strategy appeared to be what Eugene Garfield has called "obliteration by incorporation": the obliteration of the source of ideas, methods, or findings by its incorporation in currently accepted knowledge.[133]

By inventing new terms for the mechanism of ordinary benzene valence bonds (the "mesomeric" effect) and for the mechanism of activation of the organic molecule ("polarisability" caused by the "electromeric" effect), Ingold was able to claim the theory of reaction mechanisms as his (and his collaborators'), assigning to others the role of influential predecessors.[134] Like Lavoisier one hundred fifty years earlier, Ingold created a scientific language—in a "takeover of terminology"—and created a new science, a new discipline. We will look at how this strategy worked in the next chapter.

The science created was distinct from physical chemistry and from classical organic chemistry but was rooted in them both. The chemical investigations carried out from roughly 1880 to 1930 in the disciplinary network of research schools and laboratory leaders, which I have called the London-Manchester school of theoretical organic chemistry, aimed at a theoretical chemistry of general laws for intramolecular dynamics. This theoretical chemistry, while largely qualitative in character, was self-consciously a mechanical chemistry. Its practitioners reintegrated into general chemical theories, from the domain of organic chemistry, physicists' theories of ions, then electrons, thereby coming full circle to early-nineteenth-century notions of the centrality of electricity, somehow, to chemistry. Chemists in the London-Manchester school did this in an intellectual and interdisciplinary environment in which physics provided an example and challenge for applying physical ideas, particularly electrical hypotheses, to long-standing chemical problems.

Chemists at the Ecole Normale Supérieure, as we have seen, referred in the 1950s, and later, to an "Anglo-Saxon" school of theoretical and physical organic chemistry. I have identified a specific "London-Manchester" school that extended from Edward Frankland to Christopher Ingold, with roots in Leipzig, Marburg, Heidelberg, and Berlin as well as in London and Manchester and extensions to Cambridge, Oxford, and Leeds. There is an obvious paradox in my term "London-Manchester" school, since there was such strife in the 1920s *between* the London laboratories of Thorpe and Ingold and the Manchester laboratories of Lapworth and Robinson. In addition, the term "London-Manchester" school is broadly inclusive of numerous laboratory research

133. See E. Garfield, "The Obliteration Phenomenon," *Current Contents*, no. 51/52, 5–7 (22 Dec. 1975), discussed by Robert Merton in foreword to Eugene Garfield, *Citation Indexing*.

134. See especially, *Structure and Mechanism in Organic Chemistry* (Ithaca: Cornell University Press, 1953; rev. ed., 1969): 63–65.

groups over some five decades working in the interests of a theoretical organic chemistry based in mechanical hypotheses, particularly electron mechanics.

Writ large, we can contrast the Parisian and London-Manchester schools on a number of grounds: their different attitudes toward model building and pictorial imagery, toward the need for abstract and rigorous explanation, toward "up-to-date" physical hypotheses, and toward pragmatism and practical applications. The English chemists, ever more intimately involved in industrial research than their French counterparts, were far more favorable than French chemists to short-term modeling devices that predicted how a reaction would "go." At the turn of the century, there was slightly more permeability and flexibility still in England than in France to the boundaries between disciplines, intellectually, institutionally, and geographically. However, in the cases we have studied, it is the very permeability in both cases that is striking in the gradual formation of new disciplines of physical chemistry, physical organic chemistry, and a theoretical chemistry of reaction mechanisms.

This is a history, like histories of other disciplinary identities, in which there were old traditions to be overcome, new ideas to promulgate, villains to subdue, and laurels for the victors. Our heroes and heroines set out to create a theoretical chemistry that would resolve the centuries-old challenge of creating a mechanical chemistry, a truly philosophical chemistry, based in principles of matter and motion that were acceptable both to chemistry and to physics. In this, there was considerable success by the mid-1930s, before the development of quantum chemistry, which only confirmed the approach of the London-Manchester school.

8

Reaction Mechanisms

Christopher Ingold and the Integration of Physical and Organic Chemistry, 1920–1950

The Career of Christopher Ingold

Like Arthur Lapworth, and unlike Thomas Lowry and Robert Robinson, Christopher Ingold was a scientist comfortable in the laboratory domains of both physical chemistry and organic chemistry. As his student, Derek Davenport, remarked in 1987, Ingold became the clear leader of the "emerging discipline" of physical organic chemistry when he "harnessed chemical kinetics to his discussions of reaction mechanisms":

> His magisterial review, "Principles of an Electronic Theory of Organic Reactions" . . . was a kind of legal brief describing the way physical organic chemistry ought to be if only God had done his work properly. Now more than fifty years later, astonishingly little needs to be changed.[1]

Like Lapworth at the beginning of his career in the 1890s, Ingold concerned himself in the 1920s with discovering general principles that would result in the unification of aliphatic and aromatic chemistry as well with the application to chemistry of fundamental physical theories. Like Lavoisier one hundred fifty years earlier, Ingold settled in the 1930s on a program of fundamental

1. Derek A. Davenport, "On the Comparative Unimportance of the Invective Effect," *Chemtech* 17 (1987): 526–531, on 529. For histories of the rise of physical organic chemistry in the Anglo-American context, see Leon Gortler, "The Physical Organic Community in the United States, 1925–1950: An Emerging Network," *JChem.Ed.* 62 (1985): 753–757; and Martin Saltzman, "The Development of Physical Organic Chemistry in the United States and the United Kingdom: 1919–1939, Parallels and Contrasts," *JChem.Ed.* 63 (1986): 588–593. Very little survives of C. K. Ingold's scientific files and correspondence, according to his son Keith U. Ingold, personal correspondence, letter of 16 July 1990.

reform of chemical theory and chemical language that he took to have revolutionary significance.

In all of this, as we have seen, Ingold not only involved himself in controversies but became a strongly controversial figure. Like so many leaders of research schools and founders of disciplines, stories and anecdotes about him became legion. He was known for clever, sometimes brutal, harshness in his published writings but "an almost Edwardian courtliness" in person.[2] He was said to have a dry sense of humor, getting past referees and the editor of the *Journal of the Chemical Society* an alphabetical section in a paper on salt hydrates ("A is the apparatus, B is the bulb, C is the condenser, D . . . .").[3] A 1968 publication is entitled *The Importance of Being Polarizable*.[4]

Ingold's colleagues and protégés frequently have noted that his later penchant for combining physical methods with chemical techniques was presaged in a youthful precocity in physics. He began serious scientific studies at Hartley University, Southampton, as an external student of the University of London. He was "better at physics than chemistry," he later recalled, but his enthusiasm for chemistry was fired by a gifted teacher, D. R. Boyd.[5]

Ingold began studies at Imperial College with the organic chemist Thorpe in 1913 in a period when Thorpe was joking that physical chemistry was "nothing but the blessed word equilibrium." Still, Thorpe and his school were much concerned with equilibrium phenomena, and Ingold found support among his mentors for his own mathematical and physical interests.[6]

Following the war, Ingold spent two years as a research chemist with Cyanide Company in Glasgow and then returned to Imperial College, where he was a lecturer in organic chemistry. In 1923, he and Hilda Usherwood married. She had completed a Ph.D. in 1922 at the University of London, and she went on to take a D.Sc. degree in 1925. Hilda Ingold continued to do chemical work until 1939, when the outbreak of the war caused major disruptions of Ingold's laboratory.[7]

2. See Davenport: "Like Swift, Ingold criticized the ideas, not the person, but one can sympathize with a victim who had trouble keeping the two separate" (1987: 530). One example of the "invective effect" is the wording of Ingold's rejection of results published by William Taylor: "The evidence recorded by Taylor . . . is shown to be worthless. . . . Regarding the first 'item of evidence' it is now shown that the represented equilibrium does not exist. . . . His method of 'observation' is analyzed and found to be illusory." Quoted from several papers by Hughes, Ingold et al., published in 1940 by Davenport (1987: 529–530).

3. C. K. Ingold, with J. N. E. Day, E. D. Hughes, and C. L. Wilson, "Hydration of Salts with Heavy Water, and Remarks on the Constitution of Salt Hydrates," *JCS* 137 (1934): 1593–1599, on 1598. Cited in Charles W. Shoppee, "Christopher Kelk Ingold, 1893–1970," *Biographical Memoirs of Fellows of the Royal Society* 18 (1972): 349–411, on 371. This is in the tradition of the Chemical "B Club."

4. C. K. Ingold, *The Importance of Being Polarizable* (Haifa: Technion, 1968).

5. Shoppee, "Christopher Kelk Ingold," 349; also see Christopher Ingold, "Education of a Scientist," *Nature* 196 (1962): 1030–1034.

6. Seddon, "The Development of Electronic Theory," 46. I am grateful to Mrs. Jennifer Seddon Curtis for making this thesis available to me.

7. See Shoppee, "Christopher Kelk Ingold," 349, 370–371. The Ingolds' three children, two

As a young man in Thorpe's laboratory at Imperial College, Ingold was said to be "friendly, helpful, purposeful, but remote."[8] When he moved to Leeds to direct his own research school, he became more extroverted, playing cricket for the Leeds University Staff Cricket Club and entertaining research students with excellent dinners, followed by what his former student, Charles Shoppee, called "brainy" games.[9] The sports metaphor showed up later in Ingold's reflections on his years at Leeds: "With a pretty good team, keen and young, and with a major and almost boundless subject opening up, we had a good time."[10]

Appointed to succeed Robinson at University College, London (UCL), in 1930, Ingold, Robinson, and Donnan, then head of the chemistry department, briefly overlapped in their tenures in London. Ingold succeeded Donnan as director of the UCL chemistry department laboratories in 1937 and remained director until his retirement in 1961. Ingold was largely responsible for the content of a new B.Sc. special degree (1945) at London which required thorough grounding in the mathematical and physical aspects of chemistry. Under Ingold, the chemistry department began sponsoring a single series of colloquia, rather than specialized series, in conformity with Ingold's view of the unity of chemistry. As described later by John H. S. Green, in a Festschrift volume for Ingold, the UCL chemistry department successfully merged physical and organic chemistry, then inorganic chemistry, into a single discipline of chemistry.[11]

The years 1939 to 1944 were spent at University College, Aberystwyth, where the chemistry department was evacuated from London. Then, as in the past few years, much of Ingold's work was carried out in collaboration with his postdoctoral research student and later colleague Edward D. Hughes.[12] One of their students, Otto Theodor Benfey, recalls that Ingold and Hughes often came round the laboratory at Aberystwyth each day asking what was new and

daughters and a son, graduated in medicine, geography, and chemistry, respectively. Edith Hilda Ingold abandoned her experimental work in chemistry in 1939 but remained active in her husband's professional work by typing his papers and the manuscripts for both editions of his book.

8. Ibid., 370.

9. Ibid.

10. Quoted in Frederick Challenger, "Schools of Chemistry in Great Britain and Ireland. IV. The Chemistry Department of the University of Leeds," *J. Royal Inst. Chem.* 77 (1953): 161–171, on 166. Challenger was Ingold's successor at Leeds, when Ingold moved to University College, London, in 1930.

11. Shoppee, "Christopher Kelk Ingold," 350, 358, 369; and J. H. S. Green, "Sir Christopher Ingold and the Chemistry Department, University College, London," in J. H. Ridd, ed., *Studies on Chemical Structure and Reactivity Presented to Sir Christopher Ingold* (London: Methuen, 1966): 265–274, on 271.

12. Hughes taught briefly in Bangor, from 1944 to 1948, where he had grown up. Ibid., 350–351, 355. See C. K. Ingold, "Edward David Hughes, 1906–1963," *Biographical Memoirs of Fellows of the Royal Society* 10 (1964): 147–182.

encouraging research students in their work. Ingold counseled the young, somewhat frustrated Benfey that one does only one-eighth of the work during the first year of a two-year research problem, but the second year brings the other seven-eighths. Taking account of Benfey's Quaker pacifism, Ingold found him a research topic unconnected with the war effort.[13]

Through the years from his laboratory studies under Thorpe until his return to London in 1944, and still afterward, the heart of Ingold's theoretical chemistry was a recognition of the complementary roles in chemical processes of three-dimensional *structure* and physicochemical *forces.* His early career is marked by an effort to sort out the directing role in reacting molecules played by electrical and polar effects, in combination with stereochemical effects studied by chemists imbued with nineteenth-century concepts of structure and valence.

The power of Ingold's synthesis of physical chemistry and organic chemistry lay in his dealing with this tension between the conceptual demands of understanding both molecular mechanics and molecular structure. Four aspects of the work of Ingold's laboratory schools at Leeds and London contributed to framing the final theoretical synthesis: (1) the demarcation of stereochemical and polar effects; (2) the concept of resonance or mesomerism; (3) the schema of substitution and elimination reaction mechanisms; and (4) aspects of physical chemistry that included kinetics, measurement of dipole moments, and molecular spectroscopy. These lines of research, compatible with emerging quantum chemistry, developed independently of quantum mechanics, guaranteeing that the Ingold system would be a set of chemical, not physical, theories. We turn now to the first of these research programs.

"Classical" Structure Chemistry versus Polar Chemistry

When Ingold originally moved to Leeds in 1924 to succeed Julius B. Cohen as professor of organic chemistry, his teaching responsibilities led him, he later reflected, "by trial and error . . . to present organic chemistry to students more rationally and less empirically than was formally the custom—as a science rather than an art."[14] At Leeds, Ingold quickly found a valuable colleague in H. M. Dawson, who had studied at Manchester with Arthur Smithells and in Germany with van't Hoff, K. Elbs, and Abegg. Thermodynamics and kinetics were Dawson's principal interests: "Dawson taught me a lot of physical chemistry in a quiet way, and I became very interested in his attempts to sort out the kinetic effects of the constitutents of electrolytic solutions," Ingold later reminisced.[15]

13. Personal communication, 11 July 1990, from O. T. Benfey.
14. Quoted in Challenger, "Schools of Chemistry," 166.
15. Ibid.

While at Leeds from 1924 to 1930, Ingold's laboratory focused on three main topics of research: (1) the nature and mechanism of orienting effects of groups in aromatic substitution (mainly nitration); (2) the study of prototropic rearrangements (shifts of H^+) and aniontropic rearrangements (shifts of anions) as the ionic mechanisms of tautomerism; and (3) the effect of polar substituents on the velocity and orientation of addition reactions to unsaturated systems. One of Ingold's students at Leeds, John William Baker, wrote a widely read book on tautomerism.[16]

What was Ingold's preparation under Thorpe for the research program he set up at Leeds in the mid-1920s? In his postgraduate work with Thorpe, Ingold already demonstrated considerable mathematical ability, in combination with a classical, geometrical way of thinking about intramolecular ''strain.'' In 1920, he appended to a paper by R. M. Beesley and Jocelyn Thorpe a mathematical analysis of the physical forces that could give rise to curved lines of valencies accounting for the ''strain'' they treated in the classical, stereochemical tradition of Baeyer's strain theory (1885).[17] In a paper published the following year, Ingold brought into consideration the space-volumes occupied by molecular substituents, and over the next years, he and Thorpe developed a valency-distortion hypothesis, later revised in the 1940s by Ingold, Israel Dostrovsky, and Hughes into an interpretation of steric hindrance.[18]

It was Ingold's paper of 1921 that Nevil Sidgwick and Erich Hückel cited in the late 1920s as a principal source for new views on strain and ring formation. Even here, as we saw in a different context, Ingold's method of citing earlier work resulted in some ill will. Writing Sidgwick in March 1928, Thorpe complained about Sidgwick having cited Ingold, stating that it was he, Thorpe, who had initiated investigations on valency deflections and that he continued that line of research, alone, after the war. But, he said, since Sidgwick's and

16. Ibid., 167; and A. N. Shimmin, *The University of Leeds: The First Half-Century* (Cambridge: Cambridge University Press, 1954): 146. On Ingold's move to UCL, J. W. Baker was appointed ''Reader in the Mechanism of Organic Reactions'' and F. Challenger was appointed ''Professor of Organic Chemistry'' at Leeds. Baker's book is *Tautomerism* (London: Routledge, 1934), with an appendix by Ingold.

17. See Shoppee, ''Christopher Kelk Ingold,'' 351; and Seddon, ''The Development of Electronic Theory,'' 32. Ingold's note (603–620) was added to R. M. Beesley and J. F. Thorpe, ''The Formation and Stability of Associated Alicyclic Systems. Pt. I. A System of Nomenclature, and Some Derivatives of Methane-II-Cyclopropane and of Methane-III-Cyclopropane,'' *JCS* 117 (1920): 591–620; also see C. K. Ingold, R. M. Beesley, and J. F. Thorpe, ''The Formation and Stability of Spiro-Compounds. Pt. I. Spiro-Compounds from Cyclohexane,'' *JCS* 107 (1915): 1080–1106.

18. See Shoppee, ''Christopher Kelk Ingold,'' 351; Seddon, ''The Development of Electronic Theory,'' 32–33. C. K. Ingold, ''The Conditions Underlying the Formation of Unsaturated and of Cyclic Compounds,'' 305–329. C. K. Ingold and J. F. Thorpe, ''The Hypothesis of Valency-Deflexion,'' *JCS* 131 (1928): 1318–1321; and C. K. Ingold, I. Dostrovsky, and E. D. Hughes, ''Mechanism of Substitution at a Saturated Carbon Atom. Pt. XXXII. The Role of Steric Hindrance: (Section G) Magnitude of Steric Effects. Range of Occurrence of Steric and Polar Effects, and Place of the Wagner Rearrangement in Nucleophilic Substitution and Elimination,'' *JCS* 149 (1946): 173–194.

Hückel's recent publications had singled out Ingold's work, "I fear, little can be done to remove the impression created when Ingold published his 1921 paper when he was a member of my research school" that Ingold should be credited for the work.[19] Jennifer Seddon (now Curtis) argues that Ingold deserved special credit: "Ingold had staged a revival of interest in stereochemistry, producing a new theory on ring formation to supersede the nineteenth-century ideas."[20]

In a paper published with Thorpe and E. A. Perren in 1922, Ingold and his co-authors speculated on the interaction of polar and spatial effects to account for the tautomerism of glutaconic substances that have strongly electronegative substituents slightly inhibiting reaction. "We must postulate some kind of polar effect superimposed on, or in some way modifying the simple spatial phenomenon."[21] This statement calls into question the accuracy of the characterization, noted in chapter 7, that the team of Thorpe and Ingold believed "that in organic chemistry we do not know much more of the fundamentals than that the carbon atom is tetrahedral in its grasp."[22]

Still, Ingold was less willing to speculate in the early 1920s about electrons and ions than his older colleagues, Lapworth and Robinson. Indeed, Ingold's misgivings about hasty speculations regarding polar effects likely are mirrored in a 1923 paper by Usherwood. She postulated that the hypothesis of activation of a hydrogen atom is the conceptual link among phenomena of tautomerism, Michael reactions, and aldol condensations but cautioned that

> the material could with interest be interpreted in the light of the polarity hypothesis, but our aim is not so much to institute propositions regarding the ultimate cause of phenomena as to effect collection of data with the minimum of hypothesis, to reduce the tendency to undergo reaction to terms of molecular structure, and hence to seek immediate rather than ultimate causes of the facts on record.[23]

But we cannot draw the conclusion that Ingold was uninterested in theory in the early 1920s, as Shorter has suggested.[24] Rather, Ingold (like Thorpe)

19. Letter from Jocelyn Thorpe to Nevil Sidwick, dated 15 March 1928, Sidgwick Papers, Folder 66, Lincoln College, Oxford.

20. Seddon, "The Development of Electronic Theory," 33.

21. Quoted in Seddon, (1972: 41), from C. K. Ingold, E. A. Perren, and J. F. Thorpe, "Ring-Chain Tautomerism. Pt. III. The Occurrence of Tautomerism of the Three-Carbon (Glutaconic) Type between a Homocyclic Compound and Its Unsaturated Open-Chain Isomeride," *JCS* 121 (1922): 1765–1789.

22. Quoted in J. Shorter, "Electronic Theories of Organic Chemistry," 63, from editor's remarks, in *Chemistry and Industry* (1924).

23. Quoted in Seddon, (1972: 43), from E. H. Usherwood [later, Ingold], in *Chemistry and Industry* 42 (1923): 1246.

24. This is the suggestion of Shorter (1987: 63): "It is probably fair to say that in 1923 he was not greatly interested in theoretical organic chemistry." Evidence of this, according to Shorter, is that Ingold did not speak and may not have been present in Cambridge at the Faraday Society Meeting. Is it possible he was on his honeymoon?

was opposed to simple electric or electronic ideas on polarity as pursued in the early 1920s by Lowry and by Lapworth and Robinson.[25] In contrast, during the period 1922–1925, Ingold and Hilda Usherwood (Ingold) published several papers dealing with steric, kinetic, and polar influences on velocity of reaction and equilibrium position. In 1925, Hilda Ingold demonstrated that polar influences can affect the velocity of reaction but not the equilibrium position, writing that perhaps this conclusion will ''aid the sorting out of superimposed polar and spatial influences which regulate organic reactions.''[26]

The study of steric effects and steric hindrance of reactivity continued to be a long-lived aspect of investigations in the Ingold school but with a change in focus from nineteenth-century concern with the ''normal'' structure of the molecule to a new concern with calculating differences in energy and entropy between the normal and the *transition* state or activated structure of the molecule. At UCL in the 1930s, these investigations were to include elaborate kinetic analyses of the hydrolysis of carboxylic esters and work on electrometric titration curves of dibasic acids, to separate out spatial and polar factors in reactivity.[27] In addition, it was not just bond forces but forces between nonbonded atoms that were considered; Ingold concluded that it is only study of the transition state that can explain those experimental results known to be contrary to inferences from classical stereochemistry, for example, the experimental result that halogen is much easier to displace in tertiary butyl-halides (e.g., $[CH_3]_3CCl$) than in neopentyl halides (e.g., $[CH_3]_3CCH_2Cl$).[28]

Conjugated Systems and Resonance

From the work with Thorpe on ring formation in halogenated glutaconic acids, Ingold became increasingly interested in the influence of a carbon ring's semiaromatic character on its reactions. In the early 1920s, following a chat with Soddy at Oxford, Ingold reflected on a lecture Soddy had given on the subject of benzene some twenty years earlier:

25. Of Lowry's ''speculations'' on the mechanisms of cyclopropane reactions in 1923, Ingold said, they ''pass quite beyond anything we have experimentally demonstrated.'' Quoted in Seddon (1972:41).

26. E. H. Ingold, ''The Correlation of Additive Reactions with Tautomeric Change. Pt. IV. The Effect of Polar Conditions on Reversibility,'' *JCS* 127 (1925): 469–475, on 470; quoted in Seddon (1972:47).

27. See Shoppee, ''Christopher Kelk Ingold,''. 353–354; e.g., C. K. Ingold, ''The Mechanisms of, and Constitutional Factors Controlling, the Hydrolysis of Carboxylic Esters. Pt. III. The Calculation of Molecular Dimensions from Hydrolytic Stability Maxima,'' *JCS* 133 (1930): 1375–1386.

28. See C. K. Ingold, I. Dostrovsky, and E. D. Hughes, ''Mechanisms of Substitution,'' *JCS* 149 (1946): 173–194; and esp. Ingold, ''Les réactions de la chimie organique (quatre conférences), *Actualités Scientifiques et Industrielles*, no. 1037 (Paris: Hermann, 1948), 32–38. These lectures were given at the Faculty of Sciences in Paris in May 1946 on the invitation of Edmond Bauer, Jean Perrin's successor at the Sorbonne and the Laboratoire de Chimie Physique.

> I should like to see a point of union between yours and Bamberger's views and the ideas I have developed that benzene has a Dewar form tautomeric with the two Kekulé individuals.[29]

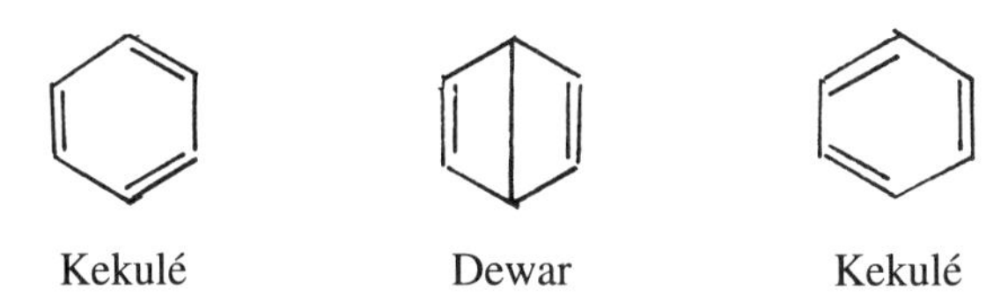

The beauty of the Dewar structure, Ingold remarked, lay in the fact that it explains quinoid transformations and aromatic tautomerism as well-established principles of aliphatic and alicyclic chemistry, that is, it provided a rationale for the unification of aliphatic and aromatic chemistry.

Ingold wrote Soddy that he was going to test the hypothesis by building up the Dewar structure by the methods used to obtain bridged-ring compounds, that is, by closing carbon rings separately.[30]

$$\begin{array}{lllll} C-C-C{=}CX & & C-C-C & & \\ \| \quad\;\; | & \rightarrow & \| \quad\; | \quad\; \| & + & XY \\ C-CY & & C-C-C & & \end{array}$$

In his first published paper on the structure of benzene in 1922, Ingold stated his aim to unify aliphatic and aromatic chemistry by studies of ring formation of unsaturated systems. He argued that chemists must accept the Dewar formulation, using a bridged phase of benzene, in order to relate the properties of aromatic compounds to the facts of aliphatic chemistry.[31]

At this time, Ingold was using classical notions of valency, including partial or oscillating valency, in the tradition of Thiele and Werner as it was now being extended in England by Flürscheim.

In contrast, Lowry, Howard J. Lucas, and Arndt all were pressing forward with the hypothesis of inner ionic charges in conjugated compounds. Arndt

29. Letter from C. K. Ingold to Frederick Soddy, undated, Soddy Papers, Ms. Eng. misc. b.179 (item 98); Bod.Oxford.

30. Undated letter from C. K. Ingold to Frederick Soddy, in Ms. Eng. misc. b.179 (item 98); Soddy Papers, Bod.Oxford. Soddy's paper is "A New Theory of Aromatic Character," read before the Oxford University Chemical Club at the Museum, 26 October 1899.

31. See Seddon, "The Development of Electronic Theory," 39. C. K. Ingold, "The Structure of the Benzene Nucleus. Pt. I. Intra-nuclear Tautomerism," *JCS* 121 (1922): 1133–1143; and "The Structure of the Benzene Nucleus. Pt. II. Synthetic formation of the Bridged Modification of the Nucleus," *JCS* 121 (1922): 1143–1153.

was focusing on the gamma pyrones, which do not exhibit the characteristic reaction of either olefins or ketones, as one would expect. Instead, they exhibit a peculiarly ''aromatic'' character.

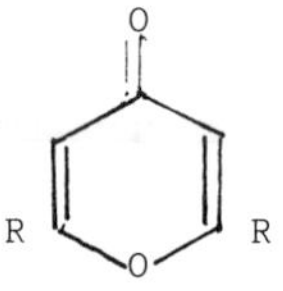

Influenced by P. Pfeiffer's 1922 paper, postulating separated charges (*Zwitterions*) within aminoacids, Arndt wrote formulas for gamma pyrone and gamma-thiopyrone as ''Zwitterion'' dipolar molecules:

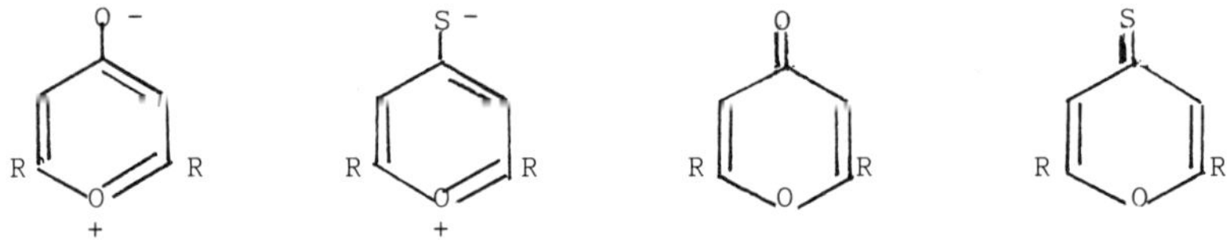

He suggested that the ionic formulas, like the nonionic formulas, ''represent formulations of extremes'' and that no bond across the ring is required. Using the hypothesis of the motions of valence electrons, as developed by Stark and Kossel, Arndt suggested the possibility of intermediate valence states (*Zwitterstufen*) as well.[32] Independently, Robinson proposed possible electronic shifts in pyrones and similar systems, but he did not state the idea of a definite ''intermediate state'' of the molecule between the ionic and uncharged formulas.[33]

By 1926, as we have seen, Ingold began to substitute notions of electron

32. Quoted in E. Campaigne, ''The Contributions of Fritz Arndt to Resonance Theory,'' *J.Chem.Ed.* 36 (1959): 336–339, on 337, from letter from Fritz Arndt to his former colleague, Campaigne. See F. Arndt, E. Scholz, and F. Nachtwey, ''Ueber Dipyrilene und über die Bindungsverhältnisse in Pyron-Ringsystem,'' *Berichte* 57 (1924): 1903–1911, on 1906; and F. Arndt, ''Gleichgewicht und 'Zwischenstufe,' '' *Berichte* 63 (1930): 2963–2966. After Arndt was dismissed from Breslau in 1933 under the infamous German ''civil service laws,'' he spent a year at UCL with Robinson. With Bernhard Eistert, he published a later article on resonance, ''Zwischenstufen,'' and electronic formulas in *ZPC* B31 (1936): 125–131. Eistert, who remained in Germany, wrote *Tautomerie und Mesomerie, Gleichgewicht und ''Resonanz''* (Stuttgart: F. Enke, 1938).

33. Arndt discusses Robinson in a letter to Campaigne, quoted on 337.

valence and electron shift for the classical affinity-valence concepts he previously had been using. In the paper written with Hilda Ingold, which Robinson called Ingold's "conversion" statement, they concluded that the theory of alternating polar charges "leads to much the same key-efficiency series as Flürscheim's, if residual affinity be interpreted as an indication of the presence of loosely held, active electrons."[34]

Linus Pauling later credited this 1926 Ingold and Ingold paper with presenting clearly the idea already broached in 1924 by Lowry, Arndt, and Pauling's Caltech colleague, Lucas,[35] that molecules in their ("real") normal state may have a structure different from any of the familiar valence-bond structures that can be imagined. Pauling's comment about Ingold and Ingold's paper apparently is a reading of their statement that the alternate noncharged and internally charged aromatic formulas express "only the direction of imaginary gross changes which actually do not at any time proceed to more than a limited (in some cases an exceedingly small) extent."[36] In 1923, Ingold and Thorpe together had noted the need for a means of expressing a molecular structure that could not be satisfactorily represented either by old-fashioned bond formulas or by new electron-pair bond formulas. But they were not using internally charged formulas.[37]

In a paper on free radicals published in 1929, Ingold and H. Burton suggested that the forces responsible for free radical stability and for the peculiarities of benzene valency arise from the delocalization of electrons permitted by alternative valency structures.[38] Ingold was influenced, he later said, by

34. C. K. Ingold and E. H. Ingold, "The Nature of the Alternating Effect in Carbon Chains. Pt. V. A Discussion of Aromatic Substitution with Special Reference to the Respective Roles of Polar and Non-Polar Dissociation; and a Further Study of the Relative Directive Efficiencies of Oxygen and Nitrogen," *JCS* 129 (1926): 1310–1328, on 1312.

35. For Lucas, see H. J. Lucas and Archibald Y. Jameson, "Electron Displacement in Carbon Compounds. I. Electron Displacement versus Alternate Polarity in Aliphatic Compounds," *JACS* 46 (1924): 2475–2482; H. J. Lucas and H. W. Moyse, "Electron Displacement in Carbon Compounds. II. HBr and 2-Pentene," *JACS* 47 (1925): 1459–1461; and H. J. Lucas, T. P. Simpson, and J. M. Carter, "Electron Displacement in Carbon Compounds. III. Polarity Differences in C-H Unions," *JACS* 47 (1925): 1462–1469.

36. Linus Pauling, "The Nature of the Theory of Resonance," 1–8, in [Sir] Alexander Todd, ed., *Perspectives in Organic Chemistry* (New York: Interscience Publishers, 1956): 6–7, regarding Ingold and Ingold (1926), 1312. Todd's volume is dedicated to Robert Robinson on the occasion of his seventieth birthday, September 13, 1956.

37. J. Thorpe and C. K. Ingold, "Quelques nouveaux aspects de la tautomérie par MM. Jocelyn Field Thorpe et Christopher Kelk Ingold," *Bull.SCF* 33 (1923): 1342–1391. In *Valence and Structure of Atoms and Molecules* (Washington, D.C.: Chemical Catalogue Company, 1923), G. N. Lewis cited this paper, after speculating that every chemical process may occur in a way to increase the net amount of conjugation. "In all such cases of conjugation it is evident that we are dealing with conditions which cannot be adequately represented either by old-fashioned bond formulae or by mere translation of the old formulae into our present theory" (153, 155–156).

38. See Shoppee, "Christopher Kelk Ingold," 355, discussing C. K. Ingold and H. Burton, "The Existence and Stability of Free Radicals," *Proceedings of the Leeds Philosophical Society*, Science Section, 1 (1929): 421ff.; also, Ingold, *Structure and Mechanism on Organic Chemistry*, 83–84.

Walter Heitler's and Fritz London's demonstration (1927) that the source of strength of the covalent bond lies in quantal resonance or the tendency of electrons to maximize their freedom from excessive localization.

As discussed in more detail in chapter 9, Erich Hückel, like Friedrich Hund and Robert S. Mulliken, was developing molecular orbital theory.[39] In a 1932 paper, Pauling analyzed molecules (like carbon dioxide, benzene, and graphite) and ions (carbonate and nitrate) in order to work out a more general and simple mathematical model than Hückel's. This led to Pauling's classic paper with George Willard Wheland in 1933. Their formulation of resonance was based on the mathematical analogy between mechanical resonance and the behavior of wave functions in quantum mechanical exchange phenomena.[40]

In 1934 Ingold defined the *chemical* term "mesomerism" to mean stable intermediate states explained *physically* by quantal resonance.[41] He stressed a preference for the word "mesomerism" over the term "tautomerism" or "resonance" on the grounds that it did not lead to the false impression that alternate structural formulas are passing into each other very rapidly, like tautomerides. If such rapid motion were going on, he argued, every molecule must spend most or all of its time in transforming itself, and the term molecular "state" simply loses its meaning.[42]

> We are here picturing the production of real states [mesomeric forms] from unreal states [i.e., the ordinary structural formulas] and not the deformation of real states by some external disturbance, as is the case in most of the physical problems to which perturbation theory is commonly applied. . . . There can be no physical separation . . . between resonance vibrations and other electronic vibrations; it follows that the unperturbed structures . . . are only of the nature of intellectual scaffolding, and that the actual state is the mesomeric state. Chemical evidence in support of these ideas is extensive.[43]

39. F. Hund, "Zur Deutung der Molekulspektren. IV," *ZP* 51 (1928): 759–795; R. S. Mulliken, "The Assignment of Quantum Numbers for Electrons in Molecules. II. Correlation of Molecular and Atomic Electron States," *Physical Review* 32 (1928): 761–772; E. Hückel, "Zur Quantentheorie der Doppelbindung," *ZP* 60 (1930): 423–456, and "Quantentheoretische Beiträge zum Benzolproblem. I. Die Elektronenkonfiguration des Benzols und Verwandter Verbindungen," *ZP* 70 (1931): 204–286.

40. Linus Pauling and G. W. Wheland, "The Nature of the Chemical Bond. V. The Quantum-Mechanical Calculation of the Resonance Energy of Benzene and Naphthalene and the Hydrocarbon Free Radicals," *J.Chem.Physics* 1 (1933): 362–374.

41. C. K. Ingold, "Mesomerism and Tautomerism," *Nature* 133 (1934): 946–947. For Pauling, see L. Pauling and G. W. Wheland, "The Nature of the Chemical Bond," 362–374; and Linus Pauling and E. Bright Wilson, Jr., *Introduction to Quantum Mechanics with Applications to Chemistry* (New York: McGraw-Hill, 1935). Werner Heisenberg's first papers on the resonance phenomenon in quantum mechanics are "Mehrkörperproblem und Resonanz in der Quantenmechanik," *ZP* 38 (1926): 411–426, and "Mehrkörperproblem und Resonanz in der Quantenmechanik. II," *ZP* 41 (1927): 239–267.

42. C. K. Ingold, "Principles of an Electronic Theory of Organic Reactions," *Chemical Reviews* 15 (1934): 225–274, on 251; reiterated, 256.

43. Ibid., 250–252.

Ingold's concern about the misunderstanding of the chemical meaning of Pauling's quantal resonance theory was well placed. London, now a refugee from Germany, was in residence at Oxford during 1933–1935. According to Paul Hoch and E. J. Yoxen, London was "virtually the only person in Oxford" other than Erwin Schrödinger, also briefly in residence there, "whom one could consult in any matter of quantum mechanics" at this time.[44] London wrote a note to Sidgwick in spring 1934 about the frequent misreading of Pauling's papers on resonance. Pauling's idea has nothing to do with the concept of moving atoms, à la Kekulé, insisted London, but demonstrates that the stationary state of a configuration of electrons with fixed nuclei cannot always be represented by just one eigenfunction corresponding to an electronic structure of the Lewis type.[45]

Pauling himself insisted on the origins of the resonance, or mesomerism, theory in the conceptions of classical organic chemistry.

> Classical structure theory was developed purely from chemical facts, without any help from physics. The theory of resonance was well on its way toward formulation before quantum mechanics was discovered. . . . It is true that the idea of resonance energy was then provided by quantum mechanics . . . but the theory of resonance in chemistry has gone far beyond the region of application in which any precise quantum mechanical calculations have been made, and its great extension has been almost entirely empirical. . . . The theory of resonance in chemistry is an essentially qualitative theory, which, like the classical structure theory, depends for its successful application largely upon a chemical feeling that is developed through practice.[46]

Ingold himself was absolutely clear in stating that the resonance thesis was necessary on "quite elementary chemical grounds."[47]

Principles for Aliphatic and Aromatic Substitution

Charles Shoppee, Ingold's student, who wrote a paper with Ingold and Thorpe in 1926 on the mechanism of tautomeric interchange,[48] later claimed

44. Paul K. Hoch and E. J. Yoxen, "Schrödinger at Oxford: A Hypothetical National Cultural Synthesis Which Failed," *Annals of Science* 44 (1987): 593–616, on 604.

45. Letter from Fritz London to Nevil Sidgwick, dated Oxford, 10 May 1934, Sidgwick Papers, L.Coll., Oxford.

46. Linus Pauling, "The Nature of the Theory of Resonance," p. 7.

47. Ingold, "Mesomerism and Tautomerism," 947.

48. C. K. Ingold, C. W. Shoppee and J. F. Thorpe, "The Mechanism of Tautomeric Interchange and the Effect of Structure on Mobility and Equilibrium. Pt. I. The Three-Carbon System," *J. Chem. Soc.* 129 (1926): 1477–1488. This apparently is the paper that triggered Robinson's letter to Jocelyn Thorpe, discussed in an earlier chapter. In the Ingold, Shoppee, and Thorpe paper, there is one reference to Robinson, namely, to the Allen, Oxford, Robinson, and Smith paper, which Ingold had read in manuscript according to Robinson. The reference to Allen, Oxford, Robinson, and Smith is paired with a reference to Ingold and Ingold, 1926, essentially giving them equal credit for priority, since Ingold notes that his paper was written before the Robinson et al. paper appeared in print.

that the seeds of the controversy between Ingold and Lapworth and Robinson lay in Ingold's work directed toward studying the benzene structure by investigating addition reactions involving carbon rings and groups like C = O and N = O.[49] As we have seen, in the 1926 paper with his wife Hilda, Ingold rejected Lapworth's theory of alternating polarities in conjugated systems but combined Robinson's fundamental principle of *ortho*, *para*-substitution by electronic displacement (in the draft copy of the Allen, Oxford, Robinson, and Smith paper) with Flürscheim's concept of alternating affinity content. The result was Ingold's "conversion" and a new theory, integrating Robinson's and Flürscheim's imagery and inventing a new symbolism and language.[50]

In his 1926 "Annual Report" on organic chemistry for the homocyclic division of the Chemical Society, Ingold generalized results from earlier papers of the year, and he distinguished two types of electronic displacement. First, the *inductive* effect (labeled *I*), adumbrated by Lewis in 1916, is due to unequal sharing of electrons in an electron pair; it is a permanently exerted mechanism analogous to electrostatic induction (designated by →). Second, the *tautomeric* effect (labeled *T*), also called the *electromeric* effect by Fry (discussed in chap. 5), is an electronic interpretation of the Thiele-Flürscheim notion of residual affinity exchange. *T* is not necessarily a permanent molecular condition but an activation mechanism resulting in the formation of a dipole molecule. It is represented by the curved arrow (↷). In 1929, Ingold reiterated this distinction between "polarization" (→) and "polarizability" (↷), emphasizing that there can be a permanent state of polarization associated with the tautomeric or electromeric effect intermediate between the formally nonpolar and dipolar forms. In this case, an electron duplet is permanently under control of three, not two, atomic nuclei. Ingold formally named this mechanism of permanent tautomeric polarization the *mesomeric* effect.[51]

Ingold specified that these effects involve only electrons and not atomic nuclei as in the phenomenon of prototropy. He stressed the significance of measurements of molecular dipole moment, the principles of which had been established in 1912 by Debye.[52] Ingold predicted that mesomeric systems would display measurable dipole effects.[53] He later said that he realized at this time that the theory of dipolar charge and intermediate form was

> too incomplete to admit of satisfactory application to the problems of velocity

49. Shoppee, "Christopher Kelk Ingold," 352.

50. Ingold and Ingold, "The Nature of the Alternating Effect," 1310–1328, esp. 1312.

51. C. K. Ingold, "[Organic Chemistry.] Pt. II. Homocyclic Division," *Annual Reports on the Progress of Chemistry* 23 (1926): 112–149; discussed by Seddon, (1972: 55); and in Ingold, "Principles of an Electronic Theory," 231–233.

52. Ingold, [Report], 1926.

53. See his later reiteration of his 1926 prediction in C. K. Ingold, *Structure and Mechanism in Organic Chemistry* (Ithaca: Cornell University Press and London: G. Bell, 1953): 105. A second edition appeared in 1969.

> and equilibrium in organic reactions as a whole; the necessary generalization [became] one of the principal tasks before the writer and his colleagues since 1926.[54]

Ingold combined his interest in kinetic studies with studies of mechanism, using competitive nitration experiments on benzene and toluene to study both percentage yields and relative velocities of reaction at the *ortho*, *meta*, and *para* positions. He self-consciously chose the competition method over straightforward kinetic methods of studying reaction rates on the grounds that intermediate steps can be masked in the latter studies.[55] The conclusion was that inductive or polarization effects are relayed mainly to the *o, p* positions with only weak effects at *m*; that tautomeric or polarizability effects are influential at *o* and *p* but not at all at *m*; that the polarizability effect is excited only at the approach of a reagent. That is, polarization controls the initial attraction of a reagent to the reacting molecule and then "polarizability" may assist but not hinder subsequent change.[56]

In explaining how all this takes place physically, Ingold invoked electric "field" effects, noting Lewis's argument in *Valence and Structure* (1923) that the inductive and tautomeric effects are relayed through space as well as through intermediate atom links. Ingold and C. C. N. Vass referred to "air-line" links in 1928, as well as to spatial orientations of the connections within the molecule.[57] As we saw above, Ingold increasingly familiarized himself with quantum mechanics in the next few years.

Busy with the move from Leeds to London in 1930, Ingold afterward was in residence at Stanford University in California in 1932, with some leisure to work out generalizations of the results already at hand. He soon published two widely read pieces: "Significance of Tautomerism and of the Reactions of Aromatic Compounds in the Electronic Theory of Organic Reactions," in the *Journal of the Chemical Society*, and the essay review, "Principles of an Electronic Theory of Organic Reactions," in *Chemical Reviews*.[58]

Perhaps Ingold was especially anxious to lay out a complete system of

54. Ingold, "Principles of an Electronic Theory," 235.

55. See C. K. Ingold, *Structure and Mechanism in Organic Chemistry*, 244–245.

56. See Seddon, "The Development of Electronic Theory," 71–72. From 1926 through 1934, Ingold wrote or co-wrote 110 of 178 published papers with 46 colleagues or junior collaborators. A bibliography of the scientific papers is given in Shoppee (1972: 385–411).

57. Ingold, "Principles of an Electronic Theory," 233–234, discussing Lewis (1923); and C. K. Ingold and C. C. N. Vass, "The Nature of the Alternating Effect in Carbon Chains. Pt. XXIII. Anomalous Orientation by Halogens, and Its Bearing on the Problem of the *Ortho-Para*-Ratio in Aromatic Substitution," *JCS* 131 (1928): 417–425.

58. C. K. Ingold, "The Principles of Aromatic Substitution from the Standpoint of the Electronic Theory of Valence," *Recueil des Travaux Chimiques* 48 (1929): 797–812; "Significance of Tautomerism and of the Reactions of Aromatic Compounds in the Electronic Theory of Organic Reactions," *JCS* 136 (1933): 1120–1127; and "Principles of an Electronic Theory of Organic Reactions," *Chemical Reviews* 15 (1934): 225–274.

reaction theory following the 1931 meeting of the fourth Solvay chemistry congress, at which Lowry, Robinson, and Ingold were all present. The topic for the meeting was ''Constitution and Configuration of Organic Molecules.'' Robinson presented a paper on the ''electrochemical'' theory of organic reaction mechanisms, remarking that he had been encouraged by colleagues to lay out the theory ''which I have developed.''

Robinson used some of Ingold's and other colleagues' terminology and results, notably, the $\delta+$, $\delta-$ notation for designating the acquisition of partial charge through electron displacement, distinguished from [+] and [-] for marking integral charges. Robinson persisted in the use of the classification ''cationoid'' and ''anionoid,'' and he explicitly criticized Ingold's use of the word ''tautomeric'' as misleading from a historical point of view.[59]

As the first to take the floor in discussion, Ingold insisted that the general theory that Robinson had outlined could be arrived at ''by another path,'' notably, Ingold's path, through the study of tautomerization, which he said was the historical origin of his theory. Ingold also insisted on the physical reality of polarization and polarizability, measurable by dipole moment and by increases or decreases in an atom's covalency. He further stressed the importance to the stability of electronic groups of the spatial distribution of electrons, as described by quantum mechanical analysis.[60] Lowry, too, noted differences with Robinson, defending his own long-held view of electrical charges possessed by the atom ''at the moment of reaction.''[61]

In his 1933 paper on ''tautomerism,'' Ingold began discussion with references to the physics of the electron, citing a 1923 paper by J. J. Thomson and a recent book by John H. Van Vleck.[62] He noted Lewis's contributions (1923) to the notion of inductive effect (→) in which electrons remain bound by their original atomic nuclei; Lowry (1923) to the notion of electromeric effect, in which there is a displacement of a duplet, shifting from one pair of atoms to another (↷)

$$\mathrm{A} \overset{\curvearrowright}{=} \mathrm{B}-\mathrm{C} \;\rightarrow\; \mathrm{A}-\mathrm{B}=\mathrm{C};$$

and Lucas (1924) for showing how the inductive effect may direct and facilitate the electromeric effect. Ingold credited the generalization of these ideas for

59. Robert Robinson, ''Quelques aspects d'une théorie électrochimique du mécanisme des réactions organiques,'' 423–489, and Discussion, 490–501, in Institut International de Chimie Solvay, *Rapports et discussions relatifs à la constitution et à la configuration des molécules organiques. Quatrième Conseil de Chimie tenu à Bruxelles du 9 au 14 avril 1931* (Paris: Gauthier-Villars, 1931): esp. 427, 436, 450.

60. Ibid., 490–495.

61. Ibid., 497.

62. C. K. Ingold, ''Significance of Tautomerism,'' reprinted in J. W. Baker, *Tautomerism*, 311–324; in reprint, 311. The references are to J. H. Van Vleck's *Electric and Magnetic Susceptibilities* (1932) and J. J. Thomson, in *Phil. Mag.* 46 (1923): 513.

aromatic substitution simultaneously to Robinson (1926) and to Ingold and Ingold (1926), claiming (disingenuously, according to Robinson), in the matter of priority, that "the second paper was communicated before the first appeared."[63]

Ingold now classified reagents as *nucleophiles* and *electrophiles*, and he included a note criticizing Lapworth's "anionoid/cationoid" classification on the grounds that the state of electrification is immaterial; the important thing is not a center's charge but whether it is an electron sink or an electron source. The activation process is one that develops either high or low electron density at the seat of reaction by virtue of polarization or polarizability. The original division into "inductive" and "tautomeric" effects is in fact too rigid, he now claimed. There "is no sharp differentation" between the two effects. "Mesomerism" or the "mesomeric effect" is "tautomeric polarization."

The explanation of "mesomerism" he argued, lies in energy relationships, namely, that there are two or more extreme forms of similar or equal energy, and the ordinary form and energy of the molecule can best be interpreted quantum mechanically. Ingold treated this idea more thoroughly in his 1937 Faraday Society article on the relation between chemical and physical theories of the source of the stability of organic free radicals.[64]

A General Theory for a New Physical Organic Chemistry

What became Ingold's classic essay defining a theory of organic reactions appeared in *Chemical Reviews* in 1934. This essay presented a theory not only of aromatic chemistry but also of aliphatic chemistry, unifying the two major branches of organic chemistry and linking them with general physical chemistry and, to a limited extent, with physics. Whereas the work on aromatic substitution had centered on the chemistry of what Ingold now called "electrophilic" (electron-seeking) reagents, emphasis increasingly now was placed on aliphatic chemistry, where (in Ingold's terminology) nucleophilic substi-

63. Ingold, "Significance of Tautomerism," pp. 312–313 in Baker, citing Lewis, *Valence*, 139; Lowry, "Studies in Electrovalency," 822–831 and 1866–1867; James Allan, Albert E. Oxford, Robert Robinson and John C. Smith, "The Relative Directive Powers," 401–411; E. H.Ingold and C. K. Ingold, "The Nature of the Alternating Effect," 1310–1328.

64. J. Seddon [Curtis] notes that this idea is similar to Robinson's purely chemical descriptive concept of the leveling out of valency (1972: 80–84). However, Fritz Arndt noted in a letter to Campaigne that while Robinson in 1925 formulated potential electronic shifts in pyrones and similar systems that could have led to the Zwitterion concept, "Robinson did not speak of a definite 'intermediate state' of the molecule" between the Zwitterion and ketone [pyrone] formula." In Arndt's view, the main point of the resonance theory is the interrelationship between valences and ionic charges; thus "Kekulé's idea, and the theories of Thiele, Weitz [?], and others cannot be regarded as anticipations of the theory of resonance." In Campaigne (1959: 337). Ingold, "The Relation between Chemical and Physical Theories of the Source of the Stability of Organic Free Radicals," *Trans.Far.Soc.* 30 (1934): 52–57.

tution and elimination dominate reaction mechanisms. Nucleophilic activity is nucleus seeking by virtue of the reagent carrying its own electron pair for bonding, for example, HO^-. Hypotheses about transition-state mechanisms could be tested by kinetic and competition studies, calculations of thermodynamic energy and entropy values, measurements of electric dipole moments, observation of optical activity, pursuit of Raman and infrared spectroscopy, and use of radioactive element tagging.

Of the investigations summarized in the 1934 essay and continued immediately thereafter, Ingold later said,

> The new work made it inescapably clear that the old order in organic chemistry was changing, the art of the subject diminishing, its science increasing; no longer could one just mix things: sophistication in physical chemistry was the base from which all chemists, including the organic chemist, must start.[65]

Ingold's strategy in the 1934 paper was to begin with physical notions of electric polarization and polarizability of molecules and submolecular groups, then to discuss the electron theory of valency. He began setting up systems of symbols by the sixth page of the essay, introducing them with statements like "XX" "may be used," "has been termed," "has been renamed," and so on, usually citing his own papers as references for the new terminology. He apportioned credit for various principles of the general theory (which, he said, have emerged "piecemeal"), using citations along the lines of the 1933 paper on tautomerism. In the essay, Ingold claims his originality (if not priority) for (1) the general theory of organic reaction mechanisms, (2) the recognition of the significance of molecular dipole moments in relation to the theory of reaction mechanisms, and (3) the concept of mesomerism, including the recognition (independently, he claims, from Gerald Branch and Linus Pauling) that the source of the energy of mesomerism lies in valency degeneracy, that is, that valency is not expressible in integral values of 1, 2, 3 . . .[66] The symbol ⌒ is superimposed on traditional formulas to denote the distributed wave function of an unlocalized electron duplet.

$R_2C = CH{-}\bar{O}$	$[R_2\overset{\frown}{C}{-}{-}CH{-}{-}\overset{\frown}{O}]^{-}$	$R_2\bar{C}{-}CH = O$
Unperturbed	Mesomeric	Unperturbed

> Thus the specific and exclusive source of energy which is necessary to determine the independence of the mesomeric effect is identified as perturbation energy

65. Quoted in Shoppee (1972: 356–357).

66. A footnote, n. 3, p. 249, in Ingold, "Principles of an Electronic Theory," states that the idea of mesomerism originating in valency degeneracy was reached independently by G. E. K. Branch "and has been used by him for several years past in teaching and in colloquia." Branch, a colleague of Lewis at Berkeley, was an undergraduate at Edinburgh who studied with Donnan at Liverpool. Branch spent some time at UCL in the 1930s, when both Donnan and Ingold were in the Chemistry Department at London.

> dependent on a degeneracy. The ultimate justification for this hypothesis is, of course, entirely *a posteriori* in character; it is derived partly from . . . physical evidence . . . and partly from the necessity for the hypothesis of the mesomeric effect in the interpretation of organic reactions.[67]

After dealing with polarization and polarizability, including an electrical "picture" of molecular activation and molecular stability, the 1934 essay turns to electron displacement, including the inductive and electromeric effects, mesomeric effect, and (a new) inductomeric effect ("polarizability" caused by induction). There is a historical assessment of analogies of these concepts with preelectronic theories. There follows a (polar) classification of atomic groups and systems, including supporting physical evidence from measurements of dipole measurement, infrared spectroscopy, and thermochemistry, a discussion of the source of the energy of mesomerism in valency degeneracy, and a (polar) classification of reagents, comparing the classification with oxidation-reduction systems and giving detailed examples of nucleophilic reagents, electrophilic reagents, and other particular reagents.

Tables for the relative inductive and electromeric strengths of various ions and molecular groups demonstrate interrelationships among electronegativity, dipolar moment, and molecular structure chemistry. Groups that are more powerful electron attractors than hydrogen (when substituted for hydrogen in a molecule) are said to exhibit negative inductive effects (-I), and groups that are less powerful electron attractors than hydrogen are said to exhibit positive inductive effects (+I). In this scheme, "less attraction than H" (which is essentially neutral) means repelling. Alkyl groups are said generally to function as feebly electron-repelling (+I) groups; radicals such as ethenyl ($CH_2 = CH$-) and phenyl (C_6H_5-) are regarded as intrinsic attractors of electrons. Anionic groups as a whole (O^-, S^-) repel electrons in comparison with neutral groups as a whole; cationic groups (R_3N^+, R_2S^+) attract electrons. (See fig. 9.)

A regular connection between the inductive effect of an atom and its position in the periodic table is adumbrated; for example, electron attraction diminishes in the series F, Cl, Br, I. Electron attraction increases with the number of the periodic group (i.e., from C to N to O) and decreases with increasing number of the periodic series (Cl to Br to I). Correspondingly, molecular dipole moment increases as the periodic group number increases for elements of the same period and, on the whole, as the period number decreases for elements of the same group. Functional groups like $C = O$, $C = N$, and $C \equiv N$ are especially susceptible to nucleophilic attack; groups like $C = C$ and $C \equiv C$ tend to be vulnerable to electrophilic attack.[68] "Now we can see

67. Ingold, "Principles of an Electronic Theory," 250–251.
68. Ibid., 237–240; also see, Edwin S. Gould, *Mechanism and Structure in Organic Chemistry* (New York: Holt, Rinehart and Winston, 1959): 207.

Electron repulsion (+*I*)

$\cdot\bar{N}R > \cdot\bar{O}$;

$\cdot\bar{O} > \cdot\bar{S} > \bar{Se}$;

$(CH_3)_3C\cdot > (CH_3)_2CH\cdot > CH_3\cdot CH_2\cdot > CH_3$

Electron attraction (−*I*)

$\cdot\overset{+}{O}R_2 > \cdot\overset{+}{N}R_3$;

$\cdot\overset{+}{N}R_3 > \cdot\overset{+}{P}R_3 > \cdot\overset{+}{As}R_3 > \cdot\overset{+}{Sb}R_3$;

$\cdot\overset{+}{O}R_2 > \cdot\overset{+}{S}R_2 > \cdot\overset{+}{Se}R_2 > \cdot\overset{+}{Te}R_2$;

$\cdot\overset{+}{N}R_3 > \cdot\overset{+-}{NO_2}$; $\cdot\overset{+-}{\overset{+-}{S}}O_2R > \cdot\overset{+-}{S}OR$; $\cdot\overset{+-}{\overset{+-}{S}}O_2R > \cdot\overset{-}{\overset{+-}{\overset{+-}{\overset{+-}{S}}}}O_3$;

$\cdot\overset{+}{N}R_3 > \cdot NR_2$; $\cdot\overset{+}{O}R_2 > \cdot OR$; $\cdot\overset{+}{S}R_2 > \cdot SR$;

$\cdot F > \cdot OR > \cdot NR_2 (> \cdot CR_3)$;

$\cdot F > \cdot Cl > \cdot Br > \cdot I$;

$:O > :NR > :CR_2$; $\vdots N > \vdots CR$;

$:O > \cdot OR$; $\vdots N > :NR > \cdot NR_2$;

$\cdot C\vdots CR > \cdot CR{:}CR_2\cdot (> \cdot CR_2\cdot CR_3)$

Figure 9. Table for Inductive Effect. From C. K. Ingold, "Principles of an Electronic Theory of Organic Reactions," Chemical Reviews, *15 (1934), p. 238.*

why the oxide-ion group on benzene orients substitution *ortho*, *para*. We can understand why in aromatic substitution "some rules suffer no exception," Ingold later wrote of the power of his theory.[69] The explanatory power of a new theoretical chemistry was firmly established.

The discussion and classification of reagents is masterful in identifying Ingold's new nomenclature and principles with more widely known oxidation-reduction and acid-base theory. The 1953 lectures at Cornell University, published as *Structure and Mechanism in Organic Chemistry*, follow this same strategy, showing how old classification schemes overlap with each other and how apparent inconsistencies disappear as old schemes are incorporated into the new one. Nineteenth-century Berzelian electrochemical dualism, revived by Lapworth and Robinson in the cationic/anionic schema, disappears into the electrophilic/nucleophilic language.

Thus, a reducing agent donates electrons, while an oxidizing agent receives them. The Brönsted-Lowry definitions of acid and base specify that

$$\text{Base} + H^+ \leftrightharpoons \text{Acid}.$$

But using Ingold's nucleophilic-electrophilic nomenclature, all "basic" reagents, organic or inorganic, are nucleophilic; and "basicity" ("affinity" for

69. Ingold, "Les réactions," 13–14.

a hydrogen nucleus or proton) may be regarded as a special case of affinity for atomic nuclei in general. Reactivity depends primarily on the presence of available electrons; the activation of a site of a reaction is due to the development of critical electron density, and thus there is no simple and general connection with the gross electrification (i.e., *charge*) of the reagent. Reagents that during reaction acquire electrons, or a share in electrons, are termed ''electrophilic.'' Brönsted-Lowry ''acids'' are ''electrophilic'' reagents that supply a proton that is bonded in one of the products of the reaction.

The ''condition of electrification is a trivial matter in comparison with the analogies of behavior on the basis of which they are classified as nucleophilic or electrophilic as the case may be.'' Thus, the categorization of reagents as cationoid or anionoid ''cannot be regarded as acceptable'' and many of the earlier controversies during 1924–1928 about aromatic substitution, Ingold argues, were wrongly laid out in terms of positive or negative polar charges.[70] Substitution in saturated compounds is most often nucleophilic and in unsaturated compounds, electrophilic (fig. 10).[71]

With respect to the nomenclature and classification system that Ingold set up, it is interesting to note the incorporation not only of traditional anthropomorphic metaphors of love and war (italicized), but also (without making too much of it) ones suggesting the modern capitalist marketplace (bold-face):

> It is a commonplace of electronic theories that chemical change is an electrical **transaction** and that reagents act by virtue of a constitutional *affinity*, either for electrons or for atomic nuclei. When, for example, an electron-seeking reagent *attacks* some center in an organic molecules . . . [reaction will occur if the center can **supply** electrons].[72]
>
> ''Sharing **economizes** electrons.''[73]

A few years later, when Ingold was inventing additional definitions and rules, he wryly apologized for yet another two new expressions: ''I hold that new words, like falsehoods, should be invented sparingly but . . . I would suggest we may as well have two and be done with it'': ''heterolytic and homolytic.''[74]

70. Ingold, ''Principles of an Electronic Theory,'' 227, 265–270; see also Ingold, *Structure and Mechanism*, esp. 200–203. ''The order in which groups arrange themselves with respect to their polarization will bear no simple relation to the order of their polarizability, and the relative importance of the two contributory effects, dependent respectively on their two independent electrical characteristics, must necessarily vary with the nature of the reaction'' (Ingold, ''Principles of an Electronic Theory,'' 228).

71. Ingold, ''Les réactions,'' 20–21. Fig. 10 is based on the useful summary in H. L. Helmprecht and L. T. Friedman, *Basic Chemistry for the Life Sciences* (New York: McGraw-Hill, 1977): 295–296.

72. Ingold, ''Principles of an Electronic Theory,'' 227.

73. Ingold, *Structures and Mechanism in Organic Chemistry*, 3–4.

74. Ingold, quoted in Davenport, ''On the Comparative Unimportance,'' 526.

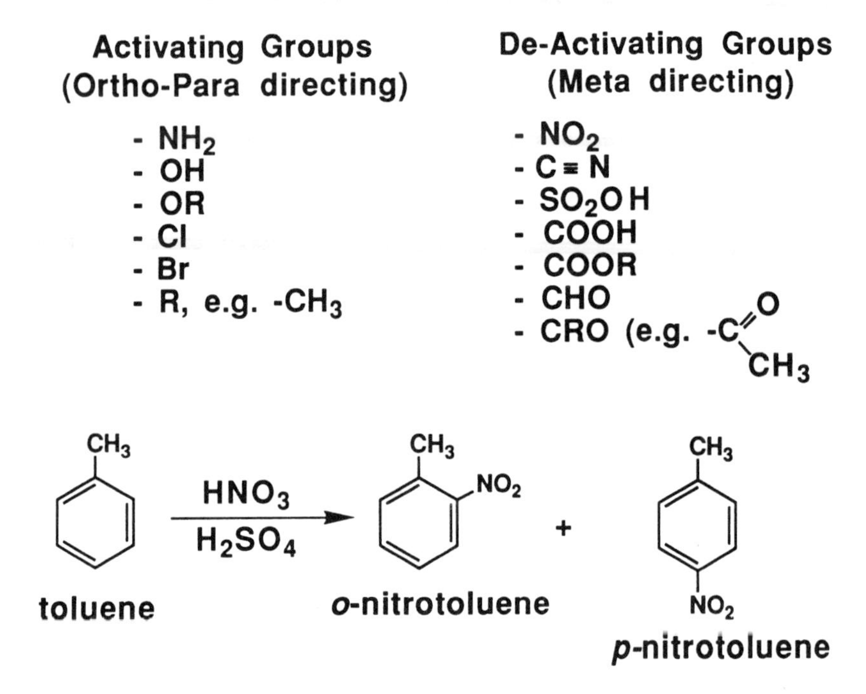

Figure 10. Electrophilic Aromatic Substitution. Drawings courtesy of Glenn Dryhurst.

$$A : B \rightarrow A + :B \text{ (heterolyis)} \qquad A : B \rightarrow A\cdot + B\cdot \text{ (homolysis)}$$

Molecules in heterolytic (polar) reactions form and break bonds by ''coordination''; and molecules in homolytic (nonpolar or free radical) reactions form and break bonds by ''colligation.''[75] (Two more new terms!) Heterolytic reactions occur mostly in solutions, usually involving ion formation and electrophilic or nucleophilic reactions; homolytic reactions occur mostly in gases and do not involve ions because less energy is required to distance the atoms into neutral radicals.[76]

Other terms that he invented include the system of classification for mechanisms of aromatic and aliphatic substitution and elimination reactions, designated S_N1, S_N2, E1, and E2. ''S'' and ''E'' refer to substitution and elimination, respectively, ''N'' to nucleophilic, and ''1'' and ''2'' to ''molecularity,'' or the number of molecules involved in a reaction step (not kinetic order, having to do with the equation for reaction rate and the concentration of reactants). Ingold first introduced some of these ideas in 1928 in a

75. Also see Ingold, *Structure and Mechanism in Organic Chemistry*, 205.
76. Ingold, ''Les réactions,'' 17–23.

paper with E. Rothstein; these ideas on substitution and elimination became more systematic in the work with Hughes during the 1930s and 1940s.[77]

In substitution reactions, a new covalent bond is formed and an old one is broken, as in the hydrolysis of tertiary-butyl bromide:

$$H\bar{O} + (CH_3)_3CBr \rightarrow (CH_3)_3COH + Br^{-} .$$

Ingold and his collaborators particularly targeted for mechanistic studies the reactions of alkyl halides, trialkylsulfonium salts, tetraalkylammonium salts, and substituted ethylene oxides with water or hydroxide ion.[78]. They distinguished two primary mechanical paths for the substitution process. In one (originally called direct displacement), the new bond is formed at the same time that the old bond is breaking; in the transition state, the incoming group and leaving group are both partly bonded to the carbon atom under attack. For example, in the hydrolysis of methyl iodide,

$$HO^{-} + CH_3I \rightarrow [HO^{\delta-} \ldots CH_3 \ldots {}^{\delta+}I] \rightarrow HO-CH_3 + I^{-}.$$

In the second type of mechanism (dissociation), the bond to the leaving group is broken before the new bond is created, the reaction proceeding through an intermediate free carbonium ion. For example, in the hydrolysis of tertiary-butyl chloride:

$$(CH_3)_3C-Cl \xrightarrow[\text{slow}]{-Cl} (CH_3)_3C^{+} \xrightarrow{+H_2O} (CH_3)_3C-OH + H^{+}.$$

In terminology that Robinson and Lapworth were to ridicule, the Ingold school described the carbonium-ion mechanism as one where the leaving group is

77. See Shoppee (1972: 359), citing C. K. Ingold and E. Rothstein, "The Nature of the Alternating Effect in Carbon Chains. Pt. XXV. The Mechanism of Aromatic Side-Chain Substitution," *JCS* 131 (1928): 1217–1221. C. K. Ingold and E. D. Hughes, "Dynamics and Mechanism of Aliphatic Substitutions," *Nature* 132 (1933): 933–934; C. K. Ingold, E. D. Hughes, and S. Masterman, "Reaction Kinetics and the Walden Inversion. Pt. I. Homogeneous Hydrolysis and Alcoholysis of *beta*-n-Octyl Halides," *JCS* 140 (1937): 1196–1201; and subsequent articles.

78. E. D. Hughes, C. K. Ingold, and S. Patel, "Influence of Poles and Polar Linkings on the Course Pursued by Elimination Reactions. Pt. XVI. Mechanism of the Thermal Decomposition of Quaternary Ammonium Compounds," *JCS* 136 (1933): 526–530; E. D. Hughes and C. K. Ingold, "Influence of Poles and Polar Linkings on the Course Pursued by Elimination Reactions. Pt. XXI. Dynamics of the Elimination of the Tert.-Butyl Group from Sulphonium Compounds," *JCS* 136 (1933): 1571–1576; J. L. Gleave, E. D. Hughes, and C. K. Ingold, "Mechanism of Substitution at a Saturated Carbon Atom. Pt. III. Kinetics of the Degradation of Sulphonium Compounds," *JCS* 138 (1935): 236–244; E. D. Hughes and C. K. Ingold," Mechanism of Substitution at a Saturated Carbon Atom. Pt. IV. A Discussion of Constitutional and Solvant Effects on the Mechanism, Kinetics, Velocity, and Orientation of Substitution," *JCS* 138 (1935): 244–255; E. D. Hughes, "Mechanism of Substitution at a Saturated Carbon Atom. Pt. V. Hydrolysis of Tert.-Butyl Chloride," *JCS* 138 (1935): 255–258.

"pulled off" by the solvent, whereas in the direct displacement reaction, it is "pushed off" by the incoming nucleophile but still with the help of the pulling action of the solvent.[79] The Ingold school argued that nucleophilic substitutions on primary carbon atoms (e.g., RCH_2-X) tend to proceed by the direct displacement (S_N2) mechanism and those on tertiary carbon atoms (R_3C—X, i.e., three R groups bonded to C—X) by dissociation (S_N1). Secondary carbon atoms (e.g., R — C(X) — R) constitute borderline cases. Tests for these could be provided by seeing whether alkyl groups (R), said to be more effective electron releasers than hydrogen, would promote the leaving of halogen (X). That is, is a trialkyl substituted methyl halide more easily hydrolyzed than R-CH_2X? And since the S_N1 mechanism was said to involve the formation of separated charge in the transition stage, is it promoted, as would be expected, by solvents of high dielectric constant, like water and formic acid? Such tests confirmed the theory.[80]

A distinction between "molecularity" and "kinetic order" was deliberately made, "Mechanism" of reaction was said to be a matter at the *molecular* level. In contrast, kinetic order is calculated from *macroscopic* quantities "which depend in part on mechanism and in part on circumstances other than mechanism."[81] The kinetic rate of a first-order reaction is proportional to the concentration of just one reactant; the rate of a second-order reaction is proportional to the product of two concentrations. In a substitution of RY by X, if the reagent X is in constant excess, the reaction is (pseudo) unimolecular with respect to its kinetic order but bimolecular with respect to mechanism, since two distinct chemical entities form new bonds or break old bonds during the rate-determining step.

Hughes, Ingold, and their co-workers also attempted to generalize about the mechanism by which elimination of substituents, rather than substitution for substituents, occurs in the aliphatic molecule, resulting in the formation of an olefin in the course of what initially was predicted to be a S_N1 mechanism.

79. See the discussion by Gould (1959: 250–254).

80. See Gould.

81. From Ingold, *Structure and Mechanism in Organic Chemistry*, 315. See Ingold, with L. C. Bateman, K. A. Cooper, and E. D. Hughes, "Mechanism of Substitution at a Saturated Carbon Atom. Pt. XIII. Mechanism Operative in the Hydrolysis of Methyl, Ethyl, Isopropyl, and Tert.-Butyl Bromides in Aqueous Solutions," *JCS* 143 (1940), 925–935.

They found that when the carbonium ion intermediate in a S_N1 reaction has one or more *beta*-hydrogen atoms, an olefin is formed.[82]

$$\underset{[\beta]}{>CH} - \underset{[\alpha]}{\overset{\overset{X}{|}}{C}} < \xrightarrow{-X^-} >CH - C^+ < \xrightarrow{-H^+} >C = C<$$

Olefin formation also occurs when the *alpha* carbon (the one attached to -X; the *beta* carbon is one position removed from *alpha*) is a secondary or tertiary alkyl carbon and more especially when an *alpha*-phenyl or *alpha*-vinyl group is present.[83] Whereas these olefin-forming reactions frequently do not require the presence of an added base, some reactions are accelerated by a base. The latter base-accelerated reactions came to be designated E2, in contrast to E1 reactions.[84]

One of the most spectacular achievements of the work on nucleophilic aliphatic substitution was the solution of the problem of the Walden inversion. In this phenomenon, first studied in detail during the 1890s by Paul Walden, a molecule is turned inside out, in the sense that the configuration of an optically active substrate is converted to its mirror image. Walden constructed a series of cycles for producing a missing enantimorph (or mirror-image isomer) from an available one making use of optical inversions.[85]

In a series of papers during 1935–1937, Hughes and Ingold demonstrated that only reactions involving breakage of a bond at an asymmetric center can lead to inversion and that every act of S_N2 substitution occurs with inversion of configuration. They also proved that in the presence of groups of "suitable polar quality and stereochemical opportunity," meaning that the carbonium-ion transition state intermediate is not shielded from the nucleophilic reagent by a bulky leaving group, the S_N1 mechanism leads to 50 percent inversion and 50 percent retention of configuration. That is, partial or complete racemization is the most characteristic stereochemical consequence of the unhindered S_N1 mechanism because the three valences of the carbonium ion in the transition state are found in a single plane.

"What Walden and Fischer confronted," concluded Ingold, "when they thought the Walden Inversion was rare, was not its rarity but the relative rarity

82. K. A. Cooper, E. D. Hughes, and C. K. Ingold, "The Mechanism of Elimination Reactions. Part III. Unimolecular Olefin Formation from Tert.-Butyl Halides in Acid and Alkaline Aqueous Solutions," *JCS* 140 (1937): 1280.

83. Ibid.

84. Each category of mechanism has very specific meaning. S_N2, for example, means that the reaction often, but not always, follows second-order kinetics, that it involves inversion of configuration because the transition state is linear, and that it has the possibility of steric retardation because of molecular congestion in the transition state. See Shoppee (1972: 359).

85. See Aaron Ihde, *The Development of Modern Chemistry,* 338.

with which the configuration is conserved."[86] Ingold and his collaborators studied the kinetics of the process by introducing radioactive ions into reaction, as well as by studying the velocity of change of optical activity.[87] In the case of inversion, for example, where *Br is a radioactive isotope chemically indistinguishable from ordinary Br,

$$ {}^{*}\mathrm{Br} + \begin{array}{c} \mathrm{C_6H_5} \\ \mathrm{H} \end{array} \!\!>\! \begin{array}{l} \mathrm{C} - \mathrm{Br} \\ \; | \\ \mathrm{CH_3} \end{array} \xrightarrow[\text{acetone}]{\text{in}} {}^{*}\mathrm{Br} - \begin{array}{c} \mathrm{C} \\ | \\ \mathrm{CH_3} \end{array} \!\!<\! \begin{array}{c} \mathrm{C_6H_5} \\ \mathrm{H} \end{array} + \mathrm{Br} \; . $$

Another triumph for the Ingold system of explanation lay in its ability now to give the mechanism for the so-called Hofmann (1851) and Saytzeff (1875) rules for olefin formation, just as it had earlier explained the Koerner (1874), Hübner (1875) and Noelting (1876) and the Crum Brown and Gibson rules for aromatic substitution.[88] The Hofmann rule specifies that when more than one olefin may form from a tetra-alkyl ammonium or trialkylsulfonium salt, the olefin bearing the smaller number of alkyl groups will predominate. The Saytzeff Rule requires that when either of a pair of olefins may result from the dehydrohalogenation of an alkyl halide, the olefin bearing the larger number of alkyl groups will often (but not always) predominate.[89]

Response to the general theory of reaction mechanisms, especially elimination reactions and nucleophilic aliphatic substitution, was by no means altogether favorable. We already have seen resistance to the sweep of Ingold's generalizations by Lapworth and Robinson, on the one hand, and by Kirrmann and Prévost, on the other. Sometimes known abroad as "the English heresy," criticisms regularly occurred especially in regard to the notion of the ionization of organic compounds into alkyl carbonium ions (now called carbocations).[90] An anonymous reviewer of a paper that Hughes submitted to the *Journal of the American Chemical Society* wrote, "This note represents views held on

86. See Shoppee (1972: 362); and Ingold (1948: 27–30). "A great part of the mystery of Walden inversions is dissipated since we have related it to different mechanisms of substitution" (Ingold [1948]: 30). Walden made his discovery in 1895 and listed more than twenty cycles of inversion and retention in *Optische Umkehrserscheinungen* (1919).

87. Ingold, "Les réactions," 27–28.

88. The first rule: negative groups (conferring or enhancing acid properties) direct a new substituent to the *meta* position; positive groups, enhancing or conferring basic properties, as well as neutral and weakly negative groups, direct new substituents into the *ortho, para* positions. The Crum Brown and Gibson Rule: X is *meta*-directing if HX can be directly oxidized to HOX; otherwise, it is *ortho, para* directing. See Ingold (1953: 224–225). Rules for orientation are given in A. F. Holleman, *Die direkte Einführung von Substituenten in den Benzolkern* (Leipzig: Veit, 1910).

89. Gould, "Mechanism and Structure, 480–481; Shoppee, "Christopher Kelk Ingold," 357.

90. See Shoppee, "Christopher Kelk Ingold," 356; and letter from O. T. Benfey to the author, 11 July 1990.

the other side of the Atlantic. It would be deplorable if they came to this country.''[91]

But by 1941, Ingold's results were incorporated into Branch and Melvin Calvin's influential textbook, *The Theory of Organic Chemistry* (1941). Americans were taking postdoctoral appointments at UCL, just as Ingold's students began taking up jobs in the United States. The first conference on organic reaction mechanisms took place at the University of Notre Dame in 1946,[92] the year that Benfey left Great Britain for a postdoctoral appointment with Louis Hammett at Columbia University.

Hammett had just recently published *Physical Organic Chemistry: Reaction Rates, Equilibria, and Mechanisms* (1940), a path-breaking textbook that played a major role in defining physical organic chemistry as a specialized field or discipline. Branch and Calvin followed with *The Theory of Organic Chemistry* in 1941 and A. Edward Remick with *Electronic Interpretations of Organic Chemistry* in 1943. Walter Hückel's ambitious two-volume *Theoretische Grundlagen der Organischen Chemie* (1931) was updated during the 1940s, unusual among organic chemists' texts in its intimate collaboration with a physicist, namely, with Erich Hückel, Walter's brother, a leader in molecular orbital calculations in the 1930s.[93]

As a student of Ingold, now working with Hammett, Benfey earlier had completed doctoral research that

> . . . aimed to lay to rest a suggestion by Hammett that one of the key arguments for a carbocation [carbonium] mechanism (the effects of specific salts) could be explained by factors not involving [these] ions. My work thoroughly convinced Hammett who in any case was quite open to the idea of organic ions.[94]

Much of the first edition of Ingold's classic 1953 book was written on the western side of the Atlantic, while he was lecturing at Cornell University during 1950–51. It surely was no accident that the title was chosen to reflect his longtime aims of unification: structure *and* mechanism. It became a fundamental text for the new field, its only rival Hammett's 1940 text that had enormous influence in the United States.[95]

Ingold's book, based on the principles outlined in 1934, addressed in detail

91. Quoted in Davenport (1987: 529).

92. See Saltzmann (1986: 590); Gortler (1985: 755, 757).

93. See Walter Hückel, *Theoretische Grundlagen der organischen Chemie*, 2 vols. (Leipzig: Akademische Verlagsgesellschaft, 1931), and later, W. Hückel, ed., *Theoretical Organic Chemistry*, 2 vols. (Office of Military Govt. for Germany, Field Information Agencies Technical, British, French, U.S., ca. 1948), a postwar product with French preface and German texts. On W. Hückel, see Ralph Oesper, ''Walter Hückel,'' *JChem.Ed.* 27 (1950): 625.

94. Personal communication from O. T. Benfey, 11 July 1990.

95. One of the oldest and most familiar quantitative relationships for relating the structure of substituted benzene derivatives to both equilibrium constants and rate constants is the ''Hammett Equation.'' See Louis Hammett, *Physical Organic Chemistry*, 184–199.

exceptions and variations from the general rules that might be encountered in the laboratory because of specific physical and chemical conditions. As Ingold said in a lecture in Paris in May, 1946, ''The idea of intramolecular electric interaction is the central theme of all our theories of reactivity.''[96] Whereas the old equation for the Arrhenius energy of activation still was ''a useful approximation'' for determining the energy necessary for production of the activated transition state, the *mechanism* for producing this transition state was to be found in a reaction site becoming a highly polar center ''which either requires electrons to be drawn away from or requires them to be driven back into, the rest of the molecule.''[97]

Electron density, kinetic velocities, and thermodynamic equilibrium all must be part of the understanding of chemical processes.[98] However, the chemical molecule is not merely the summation of physical properties. Rough approximations can be made for a molecule's overall dipole moment, for example, by adding up individual contributions due to atoms and atomic groups. But

> an accurately additive behavior would have implied the absence of intramolecular electrical interaction, and such interaction is not only a necessary property of the molecular model but also an indispensable mechanism in the theory of organic reactions.[99]

As for the relationship of quantum mechanics to classical organic chemistry, ''not only do the 'calculated' values depend heavily on empirical data but also the 'observed' values depend in part on interpretative assumptions.'' Kekulé formulas and curved arrows for electron-pair displacement still are convenient for the experimental chemist.

> Energetic attempts are being made at this time by Coulson, Daudel, Pullman and others to refine the theory of electron distribution in polycyclic aromatic systems, but it is not clear to the writer that the best compromise between simplicity and accuracy has yet been secured.[100]

After the Second World War, Ingold increasingly was to turn his attention to inorganic systems, further generalizing the scope of his mechanistic system of theoretical chemistry.[101] Others, too, were attempting to develop a mechanics for inorganic chemistry. A major text, *Electronic Structure and Chemical Binding with Special Reference to Inorganic Chemistry*, appeared in 1940. Its

96. Ingold, ''Les réactions,'' 6.
97. Ingold, *Structure and Mechanism in Organic Chemistry*, 49–50.
98. Ingold, ''Les réactions,'' 8–9.
99. Ingold, ''Principles of an Electronic Theory,'' 227.
100. Ingold, *Structure and Mechanism in Organic Chemistry*, 165, 167, 170.
101. Personal communication from O. T. Benfey, 22 September 1991.

author was Oscar K. Rice, then professor of chemistry at the University of North Carolina. In 1949, Rice published another text that was to have wide impact in the United States, this one written with the Hungarian emigré physicist Edward Teller. Their *Structure of Matter* introduced the reader to applications of quantum mechanics "useful to the chemist" and serviceable to the physicist for summarizing "topics which used to be in the forefront of his discipline only one or two decades ago."[102] It is to quantum mechanics, chemical physics, and quantum chemistry that we now turn.

102. Oscar Knefler Rice, *Electronic Structure and Chemical Binding, with Special Reference to Inorganic Chemistry* (New York: McGraw-Hill, 1940); Francis Owen Rice and Edward Teller, *The Structure of Matter* (New York: John Wiley, 1949).

Part Three

Converging Traditions and Rival Disciplines

9

Quantum Chemistry and Chemical Physics, 1920–1950

Niels Bohr's 1913 hydrogen atom paper demonstrates the traditional interest of some physicists in placing the facts and laws of chemistry within a broader framework of foundational principles laid out by physicists. During the course of the next two decades, a number of physicists who became known as quantum physicists developed physical theories and mathematical techniques that they claimed would create a mathematical and theoretical chemistry. However, few of them had much chemical knowledge beyond a general understanding of the periodic table of the elements and familiarity with the Lewis-Langmuir theory of the electron duplet and octet.

Among the physicists whose work or whose research institutes provided the strongest inspiration for mathematically trained, but chemically oriented scientists were Bohr (Copenhagen), Max Born (Göttingen), Arnold Sommerfeld (Munich), Werner Heisenberg (Leipzig), Erwin Schrödinger (Zurich), and Edwin Kemble (Harvard). While German-speaking chemists contributed surprisingly little to the emergence of quantum chemistry, German-speaking physicists played a very important role.

Friedrich Hund became friends with the young American Robert Mulliken in 1927 when Mulliken was a postdoctoral fellow at Max Born's Institute for Theoretical Physics in Göttingen. Speaking later of the beginnings of quantum chemistry, in which Mulliken was to make his reputation, Hund said that it was the German physicists Heitler and London, resident in Sommerfeld's and Schrödinger's theoretical physics institutes during 1926–27 who "began this discipline" of quantum chemistry in 1927.[1]

1. See Robert S. Mulliken, *Life of a Scientist: An Autobiographical Account of the Devel-*

Heitler and London's 1927 paper on the hydrogen molecule gained attention among both physicists and chemists. But it was not really until 1931, when Slater and Pauling independently developed methods to explain *directed* chemical valence by orbital orientation that it can truly be said that a *chemical* quantum mechanics, rather than an application of quantum mechanics to chemistry, had been created. In a study of Slater, S. S. Schweber notes the distinction between the Heitler-London-Pauling-Slater theory and the Heitler-London theory. Heitler and London successfully explained the electron-valence pair on the basis of the Goudsmit-Uhlenbeck theory of spin. Slater and Pauling explained the carbon tetrahedron. This second explanation distinguishes quantum chemistry from quantum physics.[2]

By the 1950s, there was a recognized academic field of theoretical chemistry that encompassed the subfields or disciplines of physical chemistry, physical organic chemistry, quantum chemistry, and chemical physics. In this chapter, we will consider the physicists who developed quantum mechanics and applied it to molecules in the 1920s, the physicists and chemists who created quantum chemistry and chemical physics in the late 1920s and 1930s, and the coming of age of quantum chemistry by around 1950.

The Application of Quantum Mechanics to Molecules in the 1920s

In 1920, Born wrote to G. N. Lewis about the reprint he had received of Lewis's 1916 paper on the "atom and the molecule." The "cubic" distribution of electrons, Born cautioned, had no general usefulness, and Lewis should look at how Born treated the problem in his new work, which he had sent Lewis.[3] Born expressed some humility about his knowledge of chemistry, confessing to Lewis a few years later, with respect to the new Lewis and Randall textbook on chemical thermodynamics, that he could not speak as well to the chemical side as to the physical side.[4]

---

opment of Molecular Orbital Theory, ed. Bernard J. Ransil (Berlin: Springer-Verlag, 1989): x, 52, referring to W. Heitler and A. F. London, "Wechselwirkung neutraler Atome und homopolare Bindung nach der Quantenmechanik," *ZP* 44 (1927): 455–472. After leaving Germany in 1939, London became professor of theoretical chemistry, then professor of chemical physics at the University of North Carolina. Heitler held positions in physics, completing his career at the University of Zurich as professor of theoretical physics. Kostas Gavroglu is writing a scientific biography of London.

2. See S. S. Schweber, "The Young John Clarke Slater and the Development of Quantum Chemistry," *HSPS* 20 (1990): 339–406, on 386. Schweber accepts the year 1927 as a "convenient birthdate for quantum chemistry" (398). The papers are John Slater, "Directed Valence in Polytatomic Molecules," *Physical Review* 37 (1931): 481–489; and Linus Pauling, "The Nature of the Chemical Bond," *JACS* 53 (1931): 1367–1400, 54 (1931): 988–1003, 3570–3582.

3. Letter from Max Born to G. N. Lewis, 27 November 1920, Lewis Papers, BL.UCB.

4. Letter from Max Born to G. N. Lewis, 13 September 1923, Lewis Papers, BL.UCB.

But Born did not doubt the power of the physicist to explain the facts and laws of the chemist. In the lectures published as *The Constitution of Matter*, Born developed a model of the distribution of valence electrons about a nucleus in the manner of Bohr (1913) and of Sommerfeld's protégé, Kossel (1916). Born wrote, "When we contemplate the path by which we have come we realize that we have not penetrated far into the vast territory of chemistry; yet we have travelled far enough to see before us in the distance the passes which must be traversed before physics can impose her laws upon her sister science."[5]

Born's essays making up *The Constitution of Matter* were intended to be an aid to those who did not have "time to study the larger work of Sommerfeld," namely, *Atomic Structure and Spectral Lines* (1915), which laid out Bohr's first quantum theory and Sommerfeld's emendations.[6] Born and Sommerfeld were to have enormous influence on the development of quantum mechanics and its application to atoms and molecules. Not only did many European, British, and American scientists come to study at their physics institutes but they each gave influential series of lectures at foreign institutions, nowhere with more influence than in the United States.

In 1925–26, Born made a trip that included Harvard and MIT on the East Coast and the University of California, Berkeley on the West Coast. He wrote to Lewis about his planned lectures:

> You will have no difficulty in understanding our new methods. They look rather complicated when they are in print, but they are easily explained with words.[7]

Born frequently showed less than high respect for chemical methods and theories, referring to the "arbitrary" "speculations" of chemists about valency and contrasting their ideas with the "idea of the atom as conceived by present-day physicists [which] is, in many vital respects, free from such arbitrariness." In a letter to John Slater in 1930, he unrepentantly hailed the ability of physicists to calculate binding energy for polyatomic molecules and "to establish the cases in which the valence theory of the chemist is reliable"[!][8]

5. Max Born, *The Constitution of Matter: Modern Atomic and Electron Theories*, trans. from 2d German ed., E. W. Blair and T. S. Wheeler (London: Methuen, 1923; 1st ed., 1920): 78.

6. Ibid., preface to 1st ed.; and Arnold Sommerfeld, *Atomic Structures and Spectral Lines*, trans. from 3d German ed., Henry Brose (London: Methuen, 1923).

7. Letter from Max Born to G. N. Lewis, 14 January 1926, written from MIT, Lewis Papers, BL.UCB. Also see, Alexi Josephine Assmus, "Molecular Structure and the Genesis of the American Quantum Physics Community 1916–1926" (Ph.D. dissertation, Harvard University, 1990): 160; and Assmus, "Molecular Structure and the Genesis of the American Quantum Physics Community, 1916–1926," *HSPS* 22 (1992): 209–231. The lectures at MIT were published under the title, Max Born, *Problems of Atomic Dynamics* (Cambridge: MIT Press, 1926). Similarly to Born, Henry Eyring told his graduate students at Princeton in the 1930s that the "unavoidable formality" of quantum mechanics "looks much worse on first reading it than it is." In Henry Eyring, *Quantum Chemistry* (New York: Wiley, 1944): iii.

8. Born, *Constitution of Matter*, 1; and "Zur Quantentheories der chemische Kräften," *ZP* 64 (1930): 729–740. Letter quoted in Schweber (1990: 385).

Sommerfeld lectured at Caltech, where students routinely read his *Atomic Structure* book in the physical chemistry taught by Richard Chace Tolman. Applying for a Guggenheim Fellowship in 1925 (so as to visit the theoretical physics institutes of Sommerfeld, Bohr, and Schrödinger), Pauling quoted from the German edition, the *Aufbau,* about the task ahead "of working out a complete topology of the interior of the atom and beyond this a system of mathematical chemistry."[9] In the third edition of the *Aufbau* (1922), Pauling found an account of Bohr's new model of the atom. He also found less attention paid to atomic models of the chemical properties of periodicity than in earlier editions. And he read Sommerfeld's challenge, defining a major goal for quantum physicists in the 1920s:

> All in all, the final result in the case of molecular models seems to be even more unfavourable than in that of atomic models. Here we can assert nothing even in the simplest case of the H_2 molecule.[10]

Bohr's hydrogen atom model of 1913 had provided inspiration to a few physicists, like Kossel, who were interested in chemical problems but to very few chemists concerned with the explanation of valence. First of all, the Bohr atom had a dynamic character that was not consistent with the static and stable characteristics of ordinary molecules. Second, Bohr's approach, as amended by Kossel, could not even account for the fundamental tetrahedral structure of organic molecules because it was based on a planar atomic model. Nor could it account for "homopolar" or covalent bonds, because the radii of the Bohr orbits were calculated on the basis of a Coulombic force model. Although Bohr discussed H_2, HCl, H_2O, and CH_4, physicists and physical chemists mainly took up the problem of H_2, which seemed most amenable to further treatment.[11]

Among chemists, the American physical chemist Langmuir was one of the few who expressed immediate interest in Bohr's paper. This was largely because of Bohr's calculation of the heat of formation of the H_2 molecule, a problem on which Langmuir was working at the General Electric Research

9. Linus Pauling, in interview with John Heilbron, March 27, 1964, at Pasadena, p. 12 of first session and p. 3 of second session. SHQP, Berkeley. On Pauling's graduate education, 1922–1925, at Caltech, see Judith R. Goodstein, *Millikan's School: A History of the California Institute of Technology* (New York: Norton, 1991): 180–185. Sommerfeld's statement is in the *Atombau und Spektrallinien*, 3d ed. (Braunschweig: Teubner, 1922): 88.

10. Quotation from Sommerfeld, *Atomic Structures*, 79. On p. vi, Sommerfeld notes that in the chapter on the natural system of the elements, "the former discussions of molecular models and atomic volumes have been thoroughly pared down." At Harvard in 1925, Edwin Kemble's course in quantum mechanics included the second edition of Sommerfeld's *Aufbau*; according to Schweber (1990: 348). Similarly, Joseph Hirschfelder was reading it as an undergraduate at Yale; according to Hirschfelder, "My Adventures in Theoretical Chemistry," *Ann.Rev.P.Chem.* 33 (1983): 1–29, on 4.

11. See the analysis in Helge Kragh, "Bohr's Atomic Theory and the Chemists, 1913–1925," *Rivista di storia della scienza* 2 (1985): 463–485.

Laboratory. During the period 1919–1922, Langmuir developed what he called a ''deductive chemistry'' using the electron-pair theory of valency and the quantum hypothesis. However, physicists rejected the premises and methodology of Langmuir's theory, which proposed the existence of a ''quantum force'' to counterbalance Coulombic attraction and which used the notion of principal quantum number but deduced positions of equilibrium rather than quantum jumps for electrons.[12]

Niels Bjerrum focused attention on the fact that molecular band spectra provide evidence for the existence of quanta of energy. A postdoctoral fellow in Perrin's and Nernst's laboratories, Bjerrum developed a quantum theory during 1911–1914 for molecules, not atoms, using a classical model of a vibrating rotator, modified by quantization of energy. This model, rooted in the original quantum theory, was the basis in the 1920s for Bohr's theory that could be applied both to atomic and molecular spectra by adding a frequency condition to Bjerrum's equations.[13]

Following Sommerfeld's proposal of elliptical electron orbits in 1915, Bohr amended his original theory, which had included only circular orbits.[14] A 1922 paper in *Zeitschrift für Physik* outlined the ''Aufbauprinzip'' by which electrons are fed into atomic subshells. There was a neat correlation between periodic groups containing 2, 8, 8, 18, 18, and 32 elements and electron levels containing 2, 8, up to 18, and up to 32 electrons in the so-called K, L, M, and N shells. The subshells, which Bohr originally arranged in groups of (2),(4, 4), (6, 6, 6), and (8, 8, 8, 8), he rearranged to (2), (2, 6), (2, 6, 10), and (2, 6, 10, 14). Bohr incorporated the spectral-line characteristics designated ''strong,'' ''principal,'' ''diffuse,'' and ''fundamental'' into ''s,'' ''p,'' ''d,'' and ''f'' symbols for the subshells.[15]

Slater later recalled first reading Bohr's 1922 paper as a student completing his doctoral dissertation at Harvard.

> The whole picture of the electrons in an atom was illuminated in a flash by

12. See Kragh, ''Bohr's Atomic Theory,'' 470, 477–481.

13. On Bjerrum, see Assmus, ''Molecular Structure,'' 54–65, 68–73; and Assmus, ''The Molecular Tradition,'' esp. 217–231. Niels Bjerrum, ''On the Infrared Spectra of Gases. III. The Configuration of the Carbon Dioxide Molecule and the Laws of Intramolecular Forces'' (1914), 42–55, in N. Bjerrum, *Selected Papers* (Copenhagen: E. Munksgaard, 1949). Independently, Kemble worked along these lines at Harvard, so that the theory often is called the Kemble-Bjerrum theory for molecules (Assmus, 1990: 109); see E. C. Kemble, R. T. Birge, W. F. Colby, F. W. Loomis, and L. Page, *Molecular Spectra in Gases*, Report of the Committee on Radiation in Gases, *Bulletin of the National Research Council* 11 (1926): 16–17.

14. N. Bohr, ''Ueber die Serienspektra der Elemente,'' *ZP* 2 (1920): 423–469; ''Atomic Structure,'' *Nature* 107 (1921): 104–107; and ''The Structure of the Atom and the Physical and Chemical Properties of the Elements,'' 61–126, in *The Theory of Spectra and Atomic Constitution* (Cambridge, 1922).

15. See the discussion of quantum mechanics in Aaron Ihde, *The Development of Modern Chemistry*, 541–545. Also see discussion of the importance of Edmund C. Stoner's work done in the 1920s relating spectroscopic data and periodic properties, in John Servos, *Physical Chemistry from Ostwald to Pauling*, 285–286.

> Bohr's theory of the periodic system of the elements. . . . The physicist who did not live through that period can hardly imagine the excitement felt by a new graduate student, picking up Number 1 of Volume 9 of the *Zeitschrift* and there reading for the first time the complete explanation of the periodic system of the elements.[16]

Lewis appropriated Bohr's new atom to try to unify the physical and chemical atom. If the Bohr-Sommerfeld orbits are in fixed positions and orientations, "they may be used as the building stones of an atom which has an essentially static character."[17] Bohr's dynamic theory works for the chemist, Lewis wrote, if the "average" position of an electron in a Bohr-Sommerfeld orbit is taken to correspond to the fixed position of the electron in Lewis's static chemical model. The outermost shell of electrons constitutes the "valence" electrons, and the remaining electrons constitute the "kernel."[18]

Lewis recognized that nothing in the quantum theory yet accounted for electron pairing in the valence bond, but otherwise he was optimistic. The MIT mathematical physicist Edwin Bidwell Wilson did not share the optimism, writing to Lewis "You seem to think that there is not now much difference between Bohr's point of view and that of the chemists. I very much doubt this."[19]

In 1925, Wolfgang Pauli gave chemists what they wanted from the physicists: a physical principle underlying electron-pair valency. Pauli built on the fact that in addition to the continuous, line, and band spectra, there is a fine structure of doublets, triplets, and multiple lines, some of which are split in a magnetic field (Zeeman effect). Three quantum numbers had been proposed, based on spectral lines and inferences about electron energy levels: a principal quantum number to specify energy level of the atom; an azimuthal quantum number to specify the angular momentum of electrons moving elliptically; and an inner or magnetic quantum number to express the orientation of the plane of the electron's orbit in a magnetic field.[20]

Pauli proposed the use of a fourth quantum number, which could have two values, thereby explaining why it is that electrons with identical energies behave differently in a strong magnetic field. If it is assumed that no two electrons in an atom may occupy the same atomic state, meaning that no two electrons can have the same four quantum numbers, then there might be two, but no more than two, *s* electrons for each principal quantum number. Six different

16. Quoted in Schweber (1990: 351).

17. See G. N. Lewis, *Valence and the Structure of Atoms and Molecules* (Washington, D.C.: Chemical Catalogue Co., 1923): 56–57.

18. Ibid., 58.

19. Letter from E. B. Wilson to G. N. Lewis, dated 10 June 1924, Lewis Papers, BL.UCB.

20. See Servos, *Physical Chemistry from Ostwald to Pauling*, 285; and for a more thorough but general treatment, Max Jammer, *The Conceptual Development of Quantum Mechanics* (New York: McGraw-Hill, 1966).

p electrons would be possible, and there might be ten of the *d* and fourteen of the *f*.[21]

In late fall 1925, the Dutch physicists G. Uhlenbeck and Samuel Goudsmit gave a physical interpretation to Pauli's postulate of a fourth quantum number. The electron, they proposed, may spin in one of two directions. In a given atom, a pair of electrons having three identical quantum-number values must have their spin axes oriented in opposite directions, and if paired oppositely in a single orbital, they neutralize each other magnetically.[22]

Lewis and many other chemists saw in the Pauli exclusion principle and the Uhlenbeck-Goudsmit spin hypothesis firm physical support for the chemical valence theory of the electron pair. In fact, the Pauli exclusion principle has to be postulated within the physics of the quantum theory.[23] Accepting the Nobel Prize in physics for 1945, Pauli expressed regret that the principle cannot be derived *ab initio*.

> In my original paper I stressed the circumstance that I was unable to give a logical reason for the exclusion principle or to deduce it from more general assumptions. . . . Of course in the beginning I hoped that the new quantum mechanics, with the help of which it was possible to deduce so many half-empirical formal rules in use at the time, will also rigorously deduce the exclusion principle.[24]

Pauli and other quantum physicists were confident about the principle because it was so beautifully consistent with chemical and spectroscopic facts and theory.

During summer and fall 1925, Heisenberg, then a young man, created a "matrix mechanics" based in the algebra of matrices to formulate a quantum mechanics for the electron. About the same time, the thirty-eight-year-old Schrödinger drew on Louis de Broglie's recent doctoral dissertation at Paris to develop a "wave mechanics" for the electron, building on analogies with classical wave theory.[25] De Broglie presented his dissertation in 1923, proposing that electrons exhibit wavelike characteristics, just as light waves ex-

21. Servos, *Physical Chemistry from Ostwald to Pauling*, 286–287; Ihde, *Development of Modern Chemistry*, 542. See Wolfgang Pauli, "Ueber den Einfluss der Geschwindigkeitsabhängigkeit der Elektronenmasse auf den Zeemaneffekt," *ZP* 31 (1925): 373–386.

22. G. E. Uhlenbeck and S. Goudsmit, "Zuschriften und vorläufige Mitteilungen," *Naturwissenschaften*, 13 (1925): 953; and "[Letters to the Editor:] Spinning Electrons and the Structure of Spectra," *Nature* 117 (1926): 264–265. This "letter" is followed by an untitled letter from Niels Bohr (p. 265). *Nature* 117 (1926): 264.

23. Farrington Daniels and Robert A. Alberty, *Physical Chemistry*, 2d ed. (New York: John Wiley, 1961): 471.

24. Quoted from R. Kronig and W. F. Weisskopf, eds., *W. Pauli: Collected Scientific Papers* (New York: John Wiley, 1964), II: 1085, in Peter Joseph Hall, "The Pauli Exclusion Principle and the Foundations of Chemistry," *Synthèse* 69 (1986): 267–272, on 270.

25. See the biography of Schrödinger by the Australian physical chemist Walter Moore, *Schrödinger: Life and Thought* (New York: Cambridge University Press, 1991).

hibit particlelike characteristics.[26] The most important difference between light and electrons was that the light quantum travels at the speed of light, but the electron cannot reach that limit. De Broglie's thesis director, Paul Langevin, resisted de Broglie's interpretation until Einstein persuaded him that it had merit.[27]

With the development of Schrödinger's equation to describe the dynamics of the electron, many physicists and chemists felt that a radical break with traditional physics was being averted. In lectures at Stanford University and at the University of Chicago, Slater, for example, compared the state of "mechanics up to now . . . [to] the state of optics before the discovery of interference and such things."[28]

In a line of reasoning that many of the younger quantum physicists regarded as reactionary, Schrödinger built his treatment of the electron on the well-understood mathematical techniques of wave equations as partial differential equations involving second derivatives. Schrödinger's equation for stationary electron states, as written in the *Annalen der Physik* in 1926, took the form

$$\nabla^2 \psi(x,y,z) + (8\pi^2 m/h^2) \{E - V(x,y,z)\} \psi(x,y,z) = 0,$$

where (∇^2) is the Laplace operator, m is the mass of the electron, h is Planck's constant, E is the total energy of the particle, and V is the electron's potential energy in a field of force. The function ψ was called the "field scalar" by Schrödinger, taken to be the amplitude of the electron's matter wave. The solutions of Schrödinger's equation are such that $\int |\psi(x,y,z)|^2$ is finite, and the values of E, called eigenvalues, correspond to the possible energies of an electron in nonradiating states in the potential V.[29]

But in fact, ψ (or the more general time-dependent wave function Ψ, which is the product of two functions, one involving the time alone and the other the coordinate alone) was difficult to interpret physically, because the idea of the

26. John Slater later expressed regret, even anger, that Niels Bohr deterred him from the same path of reasoning while he was a postdoctoral fellow in Copenhagen in 1923. See Schweber (1990: 351–352).

27. See Louis de Broglie's remarks in an interview with Thomas Kuhn et al., January 7, 1963, Paris, no. 1 of 2 interviews, 6–7. The French physical chemists Jules Guéron and Michel Magat have remarked on the fact that no French physicist or chemist took part in the "swift creation of the quantum theory of the chemical bond and the quantum description of chemical reactions." Jules Guéron and Michel Magat, "A History of Physical Chemistry in France," *Ann.Rev.P.Chem.* 22 (1971): 1–23, on 7.

28. Quoted in Schweber (1990: 357–358).

29. Erwin Schrödinger, "Quantisierung als Eigenwertproblem (erste Mitteilung)," *Ann.Phys.* 79 (1926): 361–376. See discussion in Helge Kragh, *Dirac: A Scientific Biography* (Cambridge: Cambridge University Press, 1990): 30–32; also Emilio Segrè, *From X-rays to Quarks: Modern Physicists and Their Discoveries* (San Francisco: Freeman, 1980): 160–164; and Linus Pauling and E. Bright Wilson, Jr., *Introduction to Quantum Mechanics with Applications to Chemistry* (New York: McGraw-Hill, 1935; republished Dover, 1985): 50–57. The symbol Δ sometimes is used instead of ∇. ∇^2 is read as "del squared."

electron wave was much more abstract than that of sound waves or electromagnetic waves, where the analogous equation determines the pressure or electric field strength.[30] However, since in optics the square of Ψ is proportional to the intensity of light, the square of Ψ in Schrödinger's equation could be taken to be a measure of electron intensity, or density, about the atom nucleus.

Pauling, who previously had heard Sommerfeld lecture on the old quantum theory at Caltech, now heard Sommerfeld lecture in Munich on the new wave mechanics. Pauling instantly liked the physicists' new interpretation of the electron. The hydrogen atom could be thought of as

> a nucleus embedded in a ball of negative electricity—the electron distributed through space. . . . The atom extends to infinity; the greater part of the atom, however, is near the nucleus—within 1 or 2 Angstroms.[31]

About the same time, Born, who had been lecturing in the United States during 1925–26, developed a probabilistic interpretation of the motions of the quantized electron. Now $|\psi(x,y,z)_2|\ d\tau$ was the probability of finding the electron in a volume element $d\tau$ at coordinates x,y,z. Born's notion was based on a classical and visual conception of particles, consistent with the positions of atoms established in x-ray crystallogaphy, as discussed in his book, *Dynamik der Krystallgitter*.

While some younger physicists, like Born's student Werner Heisenberg and Ralph Fowler's student P. A. M. Dirac, were not sanguine about Born's interpretation, it pleased chemists like Lewis, who earlier had been willing to think about the "average" position of an electron in its orbit, so as to reconcile Bohr's dynamic atom with Lewis's static atom. For Pauling, it was a natural step to take ψ^2 to be the probability distribution function for an electron's position in space.[32]

While grappling with the relationship between his matrix mechanics and Schrödinger's wave mechanics, Heisenberg wrote a paper in 1926 which drew another analogy from classical mechanics to quantum mechanics, also giving the spectra of *para* and *ortho* helium, corresponding to the singlet and triplet states, respectively, a quantum mechanical explanation.[33] In classical mechanics, the phenomenon termed "resonance" occurs when two systems, capable of sustaining similar oscillations, are coupled or allowed to interact to some degree, thereby perturbing each other. The coupled system will also have har-

30. See Daniels and Alberty, *Physical Chemistry*, 462–463.

31. Linus Pauling, "The Sizes of Ions and the Structure of Ionic Crystals," *JACS* 49 (1927): 766, quoted in Servos (1990: 283).

32. See Yuko Abe, "Pauling's Revolutionary Role in the Development of Quantum Chemistry," *Historia Scientiarum*, no. 20 (1981): 107–124, on 115–116.

33. Werner Heisenberg, "Mehrkörperproblem und Resonanz in der Quantenmechanik," *ZP* 38 (1926): 411–426.

monic oscillations (normal modes), one having a reduced and the other an increased frequency. Heisenberg applied this classical insight to the wave function of two electrons to explain the stability of the two-electron helium atom.[34] Heisenberg by no means had a classical attitude toward the wave mechanics that was an alternative to his matrix mechanics. He resisted efforts to make the new theory "visualizable," warning that one's ordinary intuitions do not work in the atomic domain.[35]

At the same time that Heisenberg was formulating his approach to the helium system, Born and Oppenheimer indicated how to formulate a quantum mechanical description of molecules that justified approximations already in use in treatment of band spectra. The theory was worked out while Oppenheimer was resident in Göttingen and constituted his doctoral dissertation. Born and Oppenheimer justified why molecules could be regarded as essentially fixed particles insofar as the electronic motion was concerned, and they derived the "potential" energy function for the nuclear motion. This approximation was to become the "clamped-nucleus" approximation among quantum chemists in decades to come.[36]

About the same time, Douglas Hartree, along with other members of the informal club for theoretical physics at Cambridge University called the Del-Squared Club, began studying approximate methods to describe many-electron atoms. Hartree developed the method of the self-consistent field, which was improved by Vladimir Fock and Slater in early 1930, so as to incorporate the Pauli principle ab initio.[37] Dirac, another Del-Squared member, published a paper in 1929 which focused on the exchange interaction of identical particles. This work became part of what soon became called the Heisenberg-Dirac approach.[38]

Two sets of papers in 1927 now provided successful treatments of the hydrogen molecule. One was written by Heitler and London, in residence at Schrödinger's institute in Zurich. The other was by Hund, Born's assistant in

34. See C. K. Ingold's treatment of the subject, in *Structure and Mechanism in Organic Chemistry*, 16–21.

35. Werner Heisenberg, "Quantenmechanik," *Naturwissenschaften* 14 (1926): 989–994.

36. Of the Born-Oppenheimer paper, David Dennison commented to Thomas Kuhn that it validated things that people were doing anyway and that he "took this for granted." In interview of David Dennison with Thomas Kuhn, January 28, 1964, Ann Arbor, Michigan, no. 2 of 3 sessions, 19. SHQP, Berkeley. Also see, Stephen J. Weininger, "The Molecular Structure Conundrum: Can Classical Chemistry be Reduced to Quantum Chemistry?" *JChem.Ed.* 61 (1984): 939–944, on 940.

37. See Kragh, *Dirac*, 10, 76. Other members of the *Del-Squared Club* included Fowler and Stoner. Also see B. Bigot and F. Volatron, "Parlez-vous chimie théorique?" *L'Actualité Chimique* (November 1984): 43–51, on 50. For a discussion of the self-consistent field, see Schweber (1990: 375–376). Hartree became professor of theoretical physics at Manchester and returned to Cambridge University after the war. For his own account of his method, see Douglas Hartree, *The Calculations of Atomic Structures* (New York: Wiley, 1957).

38. P. A. M. Dirac, "Quantum Mechanics of Many-Electron Systems," *Proc.RSL* A123 (1929): 714–733. See Kragh, *Dirac*, 76.

Göttingen. At this time, Pauling had just recently met Heitler and London in Munich, and he saw them again in Zurich. Mulliken, who as early as 1925 showed that there are multiple electronic levels in molecules similar to those in atoms, became friends with Hund at Göttingen in 1927.[39]

Heitler and London treated the H_2 molecule as two fixed hydrogen nuclei (protons) about which the electrons circulated. They assumed that (in zeroth approximation, or the lowest energy value) the wave function of each electron is centered on one of the protons, that is, that the H_2 molecule consisted of two hydrogen atoms. They showed that the two electrons, when they have antiparallel spins, tend to aggregate in the region between the two protons, thereby reducing the total energy of the system. They calculated how the energy of the system was dependent on the internuclear distance, and they demonstrated that the ground energy state was the singlet state and that the triplet state did not bind. In subsequent papers, they specifically applied these results to chemical valence theory, deducing well-known chemical facts, including the combination of hydrogen atoms to form a hydrogen molecule, the nonexistence of a diatomic helium molecule or ion, and the result that phosphorus, arsenic, tin, and bismuth may have valences of 1, 3, 5 but that nitrogen has valences only of 1 and 3.[40]

In the same year, Hund published an entirely different approach to the binding of the hydrogen molecule, based in Hartree's approach rather than Heisenberg's. Hund's work, further, was a generalization of the treatment by Oyvind Burrau of the hydrogen molecule-ion. Rather than building up resonating wave functions from individual nucleus-centered electron waves, Hund assumed that an individual electron moved in a potential field that results from all the nuclei and the other electrons present in the molecule. As Mulliken later described it, this method "regards each molecule as a self-sufficient unit and *not* as a mere composite of atoms."[41]

Physicists took some pride in what they regarded as the explanatory illumination they were shedding on chemical facts and laws. At the centenary

39. Interview of Linus Pauling with John Heilbron, 27 March 1964, 2d of 2 sessions, 7, SHQP, Berkeley; and Robert Mulliken, *Life*, 53, 61–62. On Mulliken's papers in 1925 and 1926, see Edwin C. Kemble et al., *Molecular Spectra in Gases: Report of the Committee on Radiation in Gases. Bulletin of the National Research Council* 11, pt. 3, no. 57 (December 1926): 239–240.

40. See the discussion in Schweber (1990: 381–382). W. Heitler and A. F. London, "Wechselwirkung neutraler Atome und homopolare Bindung nach der Quantenmechanik," *ZP* 44 (1927): 455–572; London, "Zur Quantentheorie der homopolaren Valenzzahlen," *ZP* 46 (1928): 455–477; Heitler, "Zur Gruppentheorie der homopolaren chemische Bindung," *ZP* 47 (1928): 835–858; and other papers.

41. Quotation from Robert Mulliken, "Spectroscopy, Molecular Orbitals, and Chemical Bonding," *Nobel Lectures. Chemistry. 1963–1970* (Amsterdam: Elsevier, 1972): 131–160, on 137. See Friedrich Hund, "Zur Deutung der Molekulspektren. IV." *ZP* 40 (1927): 742–764, 42 (1927): 93–120, 43 (1927): 805–826; and Oyvind Burrau, "Berechnung des Energiewertes des Wasserstoff-Molekel-Ions (H_2^+) im Normalzustand," *Danske Videnskabernes Selskab. Mat.-fys. Meddelser* 7 (1927): 14.

meeting of the British Association for the Advancement of Science in 1931, Debye, Hinshelwood, John Edward Lennard-Jones, Fowler, Victor Henri, Heisenberg, Born and Lawrence Bragg were participants in a symposium on ''The Structure of Simple Molecules.'' Fowler, the Cambridge mathematical physicist (and member of the Del-Squared Club), expressed the view that

> one may say now that the chemical theory of valency is no longer an independent theory in a category unrelated to general physical theory, but just a part—one of the most gloriously beautiful parts—of a simple self-consistent whole, that is of non-relativistic quantum mechanics. I have at least sufficient chemical appreciation to say rather that quantum mechanics is glorified by this success than that now ''there is some sense in valencies,'' which would be the attitude, I think, of some of my friends.[42]

It was, after all, Fowler's precocious former student, Dirac, who began his 1929 paper on the quantum mechanics of a many-electron system by saying, ''The underlying physical laws for the mathematical theory of a large part of physics and the whole of chemistry are thus completely known.''[43]

Regarding the relative methods of what had become known as the ''atomic-orbital method'' and the ''molecular-orbital method,'' Heisenberg noted in discussion that the method used by Heitler and London was not the one ''which looks most natural to the physicists,'' since they consider the interaction of two electrons in the same atom to be very large, in contrast to physicists usually neglecting this interaction as a small perturbation. The more general method is Hund's, but it does not lead in a general way to the chemical concept of valency. On this occasion, Heisenberg said what some chemists had been saying for several years: ''It seems questionable to me whether the quantum theory would have found or would have been able to derive the chemical results about valency, if it had not known them before.''[44]

In the next years, physicists came increasingly to concentrate on the mathematical development of the quantum theory, following a resolution of the apparent incompatibility between matrix mechanics and wave mechanics. Like Heisenberg, Dirac was initially disturbed by the probabilistic physicalism introduced into quantum mechanics by both Born and Pascual Jordan. The statement in 1927 of Heisenberg's ''uncertainty principle'' became the basis for a great deal of controversy about distinctions between statistical knowledge and certain knowledge, whether uncertainty lay in nature or in experimental conditions. The meaning and validity of Bohr's principle of complementarity also became a matter of controversy in 1927. Relativistic quantum mechanics, nu-

42. Ralph Fowler, ''A Report on Homopolar Valency and Its Quantum-Mechanical Interpretation,'' 26–246, in *Chemistry at the Centenary (1931) Meeting of the British Association for the Advancement of Science* (Cambridge: W. Heffer and Sons, 1932): 226.

43. P. A. M. Dirac, ''Quantum Mechanics of Many-Electron Systems,'' 714.

44. Werner Heisenberg, discussion in Dirac (1929: 247).

clear physics, and solid-state physics came increasingly to occupy the attention of the quantum physicists.[45]

Later reconstructing the history of this period, Coulson claimed that the development of wave mechanics from Schrödinger to Dirac came to a ''full stop'' about 1929 insofar as chemistry was concerned. It was one thing to deal with the simplest cases of the H_2^+ ion and the H_2 molecule. These problems hardly were comparable to methane or benzene. ''Despondency set in.''[46] Joseph Hirschfelder later recalled that there were many new techniques to be learned in the 1920s, but then the realization set in that although ''nature might be 'simple and elegant,' molecular problems were definitely more complicated. . . . At this point, the theoretical physicists left the chemist to wallow around with their messy molecules while they resumed their search for new fundamental laws of nature.''[47]

Chemists and Quantum Mechanics in the 1920s and 1930s

In 1927, Pauling returned to CalTech from postdoctoral study abroad and set out to interpret the recent results of the quantum physicists from a chemical point of view. He was now an assistant professor of theoretical chemistry, having taken an undergraduate degree in chemistry at Oregon Agricultural College in Corvallis, a doctoral degree in chemistry at Caltech, and postdoctoral fellowship residence in Munich, Copenhagen, and Zurich.

In 1927, Slater was teaching at Harvard, where he earlier had worked closely with Kemble and Percy Bridgman while taking his Ph.D. in physics. Slater had spent the 1923–24 academic year as a postdoctoral fellow in Cambridge and Copenhagen, and he would soon take time off for residence in Leipzig. Mulliken, who had been an undergraduate in chemical engineering at MIT and a Ph.D. graduate in chemistry with William Harkins at the University of Chicago, was to become an associate professor of physics at Chicago in 1928. After taking his degree in 1921, he enjoyed several periods of postdoctoral study, divided among Chicago, Harvard, London, Paris, and Germany, as well as teaching at New York University. While at Harvard, Mulliken

45. Heisenberg's principle of uncertainty (or indeterminacy) was based in the Dirac-Jordan transformation theory (see Kragh, *Dirac*, 44): P. A. M. Dirac, ''The Physical Interpretation of the Quantum Dynamics,'' *Proc.RSL* A113 (1927): 621–641; Pascual Jordan, ''Ueber eine neue Begründung der Quantenmechanik,'' *ZP* 40 (1927): 809–838; Werner Heisenberg, ''Ueber den anschaulichen Inhalt der quantentheoretischen Kinematik und Mechanik,'' *ZP* 43 (1927): 172–198. Bohr first discussed the principle of complementarity at a conference in Como in 1927; see Niels Bohr, ''The Quantum Postulate and the Recent Development of Atomic Theory,'' *Nature* 121 (1928): 580–590.

46. C. A. Coulson, ''Inaugural Lecture, Chair of Theoretical Physics, King's College, London, 1948,'' 10-p. typescript, on 3. Coulson Papers, Bod.Oxford.

47. Joseph O. Hirschfelder, ''My Adventures in Theoretical Chemistry,'' *Ann.Rev.P.Chem.* 34 (1983): 1–29, on 1.

worked with Kemble, and he and Slater occupied neighboring rooms near Harvard Square.[48]

In the 1920s, Pasadena, Chicago, and Cambridge, Massachusetts were places where there was considerable interaction among chemists, particularly physical chemists, and physicists. Pauling contrasted the state of affairs at both Berkeley and Caltech with what he found at Munich, commenting that in the mid-1920s at Caltech, ''chemists, the physical chemists, were learning as much physics and mathematics as the physicists did and they were able to take advantage of this opportunity in the way that European chemists were not.'' In contrast, at Munich, Fajans, who was professor of physical chemistry, took the attitude of most German physical chemists that ions and polarization were the key to chemical combination, at least in mineral chemistry. He later told Pauling that he never had time to learn quantum mechanics.[49]

Robert Millikan was a common link between the cross-disciplinary interests in physical science at Chicago and Caltech, just as Noyes was a link between MIT and Caltech. After completing his Ph.D., Mulliken did some postgraduate work at Chicago just as Millikan was departing for Caltech.[50] Decades later, the University of Chicago still resisted narrow specialization and continued to promote some integration of physics and chemistry. In the 1960s, chemistry was a prerequisite for introductory physics, and thermodynamics, kinetic theory, and atomic structure taught in that course were not repeated in the second year for students moving into the ''physics'' course.[51]

At Harvard, Kemble took charge of teaching advanced electromagnetism and quantum mechanics in the early 1920s, and his research group soon included the chemists Mulliken and Louis Turner and the physicists Francis Jenkins and Henry Barton, all working on band spectra and molecular spectroscopy.[52] Kemble also sought to establish cooperation with some of his MIT colleagues to do work in quantum theory, and when Slater moved from Harvard to MIT in 1930, he brought the same vision of problem-oriented, cross-disciplinary research with him which Noyes had stressed at MIT one and two decades earlier. It is no accident that the two engineering institutions, MIT and Caltech, were particularly open to cross-disciplinary work, since engineering tradition is so strongly oriented toward problem solving.[53]

48. See Anthony Serafini, *Linus Pauling: A Man and His Science* (New York: Paragon House, 1989); Schweber, ''Young John Clarke Slater,'' on Slater; as well as J. C. Slater, *Solid State and Molecular Theory: A Scientific Biography* (New York: Wiley Interscience, 1975); and Mulliken, *Life*.

49. Linus Pauling, in interview with John Heilbron, March 27, 1964, 2d of 2 sessions, 19, SHQP, Berkeley.

50. Mulliken, *Life* 32.

51. Mark G. Inghram, '' [Four Universities] University of Chicago,'' *Physics Today* (March 1968):, 44–47, on 45.

52. See Assmus, ''Molecular Structure,'' 30–32, 113–114.

53. Schweber, ''Young John Clarke Slater,'' 366, 401–402.

In retrospect, particular research orientations are striking in these and other institutions for the future development of quantum chemistry. In such places, there was a good deal of collaboration betweeen physicists and physical chemists; and there was a great deal of emphasis in both the physical chemistry and physics groups on investigations of electrolytic solutions, molecular structure, x-ray diffraction, and spectroscopic studies. Physicists at MIT and Caltech joined colleagues at Michigan, Chicago, and elsewhere in the project for mapping the molecular spectra of gases, a National Research Council project headed by Kemble at Harvard and R. T. Birge at Berkeley. Spectroscopy was to be the backbone of the new quantum chemistry in the 1920s, and American results, like the investigations of the graduate student David Dennison at Michigan, were avidly followed by European theorists.[54]

At Caltech, Pauling's earliest work was on x-ray studies and crystallography; he then turned to the relation between molecular structures and the properties of compounds, attempting to model Lewis's static electron atom into a dynamic one.[55] Fired up by his experiences in Europe, including his conversations with Heitler and London in Munich, Pauling wrote an essay review for *Chemical Reviews* on the quantum mechanics of the hydrogen molecule. He gave a nonmathematical summary of the work of Heitler and London, translating back and forth between Lewis electron-dot formulas and Heitler-London wave functions.

Pauling explained how Burrau, whose treatment in turn inspired Hund, had given an exact quantum mechanical treatment of the H_2^+ ion in which two nuclei are held together by one electron and how Heitler and London explained the stability of the more complicated two-nuclei, two-electron system that is the H_2 molecule. The old conundrum of why it is that there is attraction rather than repulsion between two hydrogen ions H^+ is explained by quantum mechanical exchange energies in analogy to mechanical resonance in classical physics.[56] In a paper published in the *Proceedings of the National Academy of Sciences*, Pauling wrote that "it has become evident that the factors mainly responsible for chemical valence are the Pauli exclusion principle and the Heisenberg-Dirac resonance phenomena."[57]

Pauling already was thinking at this time about ways to explain valence and

54. See Servos, *Physical Chemistry from Ostwald to Pauling*, 128–133, 265–274; and especially on molecular spectroscopy and quantum chemistry, see Assmus, "Molecular Structure." Assmus notes the interest of Niels Bohr, H. A. Kramers, and Wolfgang Pauli in Dennison's Ph.D. dissertation, "Molecular Structure and the Infrared Spectrum of Methane"; in Alexi J. Assmus, "The Creation of Postdoctoral Education and the Siting of American Scientific Research," MS.

55. See Yuko Abe, "Pauling's Revolutionary Role," 109–110.

56. Linus Pauling, "The Application of the Quantum Mechanics to the Structure of the Hydrogen Molecule and Hydrogen Molecule-Ion and Related Problems," *Chemical Reviews* 5 (1928): 173–213.

57. Quoted in Abe, (1981: 115) from "The Shared-Electron Chemical Bond," *Proc.NAS* 14 (1928): 359–362.

bonding in unsaturated molecules like the nitrogen oxides and benzene. Further, he, like others, wondered how to treat one of the simplest saturated organic molecules, methane. The carbon atom has six electrons, which should be distributed on the basis of quantum principles into energy states of $1s^2$, $2s^2$, $2p^2$. However, carbon has four valence electrons, suggesting that one of the $2s$ electrons has been promoted to the p energy state, so that the electron configuration in carbon is $1s^2$, $2s$, $2p_x$, $2p_y$, $2p_z$. But, if this were the case, one of the CH bonds (the $2s$ one) in CH_4 would be different from the others. It is not. Pauling later told John Heilbron that by early 1928, he thought of doing away with the distinction between $2s$ and $2p$ energy sublevels for the four shared-electron-pair bonds in methane in order to get equivalent tetrahedral valences. He mentioned the idea in a brief paragraph in his National Academy of Sciences article, but he did not yet see how to solve the problem mathematically.[58]

In the next few years, in addition to his duties at Caltech, Pauling lectured regularly at Berkeley, glad to talk with Lewis and other physical chemists there about his work. Lewis wrote his former student, Joseph Mayer, that Uhlenbeck and Sommerfeld both had been at Berkeley in spring 1929, "but the best of all by far was Pauling. . . . He gave a course three hours a week in quantum mechanics, and one of one hour a week on the size of ions and other similar problems."[59]

Mulliken, like Pauling, was keenly interested in molecular structure. His father, Samuel, was a Leipzig-trained organic chemist who taught at MIT and wrote the four-volume *Method for the Identification of Pure Organic Compounds*. Following in his father's footsteps, but distinguishing himself from his father, Robert Mulliken specialized in physical chemistry at the University of Chicago, doing research on isotope enrichment and aiming to study with Ernest Rutherford at Cambridge. Instead, the Rockefeller Foundation required him to take his postdoctoral fellowship to Harvard, where Mulliken's expertise was useful in looking for isotope effects in band spectra, a project funded by the National Research Council.[60]

Following conversations in Göttingen with Hund about his ideas for a quantum mechanical treatment of multiple-electron systems, Mulliken published a paper in 1928 in the *Physical Review*, the primary American physics journal, which welcomed papers on molecular electronic structure. As soon as he had

58. Linus Pauling in interview with John Heilbron, March 27, 1964, 2d of 2 sessions, 15–16, SHQP, Berkeley.

59. Letter from G. N. Lewis to Joseph Mayer, 3 May 1929, Lewis Papers, BL.UCB.

60. Mulliken, *Life*; and Assmus, "Molecular Structure," 30–31, 118–120. See Edwin Kemble, R. Birge et al., eds., *Molecular Spectra in Gases. Bulletin of the National Research Council* 11 (1926).

seen Hund's method, Mulliken later recalled, "I was of course immediately converted to doing it."[61]

Like Hund, Mulliken developed the basic Schrödinger equation in the direction of establishing the electron charge density resulting from a combination of the attractions of two or more nuclei and the averaged repulsions of other electrons in the system. This is a method that favors some particular region of space and disfavors others. In contrast to the Heitler-London method, it overemphasizes, rather than underemphasizes, the ionic character of a molecule. For example, for the H_2 molecule, Hund's wave function equation assumes that it is just as probable to have two electrons around the same nucleus as to have one electron around each nucleus. For a molecule made up of identical nuclei, this treatment is a considerable exaggeration of the ionic character of the molecule.

Building on his own earlier work as well as on Hund's, Mulliken assigned individual quantum numbers to electrons and fed the electrons into molecular energy levels in analogy to Bohr's use of the Aufbauprinzip for atomic energy levels. Following a suggestion by Birge at Berkeley, Mulliken designated the molecular electronic energy states by the symbols *s,p*, and *d*. The same year (1928), Hund proposed instead the symbols Σ, Π, and Δ to more clearly distinguish the energy states of molecular electrons from Bohr's atomic electrons, and in 1931, Hund began using the symbols σ and π.[62]

Mulliken, Hund, and Erich Hückel were all in Leipzig in 1930, when Mulliken was there on a Guggenheim Fellowship. In that year, Hückel began applying Hund's basic method to molecules with double bonds and to benzene and aromatic molecules.[63] Hückel was a theoretical physicist who had just returned from London and Copenhagen, where he was on a Rockefeller Foundation stipend. Both Erich Hückel and his brother Walter, who was one year older, had studied at Göttingen. Erich took his degree in physics with Debye and had spent time in Zurich as a Privatdozent; Walter took his doctoral degree in organic chemistry with Adolf Windaus and began teaching at Fribourg.

In 1931, Walter Hückel published the first edition of his *Theoretische Grundlagen der Organischen Chemie*, which was absolutely up to date on applications of physical theory to chemistry, including quantum mechanics, thanks to his brother. The second volume of this two-volume work was entirely devoted to "knowledge of the theoretical tools, which the new development of

61. Robert Mulliken, in interview with Thomas Kuhn, February 1, 1964, 2d of 2 sessions, 3, SHQP, Berkeley.

62. Robert Mulliken, "Spectroscopy, Molecular Orbitals, and Chemical Bonding," 133–138, 141–142.

63. Erich Hückel,"Quantentheorie der Doppelbindung," *ZP* 60 (1930): 423–456. Hund was at Heisenberg's Institute of Theoretical Physics and Peter Debye was at the nearby Institute of Physical Chemistry in 1930, according to Mulliken in an interview with Thomas Kuhn, February 1, 1964, 2d of 2 sessions, SHQP, Berkeley.

physics and physical chemistry is bringing for a deeper penetration into problems than pure chemical progress is likely to accomplish.''[64]

With his special family interest in organic molecules, Erich Hückel tackled the classical problem of explaining the alternating double bond system typified by butadiene and benzene, concentrating on what Mulliken was to call ''unsaturation electrons'' and what Hückel called ''electrons of the second kind.''[65] Rather than localizing the unsaturation electrons in particular valence links, Hückel considered how these electrons could be added to a structure composed of the nuclei and electrons that form the saturated bonds. He set up a series of molecular levels of increasing energy into each of which two electrons of opposite spin may be put, showing that an even number of unsaturation electrons, namely, six in the case of benzene, confers special stability in a way closely related electronically to the inertness of the rare gases.

Hückel's application of this approach to the aromatic compounds gave new confidence to those physicists and chemists following up on the Hund-Mulliken analysis. It was regarded by many people as the simplest of the quantum mechanical valence-bond methods based on the Schrödinger equation.[66] Hückel's was part of a series of applications of the method of linear combination of atom wave functions (atomic orbitals), a method that Felix Bloch had extended from H_2^+ to metals in 1928 and that Fowler's student, Lennard-Jones, had further developed for diatomic molecules in 1929. Now Hückel extended the method to polyatomic molecules.[67]

At first chemists found themselves more comfortable with the method of building up molecules from individual atoms than piling electrons into a system of nuclei already in place. There were occasionally harsh words about rival methods. ''Pauling,'' said Mulliken, ''made a special point of making everything sound as simple as possible and in that way making it very popular.'' In Mulliken's view, Pauling's approach delayed chemists' ''true'' understanding.[68]

Slater attempted to come up with a compromise between the two systems. His early interests ran in the direction taken a few steps ahead of him by de Broglie, Heisenberg, and Dirac. About 1929, Slater turned from his early strong interest in quantum electrodynamics to molecular spectroscopy and mo-

64. Walter Hückel, *Theoretische Grundlagender organischen Chemie*, I: iii, viii.

65. See discussion in Louis P. Hammett, *Physical Organic Chemistry*, 16–19; and E. Hückel, ''Zur Quantentheorie der Doppelbindung,'' *ZP* 60 (1930): 423–456; and E. Hückel, ''Quantentheoretische Beiträge zum Benzolproblem. I. Die Elektronenkonfiguration des Benzols und Verwandter Verbindungen,'' *ZP* 70 (1931): 204–286.

66. See B. Bigot and F. Volatron, ''Parlez-vous chimie théorique?'' *L'Actualité Chimique* (November 1984): 43–51, on 50.

67. See J. E. Lennard-Jones, ''The Electronic Structure of Some Diatomic Molecules,'' *Trans.Far.Soc.* 25 (1929): 668–682; and Robert Mulliken, ''Quelques aspects de la théorie des orbitales moléculaires,'' *JCP* 46 (1949): 497–542, 675–713, on 500.

68. Mulliken, in interview with Thomas Kuhn, 2d of 2 sessions, 17–18, SHQP, Berkeley.

lecular theory. He said of his change in focus, "It was obvious that I would never catch up with Dirac to the point of being clearly ahead of him."[69]

Just before returning to Europe in 1929, Slater generalized into an N-electron system the wave function used by Pauling in the treatment of helium in the 1928 *Chemical Reviews* essay. The title of Slater's paper, "The Self-Consistent Field and the Structure of Atoms," shows his debt to Hartree, although Slater's method turned out to be a great deal more practical than Hartree's, as well as consistent with the methods of Heitler, London, and Pauling.[70]

In Leipzig, Slater pursued the application of wave mechanics to electrons in simple molecules ("quantum chemistry") and in metals ("solid-state physics"). He wrote Percy Bridgman,

> The physical picture which I think is undoubtedly correct is that the interaction forces [in metals] are much as in diatomic molecules, determined by resonance phenomena, and I want to fit that in with the problem of many atoms.[71]

Slater developed an approach (the "determinantal method") that offers a way of choosing among linear combinations (essentially sums and differences) of the polar and nonpolar terms in the Hund-Mulliken equations to bring their method into better harmony with the nonpolar emphasis characteristic of the Heitler-London-Pauling approach in which polar terms do not figure in the wave equation.[72]

Independently of each other, Pauling and Slater worked out a quantum mechanical explanation of the directional valences characteristic of chemical molecules. They did this by proposing directional properties for the *p* wave functions and for the sp^3 wave functions resulting from "hybridization" of electron wave functions, or orbitals.[73]

Mulliken introduced the term "orbital" distinct from "*orbital* wave function" in 1932 in the second of fourteen papers carrying the general title, "Electronic Structures of Polyatomic Molecules and Valence." Mulliken defined atomic orbitals (AOs) and molecular orbitals (MOs) as *something like* the

69. Quoted in Schweber (1990: 373), from John Slater, "A Physicist of the Lucky Generation," MS, MIT Archives.

70. John Slater, "The Self-Consistent Field and Structure of Atoms," *Physical Review* 32 (1928): 339–348. See discussion in Schweber (1990: 376–377).

71. Quoted in Schweber (1990: 379).

72. See Schweber, "Young John Clarke Slater," 382–385. John Slater, "Cohesion in Monovalent Metals," *Physical Review* 35 (1930): 17–37.

73. Linus Pauling, "The Nature of the Chemical Bond. Applications of Results Obtained from the Quantum Mechanics and from a Theory of Paramagnetic Susceptibility to the Structure of Molecules," *JACS* 53 (1931): 1367–1400; also, "The Nature of the Chemical Bond. II. The One-Electron Bond and Three-Electron Bond," *JACS* 53 (1931): 3225–3237; John C. Slater, "Directed Valence in Polyatomic Molecules," *Physical Review* 37 (1931): 481–489. Also see Linus Pauling, *The Nature of the Chemical Bond* (1939); and Schweber, "Young John Clarke Slater," 386–387.

orbits of Bohr's theory.[74] The energy associated with each orbital was understood to be the energy required to take the electron entirely out of the orbital into free space.[75]

The basic idea of hybridization was that a valence bond wave function has a lower energy value if it is intermediary in character between the energy values associated with either an *s* (spherical) or a *p* (elliptical) wave function, or orbital. Pauling and Slater demonstrated that certain types of wave functions project out in characteristic directions, *p* waves, for example, represented by three dumbbell-shaped distributions or contour lines at right angles to one another and the *s* wave a distribution that is spherically shaped. Hybridization of these wave functions, or orbitals, produces electron distributions identical in kind and oriented toward the corners of a tetrahedron, rather than at right angles, that is, with C-H-C angles of 109.5° rather than 90°. If a bond angle is expected to be 90° and departs from that figure (109.5° in methane, 107° in ammonia, 104.5° in water), hybridization now could be suspected. The quantum mechanical calculation, Pauling said, ''provides the quantum mechanical justification of the chemist's tetrahedral carbon atom.''[76]

Pauling now turned in earnest to unsaturated molecules and to systems with alternating double bonds. He had long been interested in nitrogen oxides, and this subject likely was a topic of conversation with Sidgwick when Sidgwick was lecturing in the United States in 1931.[77] Pauling now extended the notion of mechanical resonance to molecules like carbon dioxide, benzene, and graphite as well as to carbonate and nitrate ions. This approach was entirely consistent with the idea of ''mesomerism'' or the ''Zwitterion,'' as discussed in the last chapter, but very different from hypotheses of oscillating atoms or tautomeric forms in equilibrium with one another.

In quantum mechanics, as we have already seen, one can approximately describe the hydrogen molecular ion as consisting of H_a^+ and H_b, or H_b^+ and H_a. Some combination of wave functions representing these two configurations is needed as an approximation of the actual state of affairs. The state of H_2^+ can then be thought of as a resonance hybrid of the two.

In applying this notion to many-electron systems, Pauling reasoned that a wave function might be set up to represent each of the possible classical valence, or electron-pair, bonds in compounds like carbon dioxide or benzene. Each equation corresponds to a combination of ionic and covalent character

74. Mulliken, *Life*, 90. On the ''orbital,'' Mulliken wrote in 1932 ''From here on, one-electron orbital wave functions will be referred to for brevity as *orbitals*. The method followed here will be to describe unshared electrons always in terms of *atomic orbitals* but to use *molecular orbitals* for shared electrons.'' In Robert Mulliken, ''Electronic Structures of Polyatomic Molecules and Valence,'' *Physical Review* 41 (1932): 49–71, on 50.

75. Mulliken, ''Spectroscopy, Molecular Orbitals, and Chemical Bonding,'' 136.

76. Pauling, ''The Chemical Bond. I,'' p. 1378; quoted in Yuko Abe (1981: 117).

77. Pauling and Sidgwick became acquainted when Sidgwick lectured at Cornell in 1931. See Colin Russell, *A History of Valency*, 301.

for a bond and to its energy content. For benzene, there are a series of alternative electronic structures for the molecule, including the alternating single and double bond structures of Kekulé and the bridged structures of Dewar. The actual electronic structure is none of these but something with elements from each.

Wave mechanics allows the calculation of the relative contribution of each bond to the structure, the relative weights depending on chemical and physical measurements. Each of the energy values for the alternate electronic structures is higher than the energy value for the molecule, and thus the actual "resonance hybrid" is the stable form because it has the lowest energy value. Pauling's result was a more general mathematical model for a molecule like benzene than Hückel's, which had been the most successful mathematical and mechanical model to date.[78]

In a paper with John Sherman, Pauling calculated the work required to rupture a carbon-carbon single bond (84.1 kg-cal.) in contrast to that required to rupture a double bond (15.2 kg-cal.) The difference, 67.1 kg-cal., was taken to be the binding energy due to the unsaturation electrons.[79] In comparing the binding energies of single bonds, double bonds, and the bonds in benzene (or in butadiene), it was clear that the benzene bonds are intermediary in character and considerably more saturated than an ordinary double bond. Of course, chemists had known this for decades.

The difference in energy between the resonance system and the energy predicted from any of the individual structures was termed the "resonance energy." Pauling and Sherman calculated the resonance energy of benzene and other resonance structures, using Slater's method,[80] and they found their values approximately confirmed by experimental measurements of heats of reaction. As Hammett noted, this had to be done with some caution, since resonance energy is a potential energy quantity. The amount of heat liberated in the addition of one mole of hydrogen to a hydrocarbon with a single double bond was found to vary from 32.8 to 26.6 kg-cal., whereas the addition of one

78. Linus Pauling, "The Nature of the Chemical Bond. III. The Transition from One Extreme Bond Type to Another," *JACS* 54 (1932): 98–1003; Linus Pauling, "Interatomic Distances in Covalent Molecules and Resonance between Two or More Lewis Electronic Structures," *Proc.NAS* 18 (1932): 293–297; Linus Pauling, "The Calculation of Matrix Element for the Lewis Electronic Structure of Molecules," *J.Chem.Physics* 1 (1933): 280–283; Linus Pauling and G. W. Wheland, "The Nature of the Chemical Bond. V. The Quantum-Mechanical Calculation of the Resonance Energy of Benzene and Naphthalene and the Hydrocarbon Free Radicals," *J.Chem.Physics* 1 (1933): 362–374; Linus Pauling and J. Sherman, "The Nature of the Chemical Bond. VI. Calculation from Thermochemical Data of the Energy of Resonance of Molecules Among Several Electronic Structures," *J.Chem.Physics* 1 (1933): 606–617; and Pauling and Sherman, "The Nature of the Chemical Bond. VII. The Calculation of Resonance Energy in Conjugated Systems," *J.Chem.Physics* 1 (1933): 679–686.

79. Linus Pauling and J. Sherman, "Nature of the Chemical Bond. VII," 606.

80. See Abe, "Pauling's Revolutionary Role," 119.

mole of hydrogen to a benzene hydrocarbon required the *absorption* of 6 to 8 kg-cal. [81]

Pauling and his colleagues used information from x-ray diffraction in crystals, x-ray and electron diffraction in gases, and infrared, Raman, and band spectra to establish internuclear distances and bond lengths. They found that they could tabulate covalent-bond radii for hydrogen, for singly bound carbon, and for doubly bound carbon, then calculate the length of a bond as a summation of the relevant radii and use an intermediary value between standard figures when a bond was taken to be a resonance hybrid.[82]

In a paper in 1932, Pauling proposed his scale of relative electronegativities, which could be used to predict the degree of ionization in a bond, with fluorine (4.0) the most electronegative atom on Pauling's scale and cesium the least electronegative (0.7). Electronegativity values were estimated from bond-dissociation energies and from ionization potentials.[83]

As we have seen, Pauling insisted on the origins of the resonance, or mesomerism, theory in the conceptions of classical organic chemistry.

> It is true that the idea of resonance energy was . . . provided by quantum mechanics, . . . but the theory of resonance in chemistry has gone far beyond the region of application in which any precise quantum mechanical calculations have been made, and its great extension has been almost entirely empirical, with only the valuable and effective guidance of fundamental quantum principles.[84]

While Pauling was working out his series of papers on the chemical bond, Slater followed up on his hybridization of electron energy levels by making refinements in the treatment of the nitrate and chlorate ions. However, he was struck by the great difficulty of computing results in complicated molecules, and he began to express a preference for Mulliken's method.[85] Slater became impatient with the ways in which chemists wanted to use quantum mechanics, "on the one hand [extending] its mathematical formalism, on the other . . . stretching the theory far beyond what was justified by its crude nature in making comparisons with experiment, with the result that most of the results obtained by the chemists since that time are suspect by the physicists."[86]

In his series of papers written between 1931 and 1935, Mulliken worked

81. Hammett, *Physical Organic Chemistry*, 22–23.

82. Ibid., 23–25, 31–33.

83. Linus Pauling, "The Nature of the Chemical Bond. IV. The Energy of Single Bonds and the Relative Electronegativity of Atoms," *JACS* 54 (1932): 3570–3582; and Pauling, *The Nature of the Chemical Bond.*

84. Linus Pauling, "The Nature of the Theory of Resonance," 7.

85. John Slater, "Note on the Structure of the Groups XO_3," *Physical Review* 38 (1931): 325–329; and "Molecular Energy Levels and Valence Bonds," *Physical Review* 38 (1931): 1109–1141. See Schweber, "Young John Clarke Slater," 388–390.

86. Quoted in Schweber (1990), from a 1950 document proposing a molecular theory group in the Physics Department at MIT.

out a classification of molecular orbitals on the basis of symmetry, expanding the treatment to double bonds, to the molecules H_2O, NH_3, CH_4, to the radicals CH_2 and NH_2, and distinguishing different ''species'' of molecular systems on the basis of planar, trigonal, and tetrahedral geometry. No doubt to Slater's dismay, he used the group theoretical methods that Slater characterized as ''Gruppenpest.''[87]

Pauling at Caltech, Mulliken at Chicago, and Slater at MIT, along with Van Vleck at Harvard, were among the most prominent of the founders of what was to become known as quantum chemistry. But the Americans were not the only leaders. Erich Hückel and Fritz Arndt's student, Bernhard Eistert, made contributions in Germany.[88] Particularly important in the next decade were Lennard-Jones, Coulson, and Hugh Christopher Longuet-Higgins in England. Lennard-Jones held the first chair of theoretical chemistry in Great Britain, appointed at Cambridge University in 1932. He secured rooms for his staff and students where they could work, talk, and drink coffee, so that the members of the group, ''like the members of a good laboratory, felt themselves to belong to a body with a loyalty of its own.''[89]

Mulliken visited Cambridge in summer 1930 and again in winter 1933. Coulson, who was taking his doctoral degree in chemistry after studies in mathematics and physics, recalled the second visit in glowing terms. Mulliken had shown his audience how to produce the double bond in ethylene by bringing two CH_2 groups together with sigma and pi orbitals. ''How neat, and in a sense how obviously satisfying it was,'' said Coulson.[90] Shortly afterward, Coulson read Van Vleck's papers on methane and ammonia in the *Journal of Chemical Physics* as well as the review article by Van Vleck and Sherman in *Reviews of Modern Physics*, arguing the complementarity of the atomic-orbital and molecular-orbital methods.[91] Coulson was hooked.

In 1937, Coulson's mentor, Lennard-Jones, wrote a series of papers on molecular orbitals applied to polyenes and aromatic molecules, discussing the variations in bond lengths in conjugated and aromatic molecules from a theoretical point of view.[92] He involved his student in the work. Coulson defined

87. See Schweber, ''Young John Clarke Slater,'' 377.

88. Bernhard Eistert, ''Zur Schreibweise chemischer Formeln und Reaktionsabläufe,'' *Berichte* 71B (1938): 237–240. Eistert took his doctorate at Breslau with F. Arndt and became assistant at the University of Bonn to the organic chemist P. Pfeiffer. From 1929 until the war, he worked with BASF in Ludwigshafen. After the war, he was professor of theoretical organic chemistry at the Technische Hochschule in Darmstadt.

89. N. F. Mott, ''John Edward Lennard-Jones (1894–1954),'' 175–184, in *Biographical Memoirs of the Royal Society*, I (1955), on 177.

90. Charles Coulson, ''Recent Developments in Valence Theory,'' page proofs, delivered at Australian symposium on ''Fifty Years of Valence Theory,'' 2, Coulson Papers, B.41.10, Bod.Oxford.

91. Ibid. See J. H. Van Vleck and J. Albert Sherman, ''The Quantum Theory of Valence,'' *Reviews of Modern Physics* 7 (1935): 167–228.

92. Mott, ''Lennard-Jones,'' 177.

the meaning of "fractional bond order" for bonds intermediate between integral values, at last putting Thiele's old idea of "partial valency" on firm theoretical footing. He showed that in butadiene, the end bonds have an order of 1.9 (almost a "pure" double bond) but that the center bond has an order of 1.2 and therefore is "more" than a "pure" single bond. In benzene, he calculated, all bonds are 1.67.

Using the molecular orbital method, Coulson showed how certain electrons in benzene, namely, the π electrons, can move over the whole molecule instead of being restricted to the region between two particular atoms.[93] Coulson, collaborating later with Longuet-Higgins and the French theoreticians Pascaline and Raymond Daudel and Alberte and Bernard Pullman, was to become a major presence in quantum chemistry. But on the whole, Coulson said, he was inclined to characterize the period from 1933 until the end of the Second World War as the "Mulliken Era."[94]

Quantum Chemistry and Chemical Physics in the 1930s and 1940s

In 1928, Van Vleck took a physics professorship at the University of Wisconsin. He had done his doctoral research at Harvard with Kemble, and he would return to Harvard in 1934, eventually to be named the Hollis Professor of Mathematics and Natural Philosophy. Van Vleck made a distinction between physicists and chemists by saying, "What the physicist observes about an atom is primarily the radiations, while the chemist measures its heat of dissociation, affinities, etc."[95]

Chemists were interested in atoms, to be sure, but they primarily concerned themselves with molecules and with the "affinities" by which atoms combine into molecules. The molecules became larger and larger, as biologically active molecules consumed more and more of chemists' interest at the very time that physicists were beginning to focus on smaller and smaller particles inside the atomic nucleus or the derivation of macroscopic properties from atomic mod-

93. On this, see Coulson's Inaugural Lecture for the chair of theoretical physics at King's College, London, 1948, typescript, Coulson Papers, no. 21, Bod.Oxford. Also, see J. E. Lennard-Jones and C. A. Coulson, "The Structure and Energies of Some Hydrocarbon Molecules," *Trans.Far.Soc.* 35 (1939): 811ff.

94. Charles Coulson, 4, "A History of Quantum Theory and Applications in Chemistry," 12-page typescript of after-dinner speech given August 16, 1971, at the Fourth Canadian Symposium on Theoretical Chemistry, British Columbia, in Coulson Papers, Bod.Oxford.

95. Schweber, "Young John Clarke Slater," 399, quoting from "The New Quantum Mechanics," *Chemical Reviews* 5 (1928): 467–507. Also Aaron J. Ihde, *Chemistry, as Viewed from Bascom's Hill: A History of the Chemistry Department at the University of Wisconsin at Madison* (Madison: University of Wisconsin Department of Chemistry, 1990): 165–166.

els.[96] Pauling's turn toward molecular biology, including his study of the structure of deoxyribonucleic acid, was typical among theoretical chemists who rode, or made, the crest of a new research wave toward the end of the 1930s.

In addition, spectroscopy, which had for many decades provided identification of pure elementary substances, increasingly became of more primary interest to chemists concerned with studying the active parts of the molecule and with following the kinetics of chemical reaction. The use of Raman and infrared spectroscopy, then nuclear magnetic spectrosopy and mass spectrometry, was to become standard for determining bond length, functional groups, binding energies, and the size and shape of molecules.[97] No aspect of chemistry changed more dramatically after the 1930s than optical and electromagnetic instrumentation.[98]

Yet, while changes in instrumentation were to affect virtually all chemists, certainly after the Second World War, the application of advanced mathematics, including quantum mechanics, also was revolutionary for chemists, including the majority who did not practice quantum chemistry or chemical physics.

In this development, the mathematics of thermodynamics had been the opening wedge. When asked how much mathematics he knew when he was a graduate student at the University of Chicago, Mulliken replied that his mathematics was at the level of the Lewis and Randall thermodynamics textbook.[99] But within just a decade, many chemists were expected to know a great deal more mathematics than that. Eyring required his graduate students to study quantum mechanics beginning in 1931.[100] Coulson put the matter bluntly: ''The plain truth is that you cannot have a deep theory without some sort of mathematics.''[101]

The effort to introduce advanced mathematics and fundamental theory into chemistry posed disciplinary difficulties for those developing quantum chemistry in the 1930s, as it did for those working on chemical kinetics and the radiation hypothesis in the 1920s (see chap. 5). Pauling, like Lewis, had dif-

96. As Emilio Segrè notes, Felix Bloch, Hans Bethe, and R.E. Peierls were among twentieth-century physicists working in the fields of both nuclear physics and solid-state physics. Measured by sheer volume of research, numbers of physicists employed, and numbers of published pages, solid-state physics is the largest field in physics today, in large part because of its industrial applications. See Segrè, *X-rays to Quarks*, 280.

97. See Gerhard Herzberg, *Infrared and Raman Spectra of Polyatomic Molecules* (New York: Van Nostrand, 1945): 1; and the earlier volume, *Molecular Spectra and Molecular Structure* (1939). Also Yakov Rabkin, ''Technological Innovation in Science,'' 31–54.

98. Henry Eyring, ''Physical Chemistry,'' *CENews* 54 (1976): 88–104, on 94.

99. Interview of Robert Mulliken with Thomas Kuhn, SHQP, Berkeley.

100. ''The good student who has mastered calculus will be able to follow the arguments. The way will be made easier by whatever he has learned of differential equations, vector analysis, group theory, and physical optics.'' Henry Eyring, *Quantum Chemistry* (New York: Wiley, 1944): iii.

101. Charles Coulson, ''Theoretical Chemistry: Past and Future,'' Inaugural Lecture, University of Oxford, February 13, 1973, ed. S. L. Altmann (Oxford: Clarendon, 1974): 9.

ficulty persuading the editor and referees of the *Journal of the American Chemical Society* to publish his papers.[102] Such needs were not ignored, however, since Pauling and Lewis were powerful voices within the scientific community.

Henry Barton, who had worked in Kemble's molecular spectroscopy research group at Harvard, was an influential scientific administrator in the early 1930s. In 1931, he became director of the American Institute of Physics (AIP), the recently founded consortium of five American physics societies. William Buffum, an officer in the Chemical Foundation, was influential in providing considerable financial support for the chemical community and for the *Journal of Physical Chemistry*, edited by the maverick Cornell University chemist Wilder Bancroft.[103]

Barton and Buffum proposed to an enraged Bancroft that he turn over his faltering journal to the AIP as a forum for mathematically oriented papers at the "borderland" of physics and chemistry. When he refused, the Chemical Foundation switched its financial support from the *Journal of Physical Chemistry* to the proposed AIP journal. Karl Compton, president of MIT and chairman of the governing board of the AIP, requested Lewis and others to serve on the advisory board for the journal that Harold Urey would edit. Compton's letter carried this statement defining the rationale for establishing the journal:

> Ever since the establishment of the Institute of Physics, we have discussed the problem of properly handling the increasing number of articles in physical chemistry which have distinct emphases on the physics side . . . [but are] . . . perhaps too mathematical for the *Journal of Physical Chemistry*, or too chemical for the *Physical Review*.[104]

The *Journal of Chemical Physics* would fulfill this need.

Urey had studied physical chemistry at Harvard before teaching at Johns Hopkins University and now at Columbia University. In the opening volume of his new journal, he announced,

> At present the boundary between the sciences of physics and chemistry has been completely bridged. Men [*sic*] who must be classified as physicists on the basis of training and of relations to departments or institutes of physics are working on the traditional problems of chemistry; and others who must be regarded as chemists on similar grounds are working in fields which must be regarded as

102. Linus Pauling in interview with John Heilbron, March 27, 1964, 2d of 2 sessions, 3, SHQP, Berkeley.

103. For details, see the fine discussion in Servos, *Physical Chemistry from Ostwald to Pauling*, 315–320.

104. Letter from Karl Compton to G. N. Lewis, August 6, 1932, G. N. Lewis Papers, BL.UCB.

> physics. . . . It seems proper that a journal devoted to this borderline field should be available to this group.[105]

While the *Journal of Chemical Physics* by no means restricted itself to quantum mechanics and quantum chemistry, it was now available as an important and instantly recognizable medium for the propagation of theories of the new quantum wave.

As we have already seen, and as Alexi Assmus has emphasized in her study of quantum physics in America, private foundations like the Chemical Foundation were instrumental in the creation in the United States of specialization in the field of molecular theory. The Rockefeller Foundation, the International Education Board (a Rockefeller trust), and the Guggenheim Foundation, as well as the National Research Council, all looked favorably on fellowship applications having to do with spectroscopy and molecular structure in the 1920s. (The Rockefeller Foundation was to refocus its principal interest toward molecular biology and medicine in the next decade, with similar effects for creation of a strong disciplinary field.) In the 1920s, 20 percent of all American physics Ph.D.'s and 5 to 10 percent of all chemistry Ph.D.'s received one to two years of postdoctoral support from the Rockefeller Foundation and the NRC. By 1932, 25 of 125 physics fellows and 33 of 128 chemistry fellows had spent their fellowship tenure in Europe, among them Pauling, Mulliken, and Slater.[106]

Why this emphasis? Schweber has portrayed Slater as a man who developed a deep feeling of both inferiority and competitiveness toward his European mentors and peers in the fields of atomic physics and quantum electrodynamics. Slater was not alone in this reaction, as Henry James made clear. Slater, like other American physicists and chemists, used his influence in Boston, New York, and Washington circles, as well as his position within his own institution, to build up American science in an area where Americans could take a competitive lead.[107] Donnan had written Lewis in 1921 that "you are making old Europe sit up some. If it wasn't for Planck, Einstein, Rutherford, and Bragg, we should be in a bad way."[108] But it was not enough for Europeans to sit up "some"; they must be made to gawk.

Schweber and others have argued that quantum chemistry was a quintessentially American discipline, with Mulliken, Slater, Van Vleck, Urey, Pauling, Edward Condon, Oppenheimer, Ralph Kronig, I. I. Rabi, Clarence Zener, David Dennison, Philip M. Morse, Eyring, John G. Kirkwood, George E.

105. See Editorial in *J.Chem.Physics* 1 (1933): and J. W. Stout, "The *Journal of Chemical Physics*: The First 50 Years," *Ann.Rev.P.Chem.* 37 (1986): 1–23.

106. Assmus, "Molecular Structure," 2, 9, 32, 143. Also, see Robert Kohler, *Partners in Science. Foundations and Natural Scientists, 1900–1945* (Chicago: University of Chicago Press, 1991).

107. Schweber, "Young John Clarke Slater," 356.

108. Letter from F. G. Donnan to G. N. Lewis, 1921, Lewis Papers, BL.UCB.

Kimball, E. Bright Wilson, H. M. James, and Rice among its principal practitioners in the 1930s.[109] By this time, even before the exodus of political refugees from Germany and Central Europe, many academic departments and research centers in American universities already were welcoming European chemists and physicists who were emigrating, fully trained, into American science.[110] Among these were women of no small distinction, for example, Maria Goeppert-Mayer, the German wife of G. N. Lewis's former student, Joseph Mayer. One of the important papers of the late 1930s was her calculation, with A. L. Sklar, of energies of benzene using the molecular orbital method.[111]

American quantum chemistry, then, was not entirely indigenous, nor was it entirely chemistry. Mayer wrote Lewis in 1930 about attending informal theoretical sessions at the recent meeting of the Physical Society.

> They were composed, over half of them, of Germans and Dutch. . . . Slater spoke about the stability of crystals and compounds and new quantum mechanics, not much of specific importance, but his talk of tetrahedral axes and coupling sounded so much like tetrahedrons and electron pairs that I felt much at home. Not only I, but Pauling, too, evidently, since he got up once and reminded them that the physicists were just coming to believe what the chemists had known for a long time.[112]

What was the distinction between quantum chemistry and chemical physics? After the *Journal of Chemical Physics* was established, it was easy to say that chemical physics was anything found in the new journal. This included molecular spectroscopy and molecular structures, the quantum mechanical treatment of electronic structure of molecules and crystals and the problem of chemical binding, the kinetics of chemical reactions from the standpoint of basic physical principles, the thermodynamic properties of substances and calculation by statistical mechanical methods, the structure of crystals, and surface phenomena.

Mulliken, who published the first articles in his series, ''Electronic Structures of Molecules and Valence,'' in *Physical Review*, was not keen initially on the new journal but nonetheless shifted his series to the *Journal of Chemical*

109. Schweber, ''Young John Clarke Slater,'' 341.

110. See Spencer Weart, ''The Physics Business in America, 1919–1940: A Statistical Reconnaissance,'' 295–358, in Nathan Reingold, ed., *The Sciences in the American Context: New Perspectives* (Washington, D.C.: Smithsonian Institution Press, 1979), and Robin E. Rider, ''Alarm and Opportunity: Emigration of Mathematicians and Physicists to Britain and the United States, 1933–1945,'' *HSPS* 15 (1984): 107–176.

111. Maria Goeppert-Mayer and Alfred L. Sklar, ''Calculations of the Lower Excited Levels of Benzene,'' *J.Chem.Physics* 6 (1938): 645–652.

112. Letter from Joseph Mayer to G. N. Lewis, May 19, 1930, dated from Johns Hopkins University, Lewis Letters, BL.UCB.

Physics beginning with his fifth installment.[113] Birge, head of the physics department at Berkeley, disliked the tendency toward specialization but agreed to serve on the editorial board of the new journal.[114]

The disciplinary titles of the practitioners of quantum chemistry and chemical physics varied. Pauling initially wanted his title at Caltech to be professor of theoretical chemistry and mathematical physics, but he accepted CalTech's dropping "mathematical physics"[115] and later preferred to be known as a chemist.[116] Slater always had his principal appointment in a physics department. Mulliken, who had taken his degree at Chicago in chemistry, returned in 1928 as associate professor of physics and retired in 1983 as professor of chemistry and physics.[117]

Many of those working in quantum chemistry saw themselves as more broadly based, rather than more specialized, in comparison to other colleagues in chemistry and physics. They tended to agree with Sidgwick's riposte to the accusation that scientists were coming to know more and more about less and less. "[This] must give great consolation to those who know practically nothing about absolutely everything."[118] Hirschfelder, for example, took a double Ph.D. at Princeton, studying with Eugene Wigner on the physics side and with Eyring and Hugh S. Taylor on the chemical side.[119]

Quantum chemists, like chemical physicists, tended to characterize their field as one in which the distinction between theoreticians and experimentalists is not so sharp.[120] They prided themselves on intellectual flexibility, and they frequently exhibited social flexibility. Perhaps for these very reasons quantum chemistry was able to establish a special niche in the United States in the 1930s, when the departmental structures of American universities were permeable and personal fiefdoms fairly rare.

On the chemical side, there was an important linkage between the development of an American, indeed, a larger Anglo-American, scientific community in quantum chemistry and the slightly earlier development in Anglo-American science of the field of physical organic chemistry. On the physical side, there was an equally important interdisciplinary relationship, namely, the

113. Robert Mulliken, "Spectroscopy, Quantum Chemistry and Molecular Physics," *Physics Today* (April 1968): 52–57, on 54.

114. Servos, *Physical Chemistry from Ostwald to Pauling*, 317; and Stout, "The Journal of Chemical Physics," 2.

115. Richard Tolman was professor of physical chemistry and mathematical physics at Caltech. See Linus Pauling in interview with John Heilbron, March 27, 1964, 2d of 2 sessions, 20, SHQP, Berkeley.

116. Ibid. Also, see his letter to Slater, in 1931, quoted in Schweber (1990: 402): "If I were to come to MIT . . . I prefer being in a chemistry department."

117. Mulliken, *Life*, 67.

118. Quoted from "Professor N. V. Sidgwick," 284–286, in *The Oxford Magazine*, Thursday, May 1, 1952, 286; Nevil V. Sidgwick Papers, L.Coll.Oxford.

119. Hirschfelder, "My Adventures," 5.

120. See Stout, "The *Journal of Chemical Physics*," 23.

one between studies in molecular spectroscopy and the development of quantum chemistry.

Given the emphasis of earlier chapters in this book, it is crucial to emphasize that the kinds of chemical problems to which the methods of quantum mechanics were extended in the 1930s were first and foremost the ones that earlier had concerned the chemists of the London-Manchester school, as discussed in chapters 7 and 8. Consequently, it was in England as well as in the United States that quantum chemistry first thrived.

Leon Gortler has studied the American institutions where the strongest schools of physical organic chemistry emerged in the 1920s and 1930s. These included Harvard, Columbia, Chicago, and Caltech, institutions where quantum chemistry was to establish some of its strongholds.[121] Pauling's older colleague, Lucas, was one of the first American chemists to recognize the value of the electronic interpretation in organic chemistry and to apply the inductive effect to reaction mechanisms of aromatic molecules. Lucas related traditional organic chemistry to modern chemical theory, and he was one of the first American organic chemists to use the theory of π electrons.[122]

When Mulliken visited England in 1948, he found a "brilliant group" of theoreticians at Oxford's Physical Chemistry Laboratory, including Hinshelwood and Longuet-Higgins, and "another good center of chemical physics" at Manchester.[123]

Oxford and Manchester, along with the colleges of the Universisty of London, were the leading centers for the application of the Robinson-Ingold schemes in everyday chemical synthesis. Longuet-Higgins left Oxford for Cambridge, succeeding Coulson there as professor of theoretical physics. Both were to end their careers as professors of theoretical chemistry, Longuet-Higgins appointed to that title in 1954 at Cambridge and Coulson in 1972 at Oxford.

In Germany, Hund and Mulliken were invited in 1930 to the Heidelberg meeting of the Bunsen-Gesellschaft to explain the quantum mechanics of the chemical bond.[124] The Hückel brothers contributed to the development of quantum chemistry but with considerably more influence outside Germany than inside. It is striking that in Hans Hellman's widely read *Einführung in die*

121. See Leon Gortler, "The Physical Organic Community in the United States," 753–757. Gortler argues that physical organic chemistry did not prosper at Berkeley because physical chemistry under G. N. Lewis was so strong there (754). Also see Martin Saltzmann, "The Development of Physical Organic Chemistry," 588–593.

122. Saul Winstein, handwritten 3-page obituary notice for Howard J. Lucas, H. J. Lucas Papers, Caltech; Pauling, speaking about Lucas, in interview with John Heilbron, 2d of 2 sessions, 1–2, SHQP, Berkeley. Also, see Saltzmann, "Development of Physical Organic Chemistry," 592.

123. *Life*, 127.

124. Hund, in Mulliken, *Life*, x.

Quantenchemie (1937), the journal of reference is the *Journal of Chemical Physics*.[125]

An important reason for this Anglo-American trend to chemists' interest in quantum chemistry, as Walter Hückel himself noted, was that a favorable reception to physical approaches, including the electronic theory of valence, was very limited among German organic chemists all through the period that Lapworth, Lowry, Robinson, and Ingold were developing a theory of reaction mechanisms using the electronic theory of valence. ''For a long time these ideas were almost unknown in Germany.''[126] In addition, Belgian and French chemists were more punitive in their attitudes toward German colleagues following the First World War than were physicists, and Germans were not invited to participate in the first three Solvay conferences in chemistry that took place during the 1920s.[127]

In France, there was very little interest in quantum chemistry during the 1920s and 1930s. There were a number of reasons for this state of affairs, including the emphasis of Perrin's laboratory on the radiation hypothesis and a lack of interest in quantum wave mechanics among leading physicists, including Perrin and Langevin, who had enormous influence. As Edmond Bauer put it, with a different emphasis, ''There is scarcely a physicist or physical chemist of my generation or the following generation who has not been the student of Perrin and Langevin.''[128]

Perrin's student, Bauer, who became director of Perrin's laboratory after the war, later claimed that he was mocked for his interest in theory. ''I think people thought theories were something like modes; people spoke of a fashion, a théorie à la mode.''[129] Shinn recently has done research confirming Bauer's memories. On the basis of his study of French physics between the wars, Shinn maintains that French physicists simply had no guiding theories in much of their work.[130] Charpentier-Morize, who studied chemistry under Prévost during these years, concurs, with the same judgment regarding chemists' attitudes toward theories. University laboratories in mineral chemistry and organic

125. Hans Hellmann, *Einführung in die Quantenchemie* (Leipzig: Deuticke, 1937). Hellmann was professor at the Karpow-Institut für physikalische Chemie in Moscow.

126. W. Hückel and F. Seel, *Theoretical Organic Chemistry*, 2 vols. (Wiesbaden: W. Klemm, 1948): introduction. Although the title is in English, all the papers are in German.

127. A few weeks before the first Solvay chemistry conference in 1922, French chemists signed a letter declining to reestablish relations among chemists of Western and Central Europe. A copy of this letter, 3 April 22, is in the Donnan Papers, UCL. On the Solvay conferences, see M. J. Nye, ''Chemical Explanation and Physical Dynamics: Two Research Schools at the First Solvay Chemistry Conference, 1922–1928,'' *Annals of Science* 46 (1989): 461–480.

128. Edmond Bauer, 22, in Henri Laugier et al., *Hommage national à Paul Langevin et Jean Perrin* (Paris: Orléans, 1948).

129. Edmond Bauer, in inteview with Thomas Kuhn and T. Kahan, January 8, 1963, 1st of 2 sessions, 13, SHQP, Berkeley.

130. Personal letter, May 1991, from Terry Shinn.

chemistry, she argues, were more concerned with immediate applications of chemical processes than with theories.[131]

Research in fundamental chemistry, physical chemistry, and chemical physics was carried out only at Paris, Nancy, and Strasbourg. Advanced physical chemistry was taught only at four universities in France immediately after the Second World War.[132] As we saw in chapter 6, the interest of Prévost and Kirrmann in ionic and electronic theories of reaction mechanisms developed later than in England and took a different turn than the Robinson-Ingold theory.[133]

Few French chemists were interested in both mathematics and theory, and with the creation of the Commissariat de l'Energie Atomique in 1946, they mostly were put to work on applied, rather than fundamental, problems in physical chemistry.[134] Daudel began a theoretical chemistry group in 1942, with the patronage of de Broglie, Frédéric Joliot, and A. Lepape, but he was hard-pressed to interest theoreticians in quantum chemistry rather than in quantum field theory or neutron and high-energy physics.[135] In the CNRS, work that would have been called quantum chemistry elsewhere was carried out in the section for applied mathematics until the 1970s, as noted earlier.[136]

A turning point for French quantum chemistry came in 1948 when Bauer organized a conference on the quantum theory of the chemical bond funded by the CNRS and the Rockefeller Foundation. This was two years after Bauer invited Ingold to lecture in Paris. For the Paris quantum chemistry conference, Coulson, Longuet-Higgins, Polanyi, R. W. Hill, and L. E. Sutton from England were invited to give papers, and Pauling and Mulliken from the United States were invited.

For many people, high points of the conference were Mulliken's lead-off presentation, a paper given later by Pauling, and exchanges between the two men. As Mulliken recalled, "I held forth and Pauling held forth."[137] The high point for Mulliken personally was a champagne party at Alberte and Bernard Pullman's apartment which others also remembered with enthusiasm.[138] Bauer

131. Micheline Charpentier-Morize, "La contribution de 'Laboratoires Propres' du CNRS," ms., 9–10.

132. These universities were Nancy, Strasbourg, Lyon, and Paris. See Guéron and Magat, "A History," 11; and Charpentier-Morize, "La contribution," 9.

133. See Bernard Pullman and Alberte Pullman, *Les théories électroniques de la chimie organique* (Paris: Masson, 1952): ix, on the long delayed research in electron theories of organic chemistry despite the earlier work of Prévost and Kirrmann.

134. Guéron and Magat, "A History," 12.

135. Ibid., and Raymond Daudel, "L'état de la chimie théorique," Guide Pour l'Exposition presentée au Palais de la Découverte (Paris: Sennac, 1949): 4.

136. Charpentier-Morize, "La contribution," 7.

137. Mulliken, in interview with Thomas Kuhn, 19; and "Colloque International de la Liaison Chimique, 12–16 avril 1948," *JCP* 46 (1949): 185–312, 497–542, 675–713. The English original may be found in D. A. Ramsay and J. Hinze, eds., *Selected Papers of Robert S. Mulliken* (Chicago: University of Chicago Press, 1975).

138. Mulliken, *Life*, 127.

soon gave the first course of lectures at the Sorbonne on the quantum theory of the chemical bond.[139]

Two years after the Paris Colloque, a group of chemists and physicists attending the meeting of the American Chemical Society in Detroit began discussing the need for a reliable table of the difficult integrals that inevitably turned up in valence-bond calculations. Mulliken was able to get funding to organize a Shelter Island conference at a Ram's Head resort. The first of the Shelter Island conferences, sponsored by the National Academy of Sciences, had taken place in 1947 around the theme of the foundations of positron theory. Although the initial plan had envisioned a mix of physicists and chemists, in the end only one chemist, Pauling, was invited, on the grounds that "it might not be a bad idea to get a chemical point of view."[140]

Mulliken's conference in 1952 was a "watershed," marking the coming of age of the discipline of quantum chemistry, which Mulliken equally freely called chemical physics.[141] Coulson wrote a friend that "all the experts will be there, though the total number is limited to twenty-five."[142] In fact, it was mostly an American group, joined by Coulson, Lennard-Jones, W. Moffit, and M. P. Barnett from England, Masao Kotani from Tokyo, and P. O. Löwdin from Uppsala.

At the conference, there was not much discussion of aromatic molecules, reflecting "a certain satisfaction with present π electron theory."[143] Topics for considerable discussion included bond order and hybridization, the forces between molecules and nonbonded atoms, the further mathematical development of key equations like the Hartree-Fock equations, and methods for arriving at convergence to a best value of the valence-bond and molecular-orbital approaches by a series of successive approximations. The conference concluded with agreement on procedures for collaboration in establishing a table of integrals. This, they agreed, is "the task now confronting the quantum mechanic."[144]

"Quantum mechanic" ? It was a fitting phrase. The "quantum mechanic," particularly in the Anglo-American tradition, was an eminently practical person. The new quantum mechanics, as it was applied to chemical problems and

139. The 155-page typescript was deposited at the Bibliothèque Nationale. Edmond Bauer, "Théorie quantique de la liaison chimique. II. La liaison chimique. Les Cours de Sorbonne. Centre de Documentation Universitaire, Paris V., 1953.

140. See Mulliken, *Life*, 136; and S. S. Schweber, "Shelter Island, Pocono, and Oldstone: The Emergence of American Quantum Electrodynamics after World War II," *Osiris*, 2d ser., 2 (1986): 265–302, on 277. John Van Vleck was present (279).

141. Mulliken, *Life*, 136.

142. Letter from Coulson to "Norman," August 1, 1951, Coulson Papers, New Bodleian Library, Oxford University.

143. Robert G. Parr and Bryce L. Crawford, Jr., "National Academy of Sciences Conference on Quantum-Mechanical Methods in Valence Theory," *Proc.NAS* 38 (June 1952): 547–553, on 548.

144. My emphasis. Ibid., 548–552.

to problems of the solid state, for example, the structure of metals, initially stayed very close to already known experimental results. Quantum mechanics ''interpreted,'' rather than ''discovered,'' chemical and physical facts.[145] Furthermore, accuracy, as Coulson argued, was not the strength of the new quantum mechanics. Its strength lay in the understanding it gave of underlying *processes.* The contours of equal probability density for electron motions in molecular hydrogen gave chemists a ''new *feel* for the covalent bond,''[146] providing ''flesh and blood to cover the bare bones that Couper and Frankland and Kekulé and Crum Brown had represented by a line.''[147]

Practitioners of quantum chemistry employed both the visual imagery of nineteenth-century theoretical chemists like Kekulé and Crum Brown and the abstract symbolism of twentieth-century mathematical physicists like Dirac and Schrödinger. Pauling's *Nature of the Chemical Bond* abounded in pictures of hexagons, tetrahedrons, spheres, and dumbbells. Mulliken's 1948 memoir on the theory of molecular orbitals included a list of 120 entries for symbols and words having exact definitions and usages in the new mathematical language of quantum chemistry.

Those who applied quantum mechanics to atoms and molecules had a wealth of chemists' data at hand: well defined bond properties including dipole moments, index of refractions, and ultraviolet absorption qualities and polarizability as well as well-defined valence properties of atoms in molecules. If one attempted to set up a wave equation for the water molecule, for example, there were 39 independent variables, reducible to 20 by symmetry considerations. But the experimental facts of chemistry implied or required certain properties that made it possible to solve equations by semiempirical methods. ''Chemistry could be said to be solving the mathematicians' problems and not the other way around,'' according to Coulson.[148]

Van Vleck, working at the borderline of quantum chemistry and chemical physics, saw things similarly.

> One must adopt the mental attitude and procedure of an optimist. . . . The optimist . . . is satisfied with approximate solutions of the wave equation. If they favor, say, tetrahedral and plane hexagonal models of methane and benzene, respectively, or a certain sequence among activation energies, or a paramagnetic molecule, he is content that these properties will be possessed by more accurate solutions. He appeals freely to experiment to determine constants, the direct calculation of which would be too difficult.[149]

145. See Mulliken, *Life*, 96.
146. Coulson, ''Inaugural Lecture, Chair of Theoretical Physics,'' King's College, London, 1948, 10-page typescript, 5. Coulson Papers, Bod.Oxford.
147. Coulson, ''Theoretical Chemistry: Past and Future,'' 16–17.
148. Charles Coulson, ''What Is a Chemical Bond?'' lecture at Basler Chemische Gesellschaft, April 26, 1951, 25-page typescript, 3–6, Coulson Papers, Bod.Oxford.
149. Van Vleck and Sherman, quoted in Schweber, ''Young John Clarke Slater,'' 404–405.

Or, as Nevil Mott put it, "If by making approximations and neglecting even large terms . . . one could account for something that had been observed, the thing to do was to go ahead and not to worry."[150]

In future, Mulliken and others would attempt to substitute ab initio methods for the earlier methods of successive approximations. With this, quantum chemistry was to become more abstract, more theoretical, than in its initial development. The new approaches seemed to express increasing confidence in a deep level of understanding, the hubris so characteristic of the quest for mathematical certainty. There were dangers in this which Coulson warned might deceive the theoretician. "One is almost tempted to say . . . at last I can almost see a bond. But that will never be, for a bond does not really exist at all: it is a most convenient fiction which, as we have seen, is convenient both to experimental and theoretical chemists."[151]

150. Nevil Mott, "Memories of Early Days in Solid State Physics," *Proc.RSL* A371 (1980): 56–66, on 57, quoted in S. T. Keith and Paul K. Hoch, "Formation of a Research School," on 37.

151. Coulson, "What Is a Chemical Bond?" 25, Coulson Papers, Bod.Oxford.

10

Conclusion

Theoretical Chemistry, Discipline-Building, and the Commensurability of Physics and Chemistry

Physics Is Chemistry/Chemistry Is Physics?

In 1930, the student Charles Coulson wrote out as the first lines of his chemistry exercise book at Clifton College in Bristol, "Physics is Chemistry. Chemistry is Physics. Laws of Conservation of energy and matter are FUNDAMENTAL LAWS. Lavoisier was the first to realize this clearly."[1] Almost sixty years later, the editors of the *Annual Review of Physical Chemistry* announced that they had entertained the suggestion that "Chemical Physics" be added to the title of the journal but rejected it, even while recognizing, they said, that the difference between the practice of physical chemistry and chemical physics is "small indeed."[2]

It is striking that the terms "physical chemistry" and "chemical physics" were ubiquitous in the early and mid-nineteenth century but that claims were not made for the constitution of "fields" or "disciplines" of physical chemistry and chemical physics in these years. By and large, the subject matter in these traditions was taught by chemists or by men who officially taught both chemistry and physics.

When Ostwald, Arrhenius, and van't Hoff announced their founding of a discipline of physical chemistry that deserved official recognition in the university curriculum and in distinctive scientific societies and journals, they defined physical chemistry as a borderline discipline but nonetheless as chemistry.

1. For quotation, Coulson Papers, MS 12, Bod.Oxford.
2. *Ann.Rev.P.Chem.* 39 (1988): Preface.

Scientists who were trained primarily in physics soon disputed this claim. For Nernst and Perrin, physical chemistry was chemical physics.[3] In his *Introduction to Chemical Physics* (1939), Slater made clear his view that it was a historical accident that physics and chemistry are separate sciences, that the field within which he situated his work was a unified chemistry and physics, and that it is called chemical physics ''[for] want of a better name, since physical chemistry is already preempted.''[4]

''Philosophical'' or theoretical chemistry was wide-ranging during most of the nineteenth century. In contrast, late-nineteenth-century physical chemists and twentieth-century physicists tended to narrow the definition of theoretical chemistry, eliminating organic structure theory and making theoretical chemistry almost exclusively physical and mathematical. An early indicator of this trend is Noyes's deletion of structure theory from the course in theoretical chemistry at MIT. A later indicator is the special issue of *Chemical Reviews* in 1991 which carries the title, ''Theoretical Chemistry, '' and begins with an introductory editorial entitled simply ''Quantum Theory of Matter.''[5]

However, by the 1920s, as we have seen, a self-conscious use of electron theory in a dynamical interpretation of the old static chemical molecule recovered dynamical theoretical foundations for organic chemists in what became the disciplinary specialization of physical organic chemistry. The theory of electron valency and organic reaction mechanisms, in particular, the theory of mesomerism, developed as a new theoretical chemistry, a little prior to wave mechanics, along a largely independent track.

In the course of the nineteenth century, then, chemistry had become a separate, distinct discipline from physics, but at the century's end, there was a reconvergence of the physical and chemical traditions, as practiced by individuals and research schools. In this chapter, we first review and illustrate elements of discipline-building in theoretical chemistry, referring to categories for disciplinary identity outlined at the beginning of this book. In the concluding section, we turn to the question of the conceptual commensurability of physics and chemistry, as illuminated by the aims of twentieth-century scientists practicing a ''philosophical'' or ''theoretical'' chemistry that some claimed to be reducible to physics.

Discipline-Building and Theoretical Chemistry

As we have seen, an important indicator of evolving disciplinary identity lies in changing citations of father figures by practicing scientists, both in the larger framework of physics and chemistry and in the finer framework within chemistry. Consider the case of Isaac Newton.

3. On Perrin's view, see chap. 5.
4. John Clarke Slater, *Introduction to Chemical Physics* (New York: McGraw Hill, 1939): v.
5. Ernest R. Davidson, ed., *Theoretical Chemistry. Chemical Reviews* 91, no.5 (July/August 1991). Note that in Slater's *Introduction to Chemical Physics*, only 13 of 522 pages focus on organic molecules.

As we saw in chapter 2, allusions to Newton by chemists in the eighteenth century and early nineteenth century can be understood as a means of legitimation for apron-coated chemists among the black-gowned philosophers of the university. It was no small achievement for chemists to establish a university identity in the philosophical faculties outside the professional schools of pharmacy and medicine. The chemist's laboratory was, after all, a far more appalling intrusion into academic halls than Robert Boyle's air pump or the Abbé Nollet's batteries of Leyden jars.

Following Lavoisier's achievements, any scientist making a claim to be a chemist took Lavoisier as a father figure, as demonstrated in this chapter's opening lines taken from Coulson's student notebook. A chemist's other ancestors might be claimed on the basis of more refined intellectual or national identity or both. For example, the physical chemist Pierre Duhem traced his lineage from both the "chemist" Lavoisier and the "mathematical physicist" Denis Poisson.[6] In general, French physical chemists tended to cite Lavoisier and Berthollet as progenitors; English physical chemists were more likely to cite Lavoisier and Boyle or Lavoisier and Dalton.

Historical introductions to chemistry courses and citations in journal articles provided ample opportunity for scientists to trace family lines to suit the discipline-building task at hand and to set up a record for later historians. Ostwald made sure to settle his name into a progeny of physical chemists in his history of electrochemistry. Later, Ingold minimized the historical role of contemporary rivals by spare citations to work well known at the time. Ingold sealed the extinction of French contributors to theories of reaction mechanisms by not mentioning them at all.

The strategy of discipline building includes narrative genres of progressive evolution and conflict resolution. The conflict genre of historical narrative is the more memorable and inspiring of the two genres, not only for group solidarity but also for conceptual clarity. Combat often is portrayed as conflict between ideas, rather than individuals, as in retrospective accounts of the development of quantum chemistry in the 1930s as a sparring match between atomic orbitals and molecular orbitals.

We have seen, for example, that members of Lespieau's school of theoretical chemistry narrated a history of battle within the field of French chemistry against the institutionally powerful intellectual disciples of Deville, Berthelot and Jungfleisch, on the one hand, and against the conceptually powerful, if institutionally weak, school of Duhem, on the other hand. They situated themselves adjacent and sometimes interior to the scientific circle of Perrin, with its focus on kinetics and the radiation hypothesis, and in opposition to the

6. On Duhem as the representative of the school of Poisson, see Bordeaux, Faculté des Sciences, Procès-verbaux du conseil, 27 November 1894.

Anglo-Saxon school, which in contrast to the phenomenalist French tradition, indulged in the use of pictorial models and premature hypotheses.

The Paris School was an extended one, with institutional bases in the Ecole Normale Supérieure, the Sorbonne, Strasbourg, Nancy, and Bordeaux. The London-Manchester school, even more than the Paris school, exemplifies the model of the extended research school. Its roots were in Leipzig, Marburg, Heidelberg, and Berlin as well as in London and Manchester, and it acquired outposts throughout Great Britain and America, including Leeds, Aberystwyth, Berkeley, Ithaca, and New York City. This is a school that harbored rival groups, each aiming at a general theory of organic reaction mechanisms. As we have seen, Ingold's group, extended to a "school" over several decades, came to dominate its rivals in the definition of a new discipline of physical organic chemistry.

We have seen elements in Ingold's success. After adhering for some time to the classical Continental traditions exemplified in the valence theories of Thiele, Werner, and Flürscheim, Ingold adopted and transformed the pictorial and mechanistic hypotheses of Lapworth and Robinson into a new unified system. The labeling of Ingold's ideas in America as the "English heresy" simply accentuated the significance of his work and associated it with a national tradition and a national style capable of revolutionary results. Ingold's name became associated with an Anglo-Saxon school, just as Dumas's name was identified with the nineteenth-century "French school" of chemistry. Prévost desired, but did not acquire, identification with a twentieth-century French school of chemistry.

The analogy of the extended family works well as a way of thinking about the role of research schools in upholding, combating, and creating traditions and in perpetuating, leaving, or developing scientific disciplines. Sometimes the family theme works quite literally, for example, in the case of Mulliken *fils* choosing to follow a different path than Mulliken *père* and then combining the two traditions of family interest, in organic chemistry and physical chemistry, by creating the field of quantum chemistry. The family analogy also fits neatly into the naturalist or evolutionary explanation of discipline building, which emphasizes competition, adaptation, and radiation.

Consider the following statement by Ingold. Reflecting on his education and career in 1962, he said,

> The scientific master ... must enter, or gather around himself, and, with his students, work within, an organization of other scientists and technicians, having knowledge and skills complementary to his own.... Scientific success goes to the strong groups, those that make themselves strong by thus harmoniously and intimately combining a maximum of complementary intellectual endowments and interests.[7]

7. [Sir] Christopher Ingold, "Education of a Scientist," *Nature* 196 (1962): 1030–1034, on 1032.

As the "master" of a research school, Ingold acquired an anecdotal reputation among his students for virtues and flaws, like most father figures and mythological heroes. Among these were his ruthlessness in print and his courtliness in person. With his wife, Hilda Ingold, he was appreciated for the family setting he provided, not just in the laboratory, where he encouraged or reprimanded students and colleagues, but for dinners and "brainy games" on convivial occasions, where he showed the same competitiveness demonstrated on the cricket field and in the research laboratory.

The Robinsons, the Lapworths, and the Ingolds all literally provided the rituals of familial settings for their students and colleagues, by working in the laboratories as husband and wife or, in the case of Mrs. Lapworth, by serving as departmental secretary for many years. (Perhaps it is no small part of the Lavoisier legend that Mme. Lavoisier was his hostess, translator, and illustrator.) Gertrude and Robert Robinson's home was observed by passing students to have a reassuring light shining in the study late at night. Lespieau's Paris laboratory, similarly, was specifically characterized by his former students as one of "familial sympathy."[8]

The father figures enlisted their student progeny in the struggles and controversies that characterized their everyday work and ambitions. Robinson told his students he would like to kick his antagonists with boot leather. Prévost incited his students to show their Anglo-American rivals a thing or two. Students constituted a younger generation to inspire and to educate. They further constituted a younger set of colleagues with whom to collaborate and share values.

In these matters, the choice of instruments, methods, and battle sites was crucial to the survival and esteem of the school and to its perpetuation or creation of a disciplinary specialty or field. Once basic techniques and knowledge were acquired within the secure setting of the school, institutional sites farther afield provided venues for practice and for conflict. These included the pages of journals and the halls of scientific societies, where results were reported and debated, but also conferences, like the one at Cambridge where Flürscheim threw down the gauntlet to Robinson, with Ingold and Lapworth serving as seconds for the principal antagonists.

The doors of a society or the pages of a journal might be closed to a different point of view or to the airing of controversy, leading to the need for a new venue for publishing claims and results. German chemists who established the *Berichte* of the Berlin Chemical Society adopted as standard usage the "2-volumes" notation, thereby closing down controversy over rival chemical notations in Germany and excluding the "1-volume" notation still used in France. It was necessary to establish the *Zeitschrift für physikalische Chemie* in 1887 because *Liebigs Annalen der Chemie* was de facto a specialized journal

8. Chap. 6.

in organic chemistry that did not welcome papers using predominantly physical instrumentation or theoretical reasoning. The *Journal of Chemical Physics* was established, similarly, in 1933 because of the inhospitability of the *Journal of Physical Chemistry* and the *Journal of the American Chemical Society* to the theoretical and mathematical approaches of chemists like Lewis and Pauling.

For organic chemists concerned with constructing a deductive or theoretical chemistry in the late nineteenth and early twentieth century, the tactics of physical instrumentation were of primary concern. Familiarity with new physical instruments was acquired through formal instruction in physics, by tacit learning from chemical colleagues, and, last of all, through systematic laboratory manuals and instruction. A notable distinction between Robinson's laboratory and Ingold's, which became clearer after Robinson ceased collaborating with Lapworth at Manchester, was the exclusion from Robinson's Oxford laboratory of physical instrumentation. Robinson's laboratory was identifiable as a homeland for organic synthesis and natural products chemistry, whereas the "École Sup" organic chemistry laboratory under Lespieau, Dupont, and then Kirrmann felt and looked different, with infrared spectrometers and polarimeters among its prominent apparatuses.

When Eyring came to Princeton in 1931, he experienced firsthand the power of classical methods in organic chemistry without the benefit of much in the way of physical instrumentation.

> We went through the usual procedure of securing an analysis of the pure compound, fixing the molecular weight by freezing point lowering [which had been a new and controversial method at the end of the nineteenth century], determining the nature of the bonds through measuring the reactivity of the molecule with various reagents, and finally examining the products formed from these reactions. . . . By first establishing the structure of the simple compounds into which a complex molecule fragments, chemists . . . determined ever more complicated structures, at the same time getting a profound understanding of chemical reactivity. This knowledge in turn was used in the synthesis of ever more intricate structure. . . . Many of the older chemists were much less specialized than chemists are today. . . . Each new instrument usually makes for more efficiency and specialization and less versatility.[9]

By the late 1950s, a revolution in instruments transformed the organic chemistry laboratory. Purity of a compound could be demonstrated by chromatography and mass spectroscopy. The mass spectrometer would give molecular weight or information from which the weight might be derived, and probably the empirical formula. Infrared spectra would show the presence of functional groups like -OH, -CH = O, or C = O. An x-ray crystallographic

9. Henry Eyring, "Physical Chemistry: The Past 100 Years," *CENews*, 54 no.15 (April 6, 1976), *Centennial. American Chemical Society, 1876 to 1976*, 88–104, on 94.

determination revealed the complete three-dimensional structure of the molecule.[10] As instruments became a common conceptual tool for both physicists and chemists, they literally became a new meeting ground as well, as the instrument helped to determine the layout of laboratories and the schedule of experiments.

The role of these instruments in turning up both new information and new questions became central to the evalution of research programs in chemistry. Still among organic chemists and among some theoretical chemists, the most essential instruments were the very chemical substances themselves: as in the nineteenth century, they remained both the objects of study and the means of study. As remarked by Eyring, an elaborate technology of using compounds to elucidate the structure of other compounds was built up during the course of the middle years of the nineteenth century. Organic chemists, indeed, most chemists, prided themselves on a ''chemical feeling'' (*chemische Gefühl*) that came from the kind of ''stink'' chemistry that Eyring was to relinquish in favor of physical instruments and mathematical explanation.[11]

Indeed, as Schweber has observed, organic chemists may very well have feared that using new techniques like x-ray analysis, which give an answer without doing the chemistry, might destroy traditional resources for new chemical insights and novelties. Alexander Todd recalled that Robert Robinson frequently said that the new physical methods would stop the development of organic chemistry. And in a lecture at the University of Oklahama, Carl Djerassi commented that natural product chemistry was more exciting in 1950 because it *was chemistry* and that now it is all black boxes, nuclear magnetic resonance spectroscopy, and so on.[12]

For theoretical organic chemistry at the turn of the twentieth century, the absolutely essential substances as instruments for resolving problems of chemical affinity and reaction mechanisms were benzene and acetoacetic acid, that is, the conjugated dienes that were exemplars, respectively, of ''dynamic tautomerism'' and ''tautomerism.'' To become established in the schools of Lespieau, Robinson, and Ingold, it was absolutely necessary to learn the properties and rules characteristic of the chemical behavior of these conjugated hydrocarbons. Knowledge built up through repetitive and flexible practice gave the chemical feeling which guided further progress, for example, in Robinson's

10. See Stanley Tarbell, ''Organic Chemistry: The Past 100 Years,'' *CENews* 54 (1976): 110–123, on 121.

11. See Kolbe and Armstrong on ''chemical feeling'' (chap. 7); Pauling, quoted in chap. 8; and Coulson, on the ''feel for the chemical bond'' (chap. 9). Claims were made as well, of course, for a ''physical feeling'' or a ''mathematical feeling.''

12. Personal letter from S. S. Schweber, 20 April 1992. Also, [Lord] Alexander Todd, ''Summing Up,'' in *Further Perspectives in Organic Chemistry. Ciba Foundation Symposium 53 (new series). To Commemorate Sir Robert Robinson and His Research* (Amsterdam: Elsevier, 1978); 203–204; and Carl Djerassi, lecture at University of Oklahoma, Norman, 19 November 1992.

choice of substances to test predictions of *ortho*, *meta*, and *para* orientations in response to Ingold's early challenges. The presence of some compounds and not others as laboratory instruments defined the identity of the research school.

A crucial, indeed, essential, part of the practice of any school was its use of language. An indicator of transitions within disciplines, or the coming into existence of a new discipline, is the codification of local language of a research school into classic literature for the field. There is perhaps no better clue to the waxing and waning of research schools and of disciplinary momentum than changes in language, including the use of analogy, metaphor, symbols, and models.

We have seen how eighteenth-and early-nineteenth-century chemists borrowed from ordinary language to construct metaphors like "affinity" and "saturation" which became conventions of chemical theory. We also have seen how chemists borrowed from the language of natural history to systematize knowledge of the properties and behavior of substances in terms of families, orders, classes, and species; or of types and structures modeled on behavioral functions and structural homologies. Nineteenth-century chemical language, and thus chemical explanation, of compound substances depended on a binomial nomenclature of genetic origins, not a mechanist vocabulary of matter and motion. We should remember, too, that Lavoisier's "modern" binomial nomenclature was controversial not only because it replaced many of the old names and theories but also because it was French. Had Lavoisier proposed a chemical nomenclature in Latin, there likely would have been less resistance, but he was bent on establishing a new chemistry as a *French science*.

If we think about other and later changes in symbols, vocabulary, and imagery, the struggle for invention and control of language, and thereby of theory, is obvious, whether in controversies over atomic symbols (spheres or letters?), molecular formulas (O = 8 or 16?), functional groups ("radicals" or "residues"?), or names of compounds ("rational" or "empirical"?).

Calling hydrogen the "comet" of the chemical universe[13] was laden with levels of meaning for a new mechanistic chemistry. The metaphor expressed a hypothesis that hydrogen is a finite particle, that it is in continuous and repetitive motion within a dynamic molecule, and that the gravitational analogy for chemical affinity is an apt one.

In contrast, Bohr's use of the word "shells" and Langmuir's use of the term "sheaths" in place of "orbits" discarded the old gravitational analogy for electron energy levels and electron motions. Mulliken's invention of the word "orbital" self-consciously fit his theory within the old physical tradition going back to Newton but simultaneously asserted the discovery of a new theory to inaugurate a "Mulliken era" in chemistry. We have specifically

13. Chap. 5.

noted stubborn adherence to the vocabulary of "synionie" that isolated Kirrmann and Prévost just as assuredly from the chemical mainstream as had Berthelot's adherence to equivalentist notation in the nineteenth century.

In contrast, Ingold made wise choices in developing the language of reaction mechanisms. For example, he adopted the visually mechanistic "curly arrows" of Robinson but rejected the cumbersome terms "anionoid" and "cationoid" in favor of the more anthropomorphic, yet theoretically inclusive, terms "nucleophilic" and "electrophilic." Although Robinson and Lapworth ridiculed Ingold's statement that the "leaving group" is "pulled off" by the solvent in the S_N1 mechanism and "pushed off" by the incoming "nucleophile" in the S_N2 mechanism, this picturesque language prevailed. So, too, did Ingold's commercial language of electron "supply" and electron "transactions," along with his aphorism that sharing "economizes" electrons.

Ingold's use of the terms "electron source" and "electron sink" linked his nucleophilic/electrophilic terminology to classical traditions in heat and electricity, and his use of the phrase "electron affinity" linked his approach to classical traditions in chemistry. He incorporated Lowry and Brönsted's proton theory of acidicity into the new system by describing basicity as "affinity" for a hydrogen nucleus or proton. Thus, Ingold associated his work with many lines and networks in the framework of chemical practice, translating old ideas and terms into a new chemistry, much as Lavoisier and his colleagues had done in their chemical revolution of the late eighteenth century.

In her study of the development of molecular physics in the United States, Alexi Assmus has noted how Americans, in particular, the group led by Kemble at Harvard, Dennison at Michigan, and Birge at Berkeley, asserted leadership in a new disciplinary field by standardizing the terminology for molecular spectroscopy in the mid-1920s.[14] This was Ingold's strategy in the field of organic reaction mechanisms within the discipline of physical organic chemistry. It also was Mulliken's strategy in quantum chemistry, when he presented his Paris colleagues with a list of 120 terms and symbols constituting an elementary grammar for quantum chemistry.

A common language unifies a community of practitioners, at the same time that it sets them apart from other communities. As the chemist Mayer wrote Lewis about Slater's lecture at the Baltimore meeting of the American Physical Society, the *talk* of tetrahedral axes and coupling sounded so familiar that "*I felt much at home.*"[15] In this regard, the invention of the phrase "resonance hybrid" should be remarked as a stroke of genius. No term could better have expressed the union of the natural philosophy tradition ("resonance") and the natural history tradition ("hybrid") in the theory of the electronic constitution of benzene. Both physicists and chemists felt at home.

14. Alexi Assmus, "Molecular Structure," 119–120; and conversation of August 2, 1991.
15. See chap. 9.

Slater and leaders in these new fields not only gave talks and published papers, but also wrote textbooks and monographs, some of which became classic scientific texts. The classics of physical chemistry were to include Perrin's *Les Principes* (1903) and *Les Atomes* (1913) as well as texts by Gibbs, van't Hoff, Nernst, and Ostwald. Ostwald established the series still known as *Ostwalds Klassiker* to define his view of the nature and aims of science. Lewis wrote two classics in theoretical and physical chemistry, *Valence and Structure* (1923) and, with Randall, *Thermodynamics and the Free Energy of Substances* (1923).[16] In physical organic chemistry and quantum chemistry, without question the classics were Pauling's *Nature of the Chemical Bond* (1939) and, with Wilson, *Introduction to Quantum Mechanics* (1935), Hammett's *Physical Organic Chemistry* (1940), and Ingold's *Structure and Mechanism in Organic Chemistry* (1953).

In setting up a curriculum, writing textbooks, establishing journals, and founding societies, there must be some consensus among individual practitioners that they have problems or techniques in common. This consensus may result in formation of a field of specialization, without resulting in the sense of identity associated with the systematic foundation of a discipline. Some matters of judgment and value may not even be part of the published scientific literature but largely a matter of oral tradition, tacitly acquired and passed on, in the manner of craftsmanship or folk literature. Aesthetic judgments (What is a "beautiful molecule"?) are of this sort, as theoretical chemist Roald Hoffmann notes: such judgments are "subfield (organic chemistry, physical chemistry) dependent, much like the dialects, rituals or costumes of tribal groups."[17]

The more systematic form of innovation that is disciplinary formation results, in part, from the pressures of competition and overcrowding in institutional contexts of scarce resources and limited rewards. Diversification serves the interests both of individual ambition and of institutional health. In some cases, diversification may lead to increased antagonism between principal groups or schools, as when, for example, the first chair of physical chemistry in France was established at Nancy at the expense of organic chemistry. We remember, too, Barkan's argument that physical chemistry was used by inorganic chemists as a means of reintroducing their field into institutional settings long dominated by organic chemistry.

In addition, the formation of some disciplinary fields is favored at particular times because of larger economic and political factors. Most of the new disciplinary sciences that were given university status in the late nineteenth century and early twentieth century were oriented toward the so-called applied

16. Lewis and Randall's *Thermodynamics and the Free Energy of Chemical Substances* used a notation different from Continental notation, necessitating a standardization in notation by a committee set up jointly by the Faraday Society, the Chemical Society, and the Physical Society. See Slater, *Introduction*, v.

17. Roald Hoffmann, "Molecular Beauty," *American Scientist* 74 (1988): 389–391.

sciences in industry, agriculture, and medicine, as part of national concerns for upgrading economic competitiveness. As we have seen, this trend favored physical chemistry and physical organic chemistry but not ''theoretical chemistry'' as recognized new disciplines. It is not surprising, then, that the first chairs of *theoretical* chemistry in England and France were at Cambridge University and at the Ecole Normale Supérieure, where the philosophical tradition was old and venerable, rather than at the provincial universities, where the applied and engineering sciences were fast multiplying. At an institution like Caltech, which was by nature an engineering institution, the designation of Pauling as a ''theoretical chemist'' brought academic prestige to the school initially known as Throop's Institute.

Of course, the constitution of new scientific disciplines has served the interests of entrepreneurs. While scientists themselves can be characterized as entrepreneurs acting out of self-interest within the academic setting,[18] there also are more traditional commercial interests at stake. The production of useful chemicals is an obvious commercial payoff. But so, too, is the publication of chemical books and journals and the provision of laboratory instrumentation. The greater the diversification within scientific fields, the more interests of all kinds could be served. Money was to be made, as Griffin found in the early nineteenth century, in the marketing of laboratory equipment for teaching and research. Similarly, there were profits in providing every chemical laboratory in America with an infrared spectrometer, an instrument that solidified the identity of physical organic chemistry by the mid-1950s. In turn, the technical capabilities of the instrument, whether in x-ray diffraction, infrared spectroscopy, or nuclear magnetic resonance spectrosopy, reshaped the boundaries of the disciplinary field.[19]

Our history has shown us how some leaders of research groups aspire to do more than direct their workers in the straightforward applications of disciplinary practice that we have learned to call ''normal science.''[20] Rather, choosing to address the unsolved problems that are part of the disciplinary core, the group reaches out to incorporate ideas, techniques, and materials from specialties and disciplines other than their own. As Hull has reminded us from an evolutionary point of view, innovation is favored not by isolation but by a mixing of information. As he puts it, the exchange of genetic material is more important than its replication.[21]

18. See Bruno Latour and Steve Woolgar, *Laboratory Life: The Construction of Scientific Facts*, 2d ed. (Princeton: Princeton University Press, 1986): 187–233.

19. See John D. Roberts, *Nuclear Magnetic Resonance* (New York: McGraw Hill, 1959), and John D. Roberts, *The Right Place at the Right Time* (Washington, D.C.: American Chemical Society, 1990): esp. 149ff.

20. The practice of ''obliteration by incorporation'' might lead me not to mention Thomas Kuhn's book, *The Structure of Scientific Revolutions*. Here, as in other parts of this section, I am particularly mindful that Kuhn made many of these arguments thirty years ago.

21. David Hull, *Science as Process, 22.*

Applying this idea to our history, we remember that physical chemistry has employed disciplinary methods and aims taken from both physics and chemistry and that its practice has been one of striking epistemological pluralism. Laboratory investigations aimed at understanding chemical reaction mechanisms profited from the selection of mechanical and kinetic hypotheses from physics that transformed the static molecule of classical organic chemistry into the dynamic molecule of physical organic chemistry.

Ingold exemplifies a research leader who combined information and approaches from disparate sources, reformulated questions and answers, and extended his influence widely through personal contacts and the activities of his students and collaborators. Publication was important as a means of exchanging information and exerting influence, but so, too, was personal presence. In contrast to the internationalism of the London-Manchester school, the combined intellectual and physical isolation of members of the Paris school in the 1920s and 1930s demonstrates clearly how isolation can lead to a dead end.

The union of physical chemistry with organic chemistry resulted in a larger discipline, conceptually and methodologically, than constituted by either discipline alone. To be sure, it is difficult to avoid speaking of physical chemistry, physical organic chemistry, and quantum chemistry as subfields of chemistry or specializations within the broader field or network of theoretical chemistry. Yet we cannot help but conclude that it is somewhat misleading to characterize these fields as ''subdisciplines'' of chemistry. As is the case for theoretical chemistry as a whole, their nature is in fact interdisciplinary, with their practitioners undertaking training and apprenticeship in mathematics, physics, and chemistry and, after the Second World War, in computer science. Here, as discussed at the outset of this book, Pierre Bourdieu's notion of the intellectual ''field'' is helpful.

The inclusive nature of theoretical chemistry as an interdisciplinary science is demonstrated in the Nobel Prize-winning work of Roald Hoffmann in his collaboration with Robert B. Woodward. The theoretical work was a by-product of Woodward's long efforts toward the synthesis of vitamin B_{12}. The ''Woodward-Hoffmann rules'' represent an extension of the quantum mechanical description of chemical bonding to the *course* and *stereochemistry* of organic reactions, taking into account the symmetry of the molecular orbitals in the excited and nonbonding states. With the restriction that the reaction must be a concerted process, the rules allow a prediction as to whether a given reaction will proceed thermally, via nonbonding orbitals, or photochemically, via excited electronic states. A prediction about the stereochemistry of the product also is achieved.[22] Here we have a striking example of theoretical chemistry, in which we clearly see its interdisciplinary origins.

22. See discussion in Stanley Tarbell, ''Organic Chemistry,'' 122, referring to R. Hoffmann and R. B. Woodward, ''Stereochemistry of Electrocyclic Reactions,'' *JACS* 87 (1965): 395–397;

To distinguish specialties and disciplines we may call on the concept of identity, which suggests that disciplinary identity is developed and defined by a network of elements having to do with the construction (or rejection) of what we have called genealogy and historical mythology, classic literature, common practices, formal institutions and sense of homeland, external recognition, and shared values and unsolved problems. The research group and the extended research school are the working units within modern disciplinary domains. Different schools operate within slightly different traditions, determined by local customs. With some modification of the meaning of ''research group'' and ''research school,'' this kind of analysis may be applied to early modern science as well as to modern science.

We turn now to a review and analysis of the shared epistemological values and problems that have spurred the efforts of the philosophical and theoretical chemists. We ask in conclusion whether the achievements of theoretical chemistry by the mid-twentieth century led its practitioners to identify their discipline with the aims of mechanical natural philosophy and modern physics.

Theoretical Chemistry and the Distinctiveness of Chemistry

In the course of his chemical lectures at the Ecole Polytechnique in 1794, Fourcroy distinguished the theoretical, practical, historical, and applied aspects of chemistry.

> I mean by the THEORETICAL part of chemistry the methodical exposition of all the FACTS which belong to this science and which, in the regular arrangement that must coordinate them as well as in the simple enunciation of the results that they furnish, presents them equally stripped of the experimental details by which they are proved.[23]

This statement of the meaning of theoretical chemistry lies within the methodological tradition of chemistry that we have termed ''positivist'' or ''exact,'' in contrast to the ''philosophical'' or ''realist'' tradition of the chemical philosophy that aimed, in the words of Humphry Davy, to ''ascertain the causes of all [chemical changes], and to discover the laws by which they are governed'' and that , according to Dumas, studies ''the material particles chemists call atoms and the forces to which these particles are submitted.''[24]

---

and other papers, in JACS 87(1965): 2046–2048, 2511–2513, 4388–4389, 4389–4390. Also see R. B. Woodward and Roald Hoffmann, *The Conservation of Orbital Symmetry* (Weinheim: Verlag Chemie, 1970).

23. Antoine de Fourcroy, *Système des connaissances chimiques, et de leurs applications aux phénomènes de la nature et de l'art*, 11 vols. (Paris: Baudoin, 1800), I: xxx–xxxi; translated and quoted by Janis Langins, ''The Decline of Chemistry at the Ecole Polytechnique, 1794–1805,'' *Ambix* 28 (1981): 1–19.

24. See chap. 3, referring to H. Davy, *Elements of Chemical Philosophy*, 1; and J. B. Dumas, *Leçons de philosophie chimique*, 1–2.

As we have seen, most chemists, like Fourcroy, did not much concern themselves with mechanical causal forces during the course of the nineteenth century, as they focused increasingly on aspects of organic chemistry in which the ideas and the language of biological organisms and taxonomical classification proved useful: types, structures, and functions instead of atoms, motions, and forces; discontinuity and irreversibility rather than continuity and reversibility; rich description and multiple explanation rather than mathematical certainty and consistent system.[25] As we have seen, the course of nineteenth-century chemistry was best served by a conventionalist epistemology that remained neutral in matters of truth and pragmatic in matters of practice.

Lewis noted in his popular book, *The Anatomy of Science* (1926), that the concept of "force" fell out of use among chemists because it purported to supply a single answer to the problem of causation.[26] The Oxford chemist Harcourt similarly and earlier observed that chemical laws differed from physical laws because they could be derived not from any single cause of natural force but from the whole *course* of chemical change.[27]

The interests and approaches of chemists and physicists reconverged toward the end of the nineteenth century, in a period when physicists began developing an energy-physics as an alternative to force-physics and chemists became interested in the material electron as a binding agent in a chemistry of space that included mobile atoms and the chemical valence bond.

For chemists, the problem of affinity, or what Meyer called variable valence, was the central problem of chemistry, one in which, Ostwald claimed, chemists made no progress while seeking to measure chemical "forces." Meyer, who often is identified with the tradition of physical chemistry and theoretical chemistry, as noted in chapter 3, was confident that the answer to affinity lay in theories of motion, not in species or types, just as Nernst later was to identify the end of affinity theory with its reduction to physical causes.

Stewart clearly stated the dilemma for organic chemists at the turn of the century. Through laboratory experience, they had learned to interpret the chemical bond in different ways. Although they might draw double bonds in formulas for diphenyl-ethylene, ethylene, and fulvene, chemists did not really take the bonds to resemble each other chemically. Chemists "knew" that there is an increase in unsaturation, or reactivity, of the double bond toward bromine or oxygen, from one of these compounds to the next. They "knew" that a bond must be looked at not as a fixed unit but as a sum of an infinite number of small forces or partial valences. This is what Polanyi later called tacit knowl-

25. See Otto Theodor Benfey, "The Concepts of Chemistry—Mechanical, Organicist, Magical or What?" *JChem.Ed.* 59 (1982): 395–398, who comments on rational coherence that "Aristotelian physics was highly rational, yet wrong" (398).

26. Gilbert N. Lewis, *The Anatomy of Science* (Washington D.C.: American Chemical Society, 1926; reprint, Books for Libraries Press, 1971): 99–102.

27. Vernon Harcourt, cited by Christine King, "Experiments with Time," 70.

edge, not ''theoretical'' knowledge, and what we have called ''chemical feeling.'' Now, Stewart counseled in 1908, ''we must go to the physicists and from them we can borrow as much of their *theory* as seems likely to help us with our own branch of science.''[28]

It is striking how open some chemists were to collaboration with physicists in the early decades of the twentieth century. We have seen detailed instances of this collaboration in our account in the foregoing chapters of the development of physical organic chemistry and quantum chemistry. Among the figures in the research schools we have analyzed, Lowry was one of the most outspoken about the need for joint endeavors between the physicists and chemists, saying in the early 1920s that ''the time is indeed ripe for chemists to state definitely how many types of valency they require the physicists to provide.''[29]

Lowry counseled colleagues at the 1923 Cambridge symposium, which he had organized, that ''the electron has come and has come to stay'' and that chemists or physicists ''or more probably a team containing representatives of both groups'' must investigate the electronic structure of molecules.[30] Similarly, Job commented a few years later at a Solvay chemistry conference in Brussels that organic chemists could not yet explain *why* certain chemical bonds are easy to break. To get an explanation, ''the task of the chemist will be to reunite for the physicist all the materials proper to him to suggest the solution of the problem.''[31]

What kind of solution was expected from physicists? As we have seen, many chemists, from Lavoisier on, expected that fundamental chemical problems would be accessible to mathematical solution, meaning not just precise quantification or geometrical explanation but algebraic formulation on mechanical principles.[32] For all the resentment of statements by Kelvin or Boltzmann that chemistry could be reduced to vortex atoms or the kinetics of atoms,[33] many nineteenth-century chemists shared Kekulé's vague presentiment

28. My emphasis. Alfred W. Stewart, *Recent Advances in Organic Chemistry*, 1st ed., 263, 266–267.

29. Thomas Lowry, ''Intramolecular Ionisation in Organic Compounds,'' *Trans.Far.Soc*, 19 (1923): 487–496, on 490.

30. Thomas Lowry, ''Applications in Organic Chemistry of the Electronic Theory of Valence,'' *Trans.Far.Soc.* 19 (1923): 485–487.

31. André Job, in discussion following Nevil Sidgwick's paper on variable valence, Institut International de Chimie Solvay, *Rapports et discussions sur des questions d'actualité. Troisième conseil de chimie. 12 au 18 avril 1928* (Paris, 1928); hereafter *Solvay III*, 412.

32. See discussion of Lavoisier in chap. 2 and 4. One day, Lavoisier wrote, it would be possible to ''know the energy of all these forces, to succeed in giving them a numerical value, to calculate them—this is the aim which chemistry must have.'' Lavoisier, *Oeuvres de Lavoisier*, II, 525; quoted in Maurice Crosland, ''The Development of Chemistry in the Eighteenth Century,'' 407.

33. William Thomson [Lord Kelvin], ''On Vortex Atoms,'' 16–17; and Ludwig Boltzmann, ''On the Necessity of Atomic Theories in Physics,'' 73–74. There seemed to be less resentment among chemists of J. J. Thomson's application of the ether vortex atom to chemistry in his 1882 book.

that some day there would be a "mathematico-mechanical" explanation for what nineteenth-century chemists and physicists called "atoms."

This did not mean there was much sympathy with Brodie's algebraic alternative for molecular models or with Pearson's more sophisticated attempt to introduce the mathematics of ether squirts into chemistry. Nor were chemists ready to give up the periodic table and pictorial theories in the daily work of the laboratory. But, like Robinson, who said that he considered his electronic *theory* of reaction mechanisms his "most important contribution to knowledge," many chemists considered theory, not chemical fact or chemical production, to be the highest aim of science.[34] And many agreed with Coulson that you cannot have deep theory without mathematics.

For all his interest in combining physics and chemistry, Lowry was not much convinced in 1925, on the eve of the breakthroughs by Heitler and London, Hund, and Erich Hückel, that the physicists' most recent mechanics had benefited chemists. No doubt, Lowry told colleagues at the second Solvay chemistry conference, the physical chemist should learn to think in terms of quanta and energy levels, but the mineral chemist and the organic chemist had not yet gained much from these latest physical theories.[35] In 1931, as we saw in chapter 6, Kirrmann still was of the opinion that the time had not yet arrived for the "mathematical stage" of chemical explanation.

Just as physicists had used chemical facts as a testing ground for the classical electromagnetic theory of the electron around 1900, so they again used chemical information in the late 1920s and early 1930s to develop quantum mechanics and electrodynamics. For chemists, there was a great deal of pleasure in seeing confirmed in mathematical physics the basic chemical facts and theories of nineteenth-century and early-twentieth-century chemical practice, including the diatomic nature of the elementary hydrogen molecule, the tetrahedral directionality of carbon valence bonding, and the centrality to chemical stability of the electron duplet. Nor were they surprised to find the chemical combining relationships worked out in classical structure theory confirmed in three dimensions by x-ray crystallography.

Quantum wave mechanics gave chemistry a new "understanding," but it was an understanding absolutely dependent on purely chemical facts already known. What enabled the theoretician to get the right answer the first time, in a set of calculations, was the experimental facts of chemistry, which, Coulson wrote, "imply certain properties of the solution of the wave equation, so that chemistry could be said to be solving the mathematicians' problems and not the other way around."[36] So complex are the possible interactions among valence electrons that one must either use an exact mathematical model of a

34. For Robinson, see chap. 7.
35. Thomas Lowry,"Le mécanisme de la transformation chimique," *Solvay II*, 135.
36. Coulson, "Inaugural Lecture," Coulson Papers, Bod.Oxford.

simplified system or a simplified model of the complete system. The adoption of "semiempirical" methods constituted a "threefold compromise between rigor, experiment and intuition." Some of the integrals in the Schrödinger equation are replaced by empirical parameters from reference compounds or other information that chemists already have at hand.[37] Once we get one right answer by choosing a convenient parameter, then "we get tolerably good answers for the others" in the series, noted Coulson. "Thus, the model is essentially the right one."[38]

Coulson's generation of quantum theoretical chemists was struck by the fact that the mathematical physics of wave mechanics did not result in fundamental breakthroughs or discoveries in chemistry. As we have seen, Mulliken claimed that his initial work in quantum mechanics "interpreted," rather than "discovered," chemical facts. Alberte Pullman commented in 1970;

> While it is certainly indispensable that theoretical chemists constantly try to improve the values of the sizes they calculate and more and more approach exact energy values . . . quantum chemistry risks giving the impression that its essential goal is reproducing by uncertain methods known results, in contrast to all other sciences whose goal is to use well-defined methods for the research of unknown truths."[39]

In characterizing the aims, methods, and values of chemistry, some chemists recently have stressed the irreducibility of chemistry to physics. Hoffmann, for instance, in reflecting on the question, has written that "most of the useful concepts of chemistry (for the chemist: aromaticity, the concept of a functional group, steric effects, and strain) are imprecise. When reduced to physics they tend to disappear."[40] Further, it is the very ambiguity of some chemical concepts, like "oxidation state," that appeals to chemists. For chemists, the oxidation state of an atom varies according to its role in a molecule, appearing to vary from, perhaps, +3 to -2 for an atom. Physicists are uneasy with this "elastic" heuristic device.[41]

37. Suckling, Suckling, and Suckling, *Chemistry through Models,* 65, 135.

38. Coulson, "Inaugural Lecture," Bod.Oxford.

39. Alberte Pullman, Introduction, in Raymond Daudel and Alberte Pullman, eds., *Aspects de la chimie quantique comtemporaire,* Colloques International de CNRS, 195 (Paris: Editions due CNRS, 1971): 13.

40. Roald Hoffmann, "Under the Surface of the Chemical Article," *Angewandte Chemie,* International Edition in English, 27 (1988): 1593–1602, on 1597. Also see D. W. Theobald, "Some Considerations on the Philosophy of Chemistry," *Chemical Society Reviews* 5 (1976): 203–213. To his list, Hoffmann more recently added oxidation state and chemical stability; in personal correspondence, letter of September 24, 1990.

41. Roald Hoffmann, personal correspondence, letter of October 16, 1990. Also, R. Hoffmann, "Nearly Circular Reasoning," *American Scientist,* 76 (1988): 182–185. And see Kurt Mislow and Paul Bickart, "An Epistemological Note on Chirality," *Israel Journal of Chemistry* 15 (1976–77): 1–6: "Thus 'chiral' and 'achiral' are used with two different connotations: When the terms are applied to a geometric model, they are sharply defined, whereas when used in conjunction with observables, they necessarily entail a certain fuzziness" (6).

Hoffmann and Laszlo have suggested that chemistry is more like music than like mathematics in its parting company with deductive rigor. To "represent" or "explain" a molecule, like camphor, for example, the chemist may call on any number of different representations.:

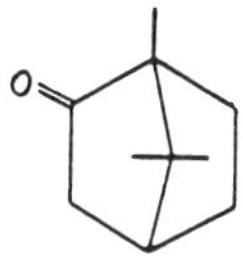

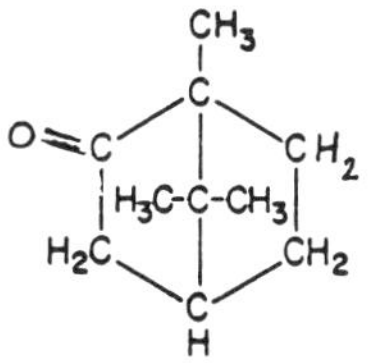

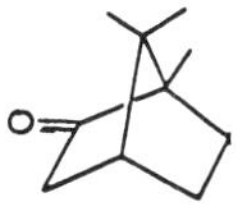

Any of these three figures, or a "ball-and-stick" model, a "space-filling model," or an electronic-distribution model, may be best for the occasion.

> Which of these representations . . . is right? Which *is* the molecule? . . . All of them are models, representations suitable for some purposes, not for others. Sometimes just the name "camphor" will do. Sometimes the formula, $C_{10}H_{16}O$ suffices.[42]

Like the woodcut suite of *Hokusai* entitled "Thirty-Six Views of Mount Fuji" (or like Claude Monet's multiple paintings of the cathedral at Rouen), there is no one rigorous answer or explanation to the "nature" of a molecule,[43] even a simple hydrocarbon like butadiene (See fig. 11.) Or as Lespieau put it at the turn of the century, in defense of nineteenth-century structural chemistry, the method of chemical science is not at all like that of mathematical science, because a chemical formula cannot be demonstrated like a theorem.[44]

Coulson described the first ten years of quantum chemists' work on the electron valence bond (roughly 1928–1938) as work spent "*escaping from the thought-forms of the physicist* [my emphasis], so that the chemical notions of directional bonding and localization could be developed."[45] Heisenberg earlier claimed that the Heitler-London treatment of the hydrogen molecule was not a characteristically *physical* approach, in contrast to Hund's more "general"

42. Roald Hoffmann and Pierre Laszlo, "Representation in Chemistry," *Angewandte Chemie, International Edition in English* 30 (1991): 1–16, on 5. These authors note that many textbooks incorrectly represent a molecule of camphor as the mirror image of the configuration given here, which is based on Hoffman and Laszlo's diagrams (n., 4).

43. Hoffmann, "Under the Surface," 1598. Also see R. B. Bernstein, Dudley R. Herschbach, and R. D. Levine, "Dynamical Aspects of Stereochemistry," *Journal of Physical Chemistry* 91 (1987), 5365–5375, on 5375.

44. See chap. 6.

45. Coulson, "Recent Developments," page proofs, 3; in Coulson Papers, Bod.Oxford.

Various Names: Bivinyl; Erythrene; Pyrrolylene; Vinylethylene; Divinyl; Biethylene; 1,3 Butadiene

Oldest Empirical Formula: CH_3
Modern Empirical Formula: C_2H_6
Molecular Weight: 54.09
Form: Colorless Gas at room temperature
Density: 0.650 gm/ml at -6°C.
Melting Point: -108.92°C.
Boiling Point: -3°C.

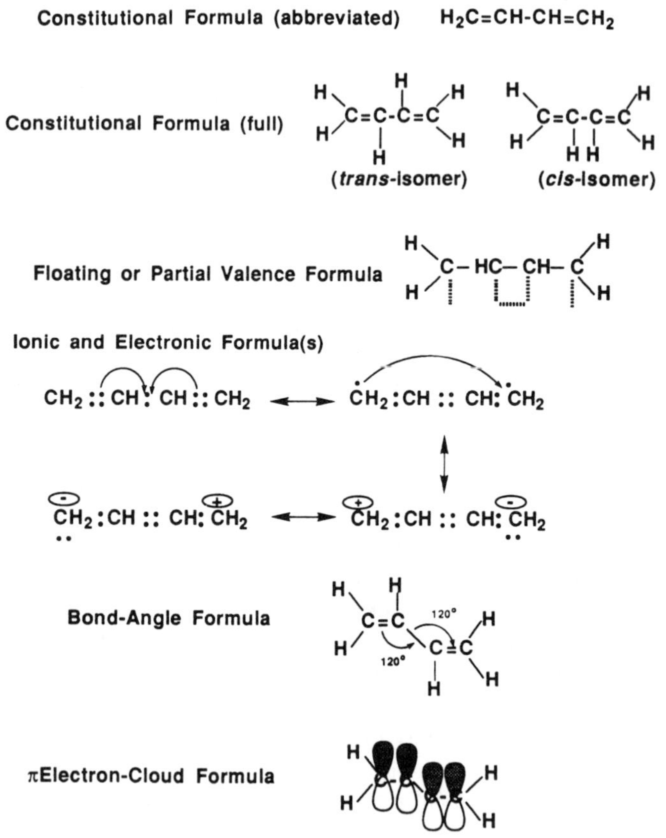

Constitutional Formula (abbreviated) $H_2C=CH\text{-}CH=CH_2$

Constitutional Formula (full)

H H H
C=C-C=C
H H H
H

(*trans*-isomer)

H H
C=C-C=C
H H H H

(*cis*-isomer)

Floating or Partial Valence Formula

H H
C– HC– CH– C
H H

Ionic and Electronic Formula(s)

$CH_2::CH:CH::CH_2$ ⟷ $\dot{C}H_2:CH::CH:\dot{C}H_2$

↕

$\overset{\ominus}{C}H_2:CH::CH:\overset{\oplus}{C}H_2$ ⟷ $\overset{\oplus}{C}H_2:CH::CH:\overset{\ominus}{C}H_2$

Bond-Angle Formula

H H
C=C 120° H
H C=C
120°
H H

πElectron-Cloud Formula

H H
H H

Figure 11. Representations of Butadiene. Drawings courtesy of Glenn Dryhurst.

and physical approach, which treated the molecule as an isolated corpuscle, neglecting electron interaction. The ongoing implications of these distinctions are reflected still in the resistance of some contemporary chemists to the quantum mechanical reduction of the chemical molecule to an isolated physical molecule.

The molecular physicist R. G. Woolley (Cavendish Laboratory at Cambridge University) and the physical chemist Hans Primas (ETH, Zurich), for instance, have reminded theoretical chemists and physicists that molecular structure is a classical idea, foreign to the principles of pioneer quantum mechanics that ''neither gives a correct nor a consistent description of molecules. . . . Quantum mechanics gives perfect predictions for all spectrosopic experiments. However, chemistry is not spectroscopy.''[46]

The problem, as Woolley addressed it, is that quantum mechanical calculations employ the fixed, or ''clamped,'' nucleus approximation (the Born-Oppenheimer approximation) in which nuclei are treated as classical particles confined to ''equilibrium'' positions. Woolley claims that a quantum mechanical calculation carried out completely from first principles, without such an approximation, yields no recognizable molecular structure and that the maintenance of ''molecular structure'' must therefore be a product not of an isolated molecule but of the action of the molecule functioning over time in its environment.[47]

The emphasis on environment and on the molecule acting in an environment is not trivial, for it lies at the heart of the conceptual aims and problems of the chemical discipline, as outlined in this book. Commenting on the difference between chemists and physicists, Mulliken suggested that the difference is that ''chemists love molecules, and get to know them individually.'' In contrast, he suggested, ''physicists are more concerned with fields of force and waves than with the individual personalities of molecule or matter.''[48] In an interview with Schweber, Wilson corroborated Mulliken's point of view, saying that for his part, ''I love my molecules.''[49]

Hoffmann likens the chemist's world of molecules to the strap hanger's world of the subway.

> I think that in their richness and variety molecules are to be compared with people. This is what I like about riding the subway in New York—the incredible range of ethnic type, physiognomy, clothes, and emotions. I see tired swarthy men, women with henna-dyed hair, people reading Korean and Russian newspapers, Caribbean blacks, a sleepy Indian girl. Angelic or rough, they're alive,

46. See the discussion in Stephen J. Weininger, ''The Molecular Structure Conundrum: Can Classical Chemistry be Reduced to Quantum Chemistry?'' *JChem.Ed* 61 (1984): 939–944. Hans Primas, ''Foundations of Theoretical Chemistry,'' 39–113, in R. G. Woolley, ed., *Quantum Dynamics of Molecules: The New Experimental Challenge to Theorists* (New York: Plenum Press, 1980); quotation on 105; and Primas, *Chemistry, Quantum Mechanics, and Reductionism* (New York: Springer Verlag, 1981). Also see, R. D. Brown, ''[Letters] Kinky Molecules,'' *Chemistry in Britain* 24 (1988), 770, contrasting the difference between chemists' and spectroscopists' quantum-mechanical treatments of structures, wherein the latter are harder to ''visualize.''

47. R. G. Woolley, ''Must a Molecule Have a Shape?'' *JCS* 100 (1978): 1073–1078; and ''Further Remarks on Molecular Structure in Quantum Theory,'' *Chemical Physics Letters* 55 (1978): 443–446.

48. R. S. Mulliken, ''Spectroscopy, Quantum Chemistry, and Molecular Physics,'' *Physics Today*, (1968): 52–57, on 54.

49. Quoted in S. S. Schweber, ''Young John Clarke Slater,'' 339–406, on 403–404.

> and in their lives are a million novels. When I open a page of *Chemical Communications* or *Angewandte Chemie*, I get a similar feeling. I recognize the molecule types (my prejudices and education determine that), but in these pages someone has pulled off something new—here a cluster of nine nickel atoms, one inside a cube of eight, there someone else has found out the curious way an NO molecule tumbles as it jumps off a metal surface to which it has been stuck. I'm a voyeur of molecules.[50]

In conclusion, we can say that if mechanics has always been an aim of scientific philosophy, then twentieth-century chemistry has revived its philosophical character, achieving a long-sought understanding of the dynamics of matter. But chemists, more than physicists, have remained self-conscious about the fit between the phenomena taking place in the laboratory and the symbols employed in the operations of explanatory mathematics. Precision, not rigor, has been characteristic of chemical methodology. Parallel representation, not single causal principle, has been characteristic of chemical explanation.

Whereas many early-twentieth-century physicists were inclined to regard conventionalism, complementarity, and indeterminacy as concessions of failure in their traditional philosophical enterprise, chemists were not surprised that a simple, ''logical'' account of the behavior of electrons and atoms, like that of molecules and people, often gives way to the inconsistencies and uncertainties of empiricism. This is a point of view they taught physicists to accept, among them, Richard Feynman:

> The fact that electrodynamics can be written in so many ways—the differential equations of Maxwell, various minimum principles with fields, minimum principles without fields, all different kinds of ways . . . was something I knew, but I have never understood. It always seems odd to me that the fundamental laws of physics, when discovered, can appear in so many different forms that are not apparently identical at first, but with a little mathematical fiddling you can show the relationship.[51]

Visual imagery, metaphorical language, constitutional formulas, and systematic classification have remained not only the legacy but the working strategy of modern theoretical chemistry, along with ''mathematical fiddling.'' Physicists and chemists alike have concerned themselves with what they call in similar language the molecule and the atom. But there is a difference in meaning, even if the language is the same.

Physicists have focused on the ''corpuscle'' and principles common to all corpuscles, whether electrons, atoms, or molecules. In contrast, the chemist

50. Roald Hoffmann, ''The Grail,'' in Roald Hoffmann and Vivian Torrence, *Chemistry Imagined* (Smithsonian Institution Press: in press).

51. Richard P. Feynman, ''The Development of the Space-Time View of Quantum Electrodynamics,'' 172–191, in *Les Prix Nobel* (Stockholm: Imprimerie Royale, 1966): 178–179.

most often has reveled not so much in the discovery or justification of uniformities in nature as in the discovery or creation of what is different and marvelous. Identifying oneself as a chemist or a physicist may hinge significantly on this different ''feeling'' about the material world, a matter of individual psychology and aesthetic preference that in the last two centuries sometimes has been characterized imperfectly as the contrast between the romantic and the classic sensibilities. The historical strategies of discipline building and the common elements of disciplinary identity rationalize and standardize values and problem sets that become commitments to one identity or another, a scientific identity or a nonscientific one, a chemical identity or a physical one.

It is difficult to resist speculating that conceptual inconsistencies in setting up the notion of disciplinary identity and resistance to the process of discipline bounding follow the same itinerary as the history of ethnic and national identities. Repeatedly, ''bridging the gap'' and ''crossing the frontier'' lead to the easing of long-accepted boundaries and established criteria of particularity. If, as we argued in the first chapter, nineteenth-century discipline formation in the scientific body was in part an analogue to nation building in the political body, then we should not be surprised at both the breakdown and the persistence of these identities with the waxing and waning of cosmopolitanism and internationalism in modern times.

Glossary

Acetoacetic Ester. Also termed ethyl acetoacetate, or $CH_3COCH_2COOC_2H_5$. As an ester, it can be hydrolyzed under certain conditions to acetoacetic acid (CH_3COCH_2COOH); as a ketone, it reacts with reagents for the carbonyl (C = O) group. Peculiarly, it also behaves like a hydroxyl (OH^-) compound. It is the prototype of the phenomenon of tautomerism, and its isomeric forms are termed the *keto* and the *enol* forms.

$$\underset{(keto)}{CH_3COCH_2COOC_2H_5} \quad \underset{(enol)}{CH_3C\underset{\underset{OH}{|}}{}\;CHCOOC_2H_5}$$

Aliphatic Compound. Saturated or unsaturated carbon compound, the structure of which may be straight, branching, or cyclic.

Alpha, Beta, Gamma Positions. Substitution or addition positions on a carbon chain, relative to a previously existing functional group:

$$\overset{\gamma}{C} - \overset{\beta}{C} - \overset{\alpha}{C} - X$$

Aromatic Compound. Fragrant and peculiarly unsaturated carbon compound, the structure of which is cyclic, with benzene as the prototype.

Conjugated System. Unsaturated hydrocarbon in which double bonds are separated by single bonds, so that single and double bonds alternate. Compounds with two double bonds are called "dienes."

Crotonic Acid. Crotonic and isocrotonic acid are geometrical isomers of the structure $CH_3CH{=}CHCOOH$, crotonic acid being the *trans* form and isocrotonic acid the *cis* form. Crotonic acid has the higher melting point (72°C, compared to 15.5°C) and is ordinarily a solid.

$$\begin{array}{ccc} H-C-CH_3 & \quad & CH_3-C-H \\ \| & & \| \\ H-C-COOH & & H-C-COOH \\ \text{(cis)} & & \text{(trans)} \end{array}$$

Isomer. One of two or more chemical compounds having the same elementary constituents (atoms) in the same proportion by weight but differing in physical or chemical properties from the other isomer(s).

Ortho, Meta, Para Positions. For a benzene ring, substitution positions on the hydrocarbon "ring" or "hexagon," relative to a previously substituted group:

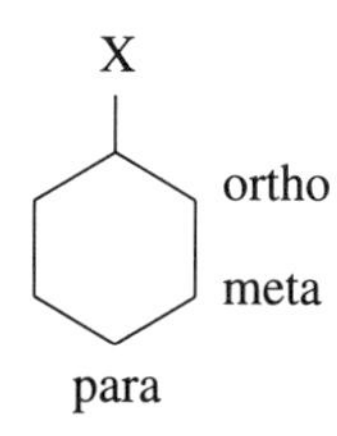

Saturated Compound. A chemical compound with completely fulfilled combining capacity; in modern terms in organic chemistry, a carbon compound containing no double or triple bonds.

Tautomer. One of two isomeric forms in equilibrium with the other.

Bibliography

Note: University theses and dissertations are included in this listing. Archives from which unpublished sources are cited are given in the list of abbreviations at the beginning of this book.

Abe, Yuko. ''Pauling's Revolutionary Role in the Development of Quantum Chemistry.'' *Historia Scientiarum* 20 (1981): 107–124.

Abegg, Richard. ''Die Valenz und das periodische System. Versuch einer Theorie der Molekular-Verbindungen.'' *Zeitschrift für anorganische Chemie* 39 (1904): 330–380.

Abir-Am, Pnina. ''The Biotheoretical Gathering, Transdisciplinary Authority and the Incipient Legitimation of Molecular Biology in the 1930s.'' *History of Science* 25 (1987): 1–70.

Alborn, Timothy L. ''Negotiating Notation: Chemical Symbols and British Society, 1831–1835.'' *Annals of Science* 46 (1898): 437–460.

Allen, James, Allen E. Oxford, Robert Robinson, and John C. Smith. ''The Relative Directive Powers of Groups of the Forms RO and RR'N in Aromatic Substitution. Pt. IV. A Discussion of the Observations Recorded in Parts I, II, and III.'' *JCS* 129 (1926): 401–411.

Allsop, C. B., and W. A. Waters. ''Thomas Martin Lowry.'' In *British Chemists*. Ed. Alexander Findlay and William Hobson Mills. London: The Chemical Society, 1947. Pp. 402–418.

Anderson, Wilda. *Between the Library and the Laboratory: The Language of Chemistry in Eighteenth-Century France*. Baltimore: The Johns Hopkins University Press, 1984.

Appel, Toby A. *The Cuvier-Geoffroy Debate: French Biology in the Decades before Darwin*. Oxford: Oxford University Press, 1987.

Armstrong, Henry Edward. ''Presidential Address of the Chemical Section.'' *BAAS Rep.* Winnipeg (1909): 420–454.

———. *The Art and Principles of Chemistry.* New York: Macmillan, 1927.

Arndt, Fritz. ''Gleichgewicht und 'Zwischenstufe.' '' *Berichte* 63 (1930): 2963–2966.

Arndt, Fritz, E. Scholz, and F. Nactwey. ''Ueber Dipyrilene und über die Bindungsverhältnisse in Pyron-Ringsystem.'' *Berichte* 57 (1924): 1903–1911.

Arrhenius, Svante. ''On the Dissociation of Substances in Aqueous Solution (1887).'' In *The Foundations of the Theory of Dilute Solution.* Edinburgh: Alembic Club, 1929.

Assmus, Alexi. ''Molecular Structure and the Genesis of the American Quantum Physics Community 1916–1926.'' Ph.D. dissertation, Harvard University, 1990.

———. ''Molecular Structure and the Genesis of the American Quantum Physics Community, 1916–1926.'' HSPS 22 (1992): 209–231.

Bachelard, Gaston. *La pluralisme cohérent de la chimie moderne.* Paris: Vrin, 1931.

———. *La formation de l'esprit scientifique.* Paris: Vrin, 1938.

———. *La psychoanalyse du feu.* Paris: Gallimard, 1938.

———. *L'activité rationaliste de la physique contemporaine.* Paris: Vrin, 1951.

Back, M. H., and Keith J. Laidler, eds. *Selected Readings in Chemical Kinetics.* Oxford: Pergamon, 1967.

Baker, J. W. *Tautomerism.* London: Routledge, 1934.

Baker, W. ''Obituaries. Lady Robinson.'' *Nature* 173 (1954): 566–567.

Baly, E. C. C., and C. H. Desch. ''The Ultra-Violet Absorption Spectra of Certain Enol-Keto-Tautomerides. Pt. II.'' *JCS* 87 (1905): 766–784.

Barkan, Diana Kormos. ''Walther Nernst and the Transition to Modern Physical Chemistry.'' Ph.D. dissertation. Harvard University, 1990.

Baumé, A., and P. J. Macquer. *Plan d'un cours de chymie expérimentale et raisonnée avec un discours historique sur la chymie.* Paris: Herissant, 1757.

Beesley, R. M., and J. F. Thorpe. ''The Formation and Stability of Associated Alicyclic Systems. Pt. I. A System of Nomenclature, and Some Derivatives of Methane-II-Cyclopropane and of Methane-III-Cyclopropane.'' *JCS* 117 (1920): 591–620.

Benezra, Claude. ''La chimie des flèches ou la notion d'agresseur et d'agressé en chimie organique.'' *L'Actualité Chimique* (1975): 21.

Benfey, Otto Theodor, ed. *Classics in the Theory of Chemical Combination.* New York: Dover, 1963.

———. ''Concepts of Time in Chemistry.'' *JChem.Ed.* 40 (1963): 574–577.

———. *From Vital Force to Structural Formulas.* Washington, D.C.: American Chemical Society, 1975.

———. ''The Concepts of Chemistry—Mechanical, Organicist, Magical or What?'' *JChem.Ed.* 59 (1982): 395–398.

Bensaude-Vincent, Bernadette. ''A Founder Myth in the History of Science? The Lavoisier Case.'' In *Functions and Uses of Disciplinary Histories. Sociology of the Sciences Yearbook.* Ed. Loren Graham et al. Dordrecht: Reidel, 1983. Pp. 53–78.

———. *A Propos de 'méthode de nomenclature chimique': Esquisse historique suivie du texte de 1787.* Paris: Centre de Documentation Sciences Humaines, 1983.

———. ''A View of the Chemical Revolution through Contemporary Textbooks: Lavoisier, Fourcroy and Chaptal.'' *BJHS* 23 (1990): 435–460.

Bernstein, D. B., Dudley Herschbach, and R. D. Levine. "Dynamical Aspects of Stereochemistry." *Journal of Physical Chemistry* 91 (1987): 5365–5377.

Berthelot, Marcellin. *Leçons sur les méthodes générales de synthèse en chimie organique*. Paris: Gauthier-Villars, 1864.

———. *La synthèse chimique*. Paris: Baillière, 1876.

———. *Essai de mécanique chimique fondée sur la thermochimie*. 2 vols. Paris: Dunod, 1879.

———. *La révolution chimique*. Paris: Alcan, 1890.

———. *Les carbures d'hydrogène*. Paris: Gauthier-Villars, 1901.

Berthelot, Marcellin, and Emile Jungfleisch. *Traité élémentaire de chimie organique*. 4th ed. 2 vols. Paris: Dunod, 1898.

Berthollet, C. L. *Researches into the Laws of Chemical Affinity*. Trans. M. Farrell. Baltimore: Philip Necklin, 1809.

Berzelius, Jöns Jacob. "Essay on the Cause of Chemical Proportions." *Annals of Philosophy* 2–3 (1813–14): 443–454, 43–52.

———. *Essai sur la théorie des proportions chimiques et sur l'influence chimique de l'électricité* [1819]. Ed. Colin Russell. Reprint ed. New York: Johnson, 1972.

Bettridge, D. "The Teaching of Chemistry in Victorian and Edwardian Times." *Royal Institute of Chemistry Reviews* 3 (1970): 161–176.

Bigot, B., and F. Volatron. "Parlez-vous chimie théorique?" *L'Actualité Chimique* (November 1984): 43–51.

Bjerrum, Niels J. *Selected Papers*. Copenhagen: E. Munksgaard, 1949.

Black, Max. *Models and Metaphors: Studies in Language and Philosophy*. Ithaca: Cornell University Press, 1962.

Bohr, Niels. "On the Constitution of Atoms and Molecules." *Philosophical Magazine* 26 (1913): 1–19, 476–502, 857–875.

———. "Ueber die Serienspektra der Elemente." *ZP* 2 (1920): 423–469.

———. "Atomic Structure." *Nature* 107 (1921): 104–107.

———. "The Structure of the Atom and the Physical and Chemical Properties of the Elements." *The Theory of Spectra and Atomic Constitution*. Cambridge: 1922. Pp. 61–126.

———. "The Quantum Postulate and the Recent Development of Atomic Theory." *Nature* 121 (1928): 580–590.

Boltzmann, Ludwig. "On the Necessity of Atomic Theories in Physics." *The Monist* 12 (1901): 65–79.

———. "Model." *Encylopedia Britannica*. 11th ed. New York: 1911. XVII: 638–640.

Born, Max. *The Constitution of Matter: Modern Atomic and Electron Theories*. Trans. E. W. Blair and T. S. Wheeler. 2d German ed. London: Methuen, 1923.

———. *Problems of Atomic Dynamics*. Cambridge: MIT Press, 1926.

Bos, H. J. M. "Mathematics and Rational Mechanics." In *The Ferment of Knowledge: Studies in Eighteenth-Century Science*. Ed. G. S. Rousseau and Roy Porter. Cambridge: Cambridge University Press, 1980. Pp. 327–355.

Bourdieu, Pierre. "Intellectual Field and Creative Project." *Social Science Information* 8 (1969): 89–119.

———. "The Social Space and the Genesis of Groups." *Social Science Information* 24 (1985): 195–220.

Bourguel, Maurice. ''Applications de l'effet Raman à la chimie organique.'' *Bull.SCF* 53 (1933): 469–505.

Bradley, John. ''On the Operational Interaction of Classical Chemistry.'' *BJPS* 6 (1955–1956): 32–42.

Bray, W. C., and Gerald E. K. Branch. ''Valence and Tautomerism.'' *JACS* 35 (1913): 1440–1448.

Brock, William H., ed. *The Atomic Debates: Brodie and the Rejection of the Atomic Theory*. Leicester: Leicester University Press, 1967.

———. ''The British Association Committee on Chemical Symbols in 1834.'' *Ambix* 33 (1986): 33–37.

———. *H. E. Armstrong and the Teaching of Science, 1880–1938*. Cambridge: Cambridge University Press, 1973.

Brock, William H., and A. J. Meadows. *The Lamp of Learning: Taylor and Francis and the Development of Science Publishing*. London: Taylor and Francis, 1984.

Brooke, John Hedley. ''Laurent, Gerhardt and the Philosophy of Chemistry.'' *HSPS* 6 (1975): 405–429.

———. ''Methods and Methodology in the Development of Organic Chemistry.'' *Ambix* 34 (1987): 147–155.

Brown, R. D. ''[Letters.] Kinky Molecules.'' *Chemistry in Britain* 24 (1988): 770.

Brunold, Charles. *Le problème de l'affinité chimique et l'atomistique: Etude du rapprochement actuel de la physique et de la chimie*. Paris: Masson, 1930.

Bud, Robert F. *The Discipline of Chemistry: The Origins and Early Years of the Chemical Society of London*. Philadelphia: University of Pennsylvania Press, 1980.

Bud, Robert F., and Gerrylynn K. Roberts. *Science vs. Practice: Chemistry in Victorian Britain*. Manchester: Manchester University Press, 1984.

Burkhardt, G. Norman. ''The University of Manchester.'' *Chemistry and Industry* (1949): 427–429.

———. ''Schools of Chemistry in Great Britain and Ireland—XIII. The University of Manchester (Faculty of Science).'' *Journal of the Royal Institute of Chemistry* 78 (1954): 448–460.

Burrau, Oyvind. ''Berechnung des Energiewertes des Wasserstoff-Molekel-Ions (H_2^+) im Normalzustand.'' *Danske Videnskabernes Selskab. Mat.-fys. Meddelser* 7 (1927): 3–18.

Burton, Milton. ''Radiation Chemistry.'' *Ann.Rev.P.Chem.* 1 (1950): 113–132.

Butterfield, Herbert. *The Origins of Modern Science, 1300–1800*. New York: Free Press, 1957.

Bykov, G. V. ''Historical Sketch of the Electron Theories of Organic Chemistry.'' *Chymia* 10 (1965): 199–253.

Cabannes, Jean. *Aniosotropie des molećules: Effet Raman, Conférences faites au Conservatoire National des Arts et Métiers les 2 et 3 mai 1930*. Paris: Hermann, 1930.

Caldin, E. F. *The Structure of Chemistry in Relation to the Philosophy of Science*. London: Sheed and Ward, 1961.

Campaigne, E. ''The Contributions of Fritz Arndt to Resonance Theory.'' *JChem.Ed.* 36 (1959): 336–339.

Canguilhem, Georges. *Etudes d'histoire et de philosophie des sciences*. Paris: Vrin, 1979.

Cannizzaro, Stanislao. *Sketch of a Course of Chemical Philosophy. Alembic Club Reprints*. Edinburgh: 1947.

Cannon, Susan Faye. *Science and Culture: The Early Victorian Period.* New York: Dawson and Science History Publications, 1978.

Cardwell, D. S. L. *From Watt to Clausius*. Ithaca: Cornell University Press, 1971.

Carrière, Justus, ed. *Berzelius und Liebig: Ihre Briefe von 1831–1845*. München: J. F. Lehmann, 1898.

Cesaro, S. Nunziante, and E. Torracca. ''Early Applications of Infra-Red Spectroscopy to Chemistry.'' *Ambix* 35 (1988): 39–47.

Challenger, Frederick. ''Schools of Chemistry in Great Britain and Ireland. IV. The Chemistry Department of the University of Leeds.'' *Journal of the Royal Institute of Chemistry* 77 (1953): 161–171.

Charpentier–Morize, Micheline. *La contribution des ''laboratoires propres'' du CNRS à la recherche chimique en France de 1939 à 1973*. No. 4. *Cahiers pour l'histoire du CNRS*. Paris: Editions du CNRS, 1989.

Chubin, Daryl E. ''State of the Field. The Conceptualization of Scientific Specialties.'' *The Sociological Quarterly* 17 (1976): 448–476.

Clark, B. N. ''The Influence of the Continent upon the Development of Higher Education and Research in Chemistry in Great Britain during the Latter Half of the Nineteenth Century.'' Ph.D. dissertation, University of Manchester, 1979.

Cloître, Michel, and Terry Shinn. ''Enclavement et diffusion du savoir.'' *Information sur les Sciences Sociales*, 25 (1986): 161–187.

Cocker, W. ''Chemistry in Manchester in the Twenties, and Some Personal Recollections.'' *Natural Product Reports. Royal Society of Chemistry* 4 (1987): 67–72.

Cohen, Ernst. *J. H. van't Hoff. Sein Leben und Wirken*. Leipzig: Akademische Verlagsgesellschaft, 1912.

Coleman, William. ''The Cognitive Basis of the Discipline: Claude Bernard on Physiology.'' *Isis* 76 (1985): 49–70.

Collie, J. N. ''A Space Formula for Benzene.'' *Trans.CS* 71 (1897): 1013–1023.

Collins, Harry. ''Stages in the Empirical Program of Relativism.'' *Social Studies of Science* 11 (1981): 3–10.

Conant, James Bryant. *Organic Chemistry. A Brief Introductory Course*. Rev. ed. New York: Macmillan, 1936.

Cooke, Josiah P., Jr. *Elements of Chemical Physics*. Boston: Little, Brown and Co., 1860.

Cooper, K. A., E. D. Hughes, and C. K. Ingold. ''The Mechanism of Elimination Reactions. Pt. III. Unimolecular Olefin Formation from Tert.-Butyl Halides in Acid and Alkaline Aqueous Solutions.'' *JCS* 140 (1937): 1280–1283.

Coulson, Charles A. ''Theoretical Chemistry: Past and Future. Lecture delivered before The University of Oxford, 23 February 1973.'' Ed. S. L. Altman. Oxford: Clarendon, 1974.

Court, S. ''The 'Annales de Chimie,' 1789–1815.'' *Ambix* 19 (1972): 113–128.

Crawford, Elisabeth. ''Arrhenius, the Atomic Hypothesis, and the 1908 Nobel Prizes in Physics and Chemistry.'' *Isis* 75 (1984): 503–522.

———. *The Beginnings of the Nobel Institution. The Science Prizes, 1901–1915*. Cambridge: Cambridge University Press, 1984.

Crosland, Maurice. *Historical Studies in the Language of Chemistry.* Cambridge: Harvard University Press, 1962.

———. "The Development of Chemistry in the Eighteenth Century." *Studies on Voltaire and the Eighteenth Century* 24 (1963): 369–441.

———. *The Society of Arcueil: A View of French Science at the Time of Napoleon I.* Cambridge: Harvard University Press, 1967.

———. *Gay-Lussac: Scientist and Bourgeois.* Cambridge: Cambridge University Press, 1978.

———. *Science Under Control. The French Academy of Sciences 1795–1914.* Cambridge: Cambridge University Press, 1992.

Crowther, James Gerald. *The Cavendish Laboratory, 1874–1974.* New York: Science History Publications, 1974.

Dagognet, François. *Tableaux et langages de la chimie.* Paris: Seuil, 1969.

———. *Ecriture et iconographie.* Paris: Vrin, 1973.

Dalton, John. *A New System of Chemical Philosophy.* 2 vols. in 3. Manchester: S. Russel, 1808; Russel and Allen, 1808; and Executors of S. Russel for G. Wilson, 1827.

Daniels, Farrington, and Robert A. Alberty. *Physical Chemistry.* 2d ed. New York: John Wiley, 1963.

Daudel, Raymond. *L'état de la chimie théorique: Guide pour l'exposition presentée au Palais de la Découverte.* Paris: Sennac, 1949.

———, and Alberte Pullman, eds. *Aspects de la chimie quantique contemporaire.* Colloques Internationaux de CNRS, #195. Paris: Editions du CNRS, 1971.

Davenport, Derek. "On the Comparative Unimportance of the Invective Effect." *Chemtech* 17 (1987): 526–531.

Davies, Peter John. "Sir Arthur Schuster, 1851–1934." Ph.D. dissertation, University of Manchester Institute of Science and Technology, 1983.

Davy, Humphry. *Elements of Chemical Philosophy.* London: Printed for J. Johnson [by W. Bulmer], 1812.

Debus, Allen G. *The Chemical Philosophy: Paracelsian Science and Medicine in the Sixteenth and Seventeenth Centuries.* 2 vols. New York: Science History Publications, 1977.

Delattre, P., and M. Thullier, eds. *Elaboration et justification des modèles: Applications en biologie.* Paris: Maloine, 1979.

Deluc, J. A. *Introduction à la physique terrestre . . . précedé de deux memoires sur la nouvelle théorie chimique.* Paris: Nyon, 1803.

Detouches, Jean-Louis. "Sur la notion de modèle en microphysique." *Synthese* 12 (1960): 176–181.

Dhar, Nil Ratan. "Coefficient de température de réactions catalytiques." Doctoral thesis, University of Paris, 1916.

———. *The Chemical Action of Light.* London: Blackie and Son, 1931.

Dirac, P. A. M. "The Physical Interpretation of the Quantum Dynamics," *Proc. RSL* A 113 (1927): 621–641.

———. "Quantum Mechanics of Many-Electron Systems." *Proc.RSL* A123 (1929): 714–733.

Dolby, R. G. A. "The Transmission of Two New Scientific Disciplines from Europe to North America in the Late 19th Century.'' *Annals of Science* 34 (1977): 287–310.

Donovan, Arthur. *Philosophical Chemistry in the Scottish Enlightenment. The Doctrines and Discoveries of William Cullen and Joseph Black.* Edinburgh: Edinburgh University Press, 1975.

———. ''Lavoisier and the Origins of Modern Chemistry.'' Ed. Arthur Donovan. *Osiris*, 2d ser. 4 (1988): 214–231.

Dubpernell, George, and T. H. Westbrook, eds. *Selected Topics in the History of Electrochemistry.* Princeton: The Electrochemical Society, 1978.

Duhem, Pierre. *Traité élémentaire de mécanique chimique fondée sur la thermodynamique.* 2 vols. Paris: Hermann, 1897.

———. ''Une science nouvelle: La chimie physique.'' *Revue Philomatique de Bordeaux et du Sud-Ouest* (1899): 205–219, 260–280.

———. *Le mixte et la combination chimique.* Paris: Naud, 1902.

———. *La théorie physique: Son objet, et sa structure.* Paris: Chevalier et Rivière, 1906.

———. *The Aim and Structure of Physical Theory.* Trans. Philip Wiener. From 2d French ed. Princeton: Princeton University Press, 1954.

Dulou, Raymond, and Albert Kirrmann. ''Le Laboratoire de Chimie de l'Ecole Normale Supérieure: Notes historiques.'' *Bull.SAENS* 54 (1973).

Dumas, Jean-Baptiste. *Leçons sur la philosophie chimique.* Ed. M. Bineau. Paris: Ebrard, 1837.

———. ''Premier mémoire sur les types chimiques.'' *Ann.Chim.Phys.* 73 (1840): 73–103.

Dupont, Georges. ''Sur la théorie électronique de la valence: Essai de représentation stéréochimique des éléments.'' *Bull.SCF* 41 (1927): 1101–1137.

———. ''Robert Lespieau, 1864–1947.'' *Bull.SCF* 16 (1949): 1–9.

Edge, David O., and Michael J. Mulkay. *Astronomy Transformed: The Emergence of Radio Astronomy in Britain.* New York: Wiley, 1976.

Eistert, Bernhard. *Tautomerie und Mesomerie: Gleichgewicht und 'Resonanz.'* Stuttgart: F. Enke, 1938.

———. ''Zur Schreibweise chemischer Formeln und Reaktionsabläufe.'' *Berichte* 71B (1938): 237–240.

Eyre, J. Vargas. *Henry Edward Armstrong, 1848–1937: The Doyen of British Chemists and Pioneer of Technical Education.* London: Butterworths Scientific Publications, 1958.

Eyring, Henry. *Quantum Chemistry.* New York: John Wiley, 1944.

———. ''Physical Chemistry: The Past 100 Years.'' *CENews* 54 (1976): 88–104.

Falk, K. George, and John M. Nelson. ''The Electron Conception of Valence.'' *JACS* 32 (1910): 1637–1654.

Feynman, Richard P. ''The Development of the Space-Time View of Quantum Electrodynamics.'' In *Les Prix Nobel.* Stockholm: Imprimerie Royale, 1966. Pp. 172–191.

Findlay, Alexander. *A Hundred Years of Chemistry.* 2d ed. London: Duckworth, 1948.

Forman, Paul, John Heilbron, and Spencer Weart, eds. *Physics ca. 1900.* V. *Historical Studies in the Physical Sciences.* Princeton: Princeton University Press, 1975.

Foucault, Michel. *Les mots et les choses: Une archéologie des sciences humaines.* Paris: Gallimard, 1966.

———. *Discipline and Punish: The Birth of the Prison.* New York: Vintage, 1979.

Fourcroy, Antoine de. *Philosophie chimique, ou, vérités fondamentale de la chimie moderne disposés dans un nouvel ordre.* Paris: Imprimerie de Cl. Simon, 1792.

———. *Système des connaissances chimiques, et de leurs applications aux phénomènes de la nature et de l'art.* 11 vols. Paris: Baudoin, 1800.

Fowler, Ralph. ''A Report on Homopolar Valency and Its Quantum-Mechanical Interpretation.'' In *Chemistry at the Centenary (1931) Meeting of the British Association for the Advancement of Science.* Cambridge: W. Heffer and Sons, 1932. Pp. 226–246.

Fox, Robert. ''The Rise and Fall of Laplacian Physics.'' *HSPS* 4 (1974): 89–136.

Frankland, Edward. ''On a New Series of Organic Compounds Containing Metals.'' *Philosophical Transactions of the Royal Society* 142 (1852): 417–444.

———. *Lecture Notes for Chemical Students: Embracing Mineral and Organic Chemistry.* London: John Van Voorst, 1866.

———. *Experimental Researches in Pure, Applied, and Physical Chemistry.* London: J. Van Voorst, 1877.

Freudenthal, Gad, ed. *Études sur Hélène Metzger.* Leiden: E. J. Brill, 1990.

Friedel, Charles. *Cours de chimie organique, professé à la Faculté des Sciences, Paris.* Paris: Faculté des Sciences, 1887.

Friedel, Robert. ''Defining Chemistry: Origins of the Heroic Chemist.'' In *Chemical Sciences in the Modern World.* Ed. Seymour Mauskopf. Philadelphia: University of Pennsylvania Press, in press.

Fruton, Joseph S. ''Contrasts in Scientific Style: Emil Fischer and Franz Hofmeister; Their Research Schools and Their Theories of Protein Structure.'' *Proceedings of the American Philosophical Society* 129 (1985): 313–370.

———. ''The Liebig Research Group, A Reappraisal.'' *Proceedings of the American Philosophical Society* 132 (1988): 1–66.

Fry, Harry Shipley. *The Electronic Conception of Valence and the Constitution of Benzene.* London: Longmans, Green, 1921.

Fudano, Jun. ''Early X-ray Research at Physical Laboratories in the United States of America, circa 1900: A Reappraisal of American Physics.'' Ph.D. dissertation, University of Oklahoma, 1990.

Garber, Elizabeth. ''Siméon-Denis Poisson: Mathematics versus Physics in Early Nineteenth-Century France.'' In *Beyond History of Science: Essays in Honor of Robert E. Schofield.* Ed. E. Garber. Bethlehem: Lehigh University Press, 1990.

Garfield, Eugene. *Citation Indexing—Its Theory and Application in Science, Technology and Humanities.* Philadelphia: ISI, 1979.

Gascoigne, Mortimer. *A Historical Catalogue of Scientific Periodicals, 1665–1900.* New York: Garland, 1985.

Gee, W. W. Haldane, Hubert Frank Coward, and Arthur Harden. ''John Dalton's Lectures and Lecture Illustrations.'' *Mem.Manchester LPS* 59, XII (1914–15). 66 pp.

Geison, Gerald L. *Michael Foster and the Cambridge School of Physiology: The Scientific Enterprise in Late Victorian Society.* Princeton: Princeton University Press, 1978.

———. "Scientific Change, Emerging Specialties, and Research Schools." *History of Science* 19 (1981): 20–37.

Gibson, C. S., and A. J. Greenaway, eds. *Faraday Lectures 1869–1928*. London: The Chemical Society, 1928.

Gooday, Graeme. "Precision Measurement and the Genesis of Physics Teaching Laboratories in Victorian Britain." *BJHS* 23 (1990): 23–51.

Goodstein, Judith R. *Millikan's School. A History of the California Institute of Technology*. New York: Norton, 1991.

Gortler, Leon. "The Physical Organic Community in the United States, 1925–1950: An Emerging Network." *JChem.Ed.* 62 (1985): 753–757.

Gould, Edwin S. *Mechanism and Structure in Organic Chemistry*. New York: Holt, Rinehart and Winston, 1959.

Graham, Thomas. *Elements of Chemistry: Including the Applications of the Science in the Arts*. 2d ed. 2 vols. London: Baillière, 1850.

Green, J. H. S. "Sir Christopher Ingold and the Chemistry Department, University College, London." In *Studies on Chemical Structure and Reactivity. Presented to Sir Christopher Ingold*. Ed. J. H. Ridd. London: Methuen, 1966. Pp. 265–274.

Greenaway, John. "Memorial Notice: William Henry Perkin." In *The Life and Work of Professor William Henry Perkin*. London: The Chemical Society, 1932. Pp. 7–38.

Gregory, Frederick. "Romantic Kantianism and the End of the Newtonian Dream in Chemistry." *AIHS* 34 (1984): 108–123.

Gren, F. A. C. *Grundriss der Naturlehre in seinem mathematischen und chemischen Theile*. Halle: 1788.

Grignard, Victor, ed. *Traité de chimie organique*. Paris: Masson, 1935.

Grimaux, Edouard. *Introduction à l'étude de la chimie: Théories et notations chimiques*. Paris: Dunod, 1883.

Groult, Martine, Pierre Louis, and Jacques Roger, eds. *Transfert de vocabulaire dans les sciences*. Paris: Editions du CNRS, 1988.

Guerlac, Henry. "Chemistry as a Branch of Physics: Laplace's Collaboration with Lavoisier." *HSPS* 7 (1976): 193–276.

Guéron, Jules, and Michel Magat. "A History of Physical Chemistry in France." *Ann.Rev.P.Chem.* 22 (1971): 1–25.

Guyton de Morveau, L. B. *Elemens de chymie, théorique et pratique*. Paris: 1777.

Hacking, Ian. *The Emergence of Probability: A Philosophical Study of Early Ideas about Probability, Induction and Statistical Inference*. Cambridge: Cambridge University Press, 1975.

Hall, Marie Boas. *All Scientists Now: The Royal Society in the Nineteenth Century*. Cambridge: Cambridge University Press, 1984.

Hall, Peter Joseph. "The Pauli Exclusion Principle and the Foundations of Chemistry." *Synthese* 69 (1986): 267–272.

Hammett, Louis P. *Physical Organic Chemistry*. New York: McGraw-Hill, 1940.

Hankins, Thomas. *Science and the Enlightenment*. Cambridge: Cambridge University Press, 1985.

Hannaway, Owen. *The Chemist and the Word*. Baltimore: Johns Hopkins University Press, 1975.

Harman, Peter M. *Energy, Force and Matter.* Cambridge: Cambridge University Press, 1982.

Harnack, Adolf. *Geschichte der Königlich Preussischen Akademie der Wissenschaften zu Berlin.* 3 vols. Berlin: 1900.

Hartley, Harold. "Schools of Chemistry in Great Britain and Ireland. XVI. The University of Oxford." *Journal of the Royal Institute of Chemistry* 79 (1955): 118–127, 176–184.

Heilbron, John L. *H. G. J. Moseley: The Life and Letters of an English Physicist, 1887–1915.* Berkeley, Los Angeles, and London: University of California Press, 1974.

———. "Rutherford-Bohr Atom." *American Journal of Physics* 49 (1981): 223–231.

Heisenberg, Werner. "Mehrkörperproblem und Resonanz in der Quantenmechanik." *ZP* 38 (1926): 411–426.

———. "Quantenmechanik." *Die Naturwissenschaften* 14 (1926): 989–994.

———. "Mehrkörperproblem und Resonanz in der Quantenmechanik. II." *ZP* 41 (1927): 239–267.

———. "Ueber den anschaulichen Inhalt der quantentheoretischen Kinematik und Mechanik." *ZP* 43 (1927): 172–198.

Heitler, Walther, and A. Fritz London. "Wechselwirkung neutraler Atome und homopolare Bindung nach der Quantenmechanik." *ZP* 44 (1927): 455–472.

Hellmann, Hans. *Einführung in die Quantenchemie.* Leipzig: Deuticke, 1937.

Helmholtz, Hermann von. "On the Modern Development of Faraday's Conception of Electricity." *JCS* 39 (1881): 277–304.

Henrich, Ferdinand. *Theories of Organic Chemistry.* Rev. 4th German ed. New York: John Wiley, 1922.

Herzberg, Gerhard. *Infrared and Raman Spectra of Polyatomic Molecules.* New York: Van Nostrand, 1945.

Hesse, Mary. *Models and Analogies in Science.* South Bend: University of Notre Dame Press, 1966.

Hiebert, Erwin N. "Nernst and Electrochemistry." In *Selected Topics in the History of Electrochemistry.* Ed. George Dubpernell and J. H. Westbrook. Princeton: Electrochemical Society, 1978. Pp. 180–200.

Hinshelwood, Cyril N. *The Kinetics of Chemical Change in Gaseous Systems.* 2d ed. Oxford: Clarendon Press, 1929.

Hirschfelder, Joseph. "My Adventures in Theoretical Chemistry." *Ann.Rev.P.Chem.* 33 (1983): 1–29.

Hobsbawm, Eric, and Terence Ranger, eds. *Invented Traditions.* Cambridge: Cambridge University Press, 1983.

Hoch, Paul K., and E. J. Yoxen. "Schrödinger at Oxford: A Hypothetical National Cultural Synthesis Which Failed." *Annals of Science* 44 (1987): 593–616.

Hoffmann, Roald. "Under the Surface of the Chemical Article." *Angewandte Chemie, International Edition in English* 27 (1988): 1593–1602.

———. "Nearly Circular Reasoning." *American Scientist* 76 (1988): 182–185.

Hoffmann, Roald, and Pierre Laszlo. "Representation in Chemistry." *Angewandte Chemie. International Edition in English* 30 (1991): 1–16.

Hoffmann, Roald, and Vivian Torrence. *Chemistry Imagined.* Smithsonian Institution Press: In press.

Hofmann, A. W. von "On the Combining Power of Atoms." *Chemical News* 12 (1865): 166–169, 175–179, 187–190.

Holmes, Frederic L. "From Elective Affinities to Chemical Equilibrium: Berthollet's Law of Mass Action." *Chymia* 8 (1962): 105–146.

———. *Claude Bernard and Animal Chemistry: The Emergence of a Scientist.* Cambridge: Harvard University Press, 1974.

———. *Lavoisier and the Chemistry of Life. An Exploration of Scientific Creativity.* Madison: University of Wisconsin Press, 1985.

———. "The Formation of the Munich School of Metabolism." In *The Investigative Enterprise: Studies on Nineteenth-Century Physiology and Medicine.* Ed. William Coleman and F. L. Holmes. Berkeley, Los Angeles, and London: University of California Press, 1988.

———. "The Complementarity of Teaching and Research in Liebig's Laboratory." Ed. Kathryn M. Olesko. *Osiris*, 2d ser., 5 (1989): 121–164.

———. *Eighteenth-Century Chemistry as an Investigative Enterprise.* Berkeley, Los Angeles, and London: University of California Press, 1989.

———. "Justus Liebig and the Construction of Organic Chemistry." In *Chemical Sciences in the Modern World.* Ed. Seymour Mauskopf. Philadelphia: University of Pennsylvania Press, in press.

Holton, Gerald. *Thematic Origins of Scientific Thought: Kepler to Einstein.* Rev. ed. Cambridge: Harvard University Press, 1988.

Hückel, Erich. "Zur Quantentheorie der Doppelbindung." *ZP* 60 (1930): 423–456.

———. "Quantentheoretische Beiträge zum Benzolproblem. I. Die Elektronenkonfiguration des Benzols und Verwandter Verbindungen." *ZP* 70 (1931): 204–286.

Hückel, Walter. *Theoretische Grundlagen der organischen Chemie.* 2 vols. Leipzig: Akademische Verlagsgesellschaft, 1931.

Hückel, Walter, and F. Seel, eds. *Theoretical Organic Chemistry.* Wiesbaden: W. Klemm, 1948.

Hufbauer, Karl. *The Formation of the German Chemical Community, 1729–1795.* Berkeley, Los Angeles and London: University of California Press, 1982.

Hull, David L. *Science as a Process: An Evolutionary Account of the Social and Conceptual Development of Science.* Chicago: University of Chicago Press, 1988.

Hund, F. "Zur Deutung der Molekulspektren. IV." *ZP* 51 (1928): 757–795.

Ihde, Aaron J. *The Development of Modern Chemistry.* New York: Harper and Row, 1964.

———. *Chemistry, as Viewed from Bascom's Hill: A History of the Chemistry Department at the University of Wisconsin at Madison.* Madison: University of Wisconsin Chemistry Department, 1990.

Inghram, Mark G. "[Four Universities] University of Chicago." *Physics Today* (March 1968): 44–47.

Ingold, Christopher K. "The Conditions Underlying the Formation of Unsaturated and of Cyclic Compounds from Halogenated Open-Chain Derivatives. Pt. I. Products Derived from alpha-Halogenated Glutaric Acids, Pt. II. Products Derived from alpha-Halogenated Adipic Acids." *JCS* 119 (1921): 305–329, 951–970.

———. "The Structure of the Benzene Nucleus. Pt. I. Intra-nuclear tautomerism." *JCS* 121 (1922): 1133–1143.

———. ''The Structure of the Benzene Nucleus. Pt. II. Synthetic Formation of the Bridged Modification of the Nucleus.'' *JCS* 121 (1922): 1143–1153.

———. ''The Principles of Aromatic Substitution from the Standpoint of the Electronic Theory of Valence.'' *Recueil des Travaux Chimiques* 48 (1929): 797–812.

———. ''The Mechanisms of, and Constitutional Factors Controlling, the Hydrolysis of Carboxylic Esters. Pt. III. The Calculation of Molecular Dimensions from Hydrolytic Stability Maxima.'' *JCS* 133 (1930): 1375–1386.

———. ''Significance of Tautomerism and of the Reactions of Aromatic Compounds in the Electronic Theory of Organic Reactions.'' *JCS* 136 (1933): 1120–1127.

———. ''Principles of an Electronic Theory of Organic Reactions.'' *Chemical Reviews* 15 (1934): 225–274.

———. ''Mesomerism and Tautomerism.'' *Nature* 133 (1934): 946–947.

———. ''The Relation between Chemical and Physical Theories of the Source of the Stability of Organic Free Radicals.'' *Trans.Far.Soc.* 30 (1934): 52–57.

———. ''Les réactions de la chimie organique (quatre conférences).'' *Actualités Scientifiques et Industrielles*, no. 1037 (1948).

———. *Structure and Mechanism in Organic Chemistry*. Ithaca: Cornell University Press, 1953.

———. ''Bernard Jacques Flürscheim.'' *JCS* (1956): 1087–1089.

———. ''Education of a Scientist.'' *Nature* 196 (1962): 1030–1034.

———. ''Edward David Hughes, 1906–1963.'' *Biographical Memoirs of Fellows of the Royal Society* 10 (1964): 147–182.

Ingold, Christopher K., I. Dostrovsky, and E. D. Hughes. ''Mechanism of Substitution at a Saturated Carbon Atom. Pt. XXXII. The Role of Steric Hindrance (Section G) Magnitude of Steric Effects. Range of Occurrence of Steric and Polar Effects, and Place of the Wagner Rearrangement in Nucleophilic Substitution and Elimination.'' *JCS* 149 (1946): 173–194.

Ingold, Christopher K., E. D. Hughes, and S. Masterman. ''Reaction Kinetics and the Walden Inversion. Pt. I. Homogeneous Hydrolysis and Alcoholysis of beta-n-octyl Halides.'' *JCS* 140 (1937): 1196–1201.

Ingold, C. K., C. W. Shoppee, and J. F. Thorpe. ''The Mechanism of Tautomeric Interchange and the Effect of Structure on Mobility and Equilibrium. Pt. I. The Three-Carbon System.'' *JCS* 129 (1926): 1477–1488.

Ingold, Christopher K., and Jocelyn F. Thorpe. ''The Hypothesis of Valency-Deflexion.'' *JCS* 131 (1928): 1318–1321.

Ingold, Christopher K., and C. C. N. Vass. ''The Nature of the Alternating Effect in Carbon Chains. Pt. XXIII. Anomalous Orientation by Halogens, and Its Bearing on the Problem of the Ortho-Para Ratio in Aromatic Substitution.'' *JCS* 131 (1928): 417–425.

Ingold, Edith Hilda. ''The Correlation of Additive Reactions with Tautomeric Change. Pt. IV. The Effect of Polar Conditions on Reversibility.'' *JCS* 127 (1925): 469–475.

Ingold, Edith Hilda, and Christopher K. Ingold. ''The Nature of the Alternating Effect in Carbon Chains. Pt. V. A Discussion of Aromatic Substitution with Special Reference to the Respective Roles of Polar and Non-Polar Dissociation; and a Further Study of the Relative Directive Efficiencies of Oxygen and Nitrogen.'' *JCS* 129 (1926): 1310–1328.

Jacques, Jean. *Confessions d'une chimiste ordinaire*. Paris: Seuil, 1981.
———. *Berthelot. Autopsie d'un mythe*. Paris: Belin, 1987.
Jaki, Stanley L. *Uneasy Genius: The Life and Work of Pierre Duhem*. The Hague: Nijhoff, 1984.
Jammer, Max. *The Conceptual Development of Quantum Mechanics*. New York: McGraw-Hill, 1966.
Job, André. "La mobilité chimique." *Rapports et discussions sur cinq questions d'actualité: Premier conseil de chimie, 21 au 17 avril 1922*. Ed. Institut International de Chimie Solvay. Paris: Gauthier-Villars, 1925. Pp. 284–319.
Johnson, Jeffrey A. "Academic Chemistry in Imperial Germany." *Isis* 76 (1985): 500–524.
———. "Hierarchy and Creativity in Chemistry, 1871–1914." Ed. Kathryn Olesko. *Osiris*, 2d ser., 5 (1989): 214–240.
Joly, A., and R. Lespieau. *Nouveau précis de chimie (notation atomique)*. 4th ed. 3 vols. Paris: Hachette, 1904–05.
Jones, Harry Clary. *The Elements of Physical Chemistry*. New York: Macmillan, 1902.
Jordan, Pascual, "Ueber eine neue Begründung der Quantenmechanik." *ZP* 40 (1927): 809–838.
Jungnickel, Christa. "Teaching and Research in the Physical Sciences and Mathematics in Saxony, 1820–1850." *HSPS* 10 (1979): 3–47.
———, and Russell McCormmach. *Intellectual Mastery of Nature: Theoretical Physics from Ohn to Einstein*. 2 vols. Chicago: University of Chicago Press, 1986.
Kargon, Robert H. *Science in Victorian Manchester: Enterprise and Expertise*. Baltimore: Johns Hopkins University Press, 1977.
———. "Model and Analogy in Victorian Science: Maxwell's Critique of the French Physicists." *Journal of the History of Ideas* 30 (1969): 423–436.
Kastler, Alfred, and Yves Noel, eds. *Hommage à Albert Kirrmann*. Paris: Ecole Normale Supérieure, 1970. 18 pp.
Keas, Michael N. "The Structure and Philosophy of Group Research: August Wilhelm von Hofmann's Research Program in London (1845–1865)." Ph.D. dissertation, University of Oklahoma, 1992.
Keith, S. T., and Paul K. Hoch. "Formation of a Research School: Theoretical Solid State Physics at Bristol, 1930–1954." *BJHS* 19 (1986): 19–44.
Kekulé, Auguste. "Ueber einige Condensationsprodukte des Aldehyds." *Liebig's Ann.* 162 (1872): 77–124, 309–320.
Kemble, E. C., R. T. Birge, W. F. Colby, F. W. Loomis, and L. Page. *Molecular Spectra in Gases*. XI. *Bulletin of the National Research Council* (1926).
Kermack, W. O., and Robert Robinson. "An Explanation of the Property of Induced Polarity of Atoms and an Interpretation of the Theory of Partial Valencies on an Electronic Basis." *JCS* 121 (1922): 427.
Kim, Mi Gyung. "Practice and Representation: Investigative Programs of Chemical Affinity in the Nineteenth Century." Ph.D. dissertation, University of California, Los Angeles, 1990.
———. "The Layers of Chemical Language. I: Constitution of Bodies v. Structure of Matter." *History of Science* 30 (1992): 69–96.
———. "The Layers of Chemical Language. II: Stabilizing Atoms and Molecules in the Practice of Organic Chemistry." *History of Science* 30 (1992): 397–437.

King, Christine. "Experiments with Time: Programs and Problems in the Development of Chemical Kinetics." *Ambix* 28 (1981): 70–82.

———. "Chemical Kinetics and the Radiation Hypothesis." *Archives for History of Exact Sciences* 30 (1984): 45–86.

Kirrmann, Albert. "Le moment électrique des molécules." *Revue Générale des Sciences* 39 (1928): 598–603.

———. "Recherches sur les aldehydes a-bromées et quelques-uns de leurs dérivés." *Annales de Chimie* 11 (1929): 223–286.

———. "Structure chimique et effet Raman." *Journal de Physique et le Radium,* ser. 7, 9 (1938): 48S.

———. *Chimie organique.* Paris: Colin, 1947.

———. "Aspects actuels de la chimie organique." *Revue Philosophique* 156 (1966): 53–58.

———. "La naissance des formules moléculaires en chimie organique." *L'Actualité Chimique* (1974): 45–49.

Klosterman, L. J. "A Research School of Chemistry in the Nineteenth Century: Jean-Baptiste Dumas and His Research Students." *Annals of Science* 43 (1985): 1–80.

Knight, David, ed. *Classical Scientific Papers: Chemistry.* New York: American Elsevier, 1968.

———. *Sources for the History of Science, 1660–1914.* Cambridge: Cambridge University Press, 1975.

———. *The Transcendental Part of Chemistry.* Folkestone, Kent: Dawson, 1978.

Knorr-Cetina, Karin D. "The Ethnographic Study of Scientific Work: Towards a Constructivist Interpretation of Science." *Science Observed.* Ed. Karin D. Knorr-Cetina and Michael Mulkay. Hollywood: Sage, 1983. Pp. 115–140.

Knott, C. G. *Life and Scientific Work of Peter Guthrie Tait.* Cambridge: Cambridge University Press, 1911.

Königsberger, Leo. Hermann von Helmholtz. Trans. Frances A. Welby. Oxford: Clarendon Press, 1906.

Kohler, Robert E., Jr. "The Origin of G. N. Lewis's Theory of the Shared Pair Bond." *HSPS* 3 (1971): 343–376.

———. *From Medical Chemistry to Biochemistry: The Making of a Biomedical Discipline.* Cambridge: Cambridge University Press, 1982.

Kolbe, Hermann. "Ueber den Zustand der Chemie in Frankreich," *Journal für praktische Chemie* 110 (1870): 173–183.

———. "Meine Betheiligung an der Entwickelung der theoretischen Chemie." *Journal für praktische Chemie* 23–24 (1881): 305–323, 353–379, 497–517; 374–425.

Kon, G. A. R., and R. P. Linstead. "Jocelyn Field Thorpe." In *British Chemists.* Ed. Alexander Findlay. London: The Chemical Society, 1947. Pp. 369–401.

Kopp, Hermann. *Geschichte der Chemie.* 4 vols. Braunschweig: Vieweg, 1843–1847.

Kossel, Walther. "Ueber Molekülbildung als Frage des Atombaus." *Ann.Phys.* 49 (1916): 229–362.

Kragh, Helge. "Julius Thomsen and Classical Thermochemistry." *BJHS* 17 (1984): 255–272.

———. "Bohr's Atomic Theory and the Chemists, 1913–1925." *Rivista di storia della scienza* 2 (1985): 463–485.

———. "The Aether in Late 19th-Century Chemistry." *Ambix* 36 (1989): 49–65.

———. *Dirac. A Scientific Biography*. Cambridge: Cambridge University Press, 1990.

Kronick, D. A. *A History of Scientific and Technical Periodicals*. 2d ed. Metuchen: Scarecrow Press, 1976.

Kronig, R., and W. F. Weisskopf, eds. *W. Pauli. Collected Scientific Papers*. New York: John Wiley, 1964.

Kuhn, Thomas S. *The Structure of Scientific Revolutions*. 2d ed. Chicago: Chicago University Press, 1970.

———. *The Essential Tension: Selected Studies in Scientific Tradition and Change*. Chicago: University of Chicago Press, 1977.

———. "Metaphor in Science." In *Metaphor and Thought*. Ed. Andrew Ortony. Cambridge: Cambridge University Press, 1979. Pp. 409–419.

Lachman, Arthur. *The Spirit of Organic Chemistry*. New York: Macmillan, 1899.

Ladenburg, Albert. *Vorträge über die Entwicklungsgeschichte der Chemie von Lavoisier zur Gegenwart*. Braunschweig: Vieweg, 1869.

———. *Theorie der aromatische Verbindungen*. Brunswick, 1876.

Laidler, Keith J. "Chemical Kinetics and the Origins of Physical Chemistry." *Arch.Ex.Sci.* 32 (1985): 43–75.

Langins, Janis. "The Decline of Chemistry at the École Polytechnique, 1794–1805." *Ambix* 28 (1981): 1–19.

Langmuir, Irving. "The Arrangement of Electrons in Atoms and Molecules." *JACS* 41 (1919): 868–934.

———. "Isomorphism, Isoterism and Covalence." *JACS* 41 (1919): 1543–1559.

———. "The Structure of Atoms and the Octet Theory of Valence." *Proc. NAS* 5 (1919): 252–259.

———. "The Structure of Molecules." *BAAS Rep.* (1922): 468–469.

———. "Modern Concepts in Physics and Their Relation to Chemistry." *Science* 70 (1929): 385–396.

Lapworth, Arthur. "The Form of Change in Organic Compounds and the Function of the alpha-meta Orientating Groups." *JCS* 79 (1901): 1265–1284.

———. "Latent Polarities of Atoms and Mechanism of Reaction, with Special Reference to Carbonyl Compounds." *Mem.Manchester LPS* 64.3 (1919–1920): 1–16.

Lapworth, Arthur, and Robert Robinson. "Remarks on some Recent Contributions to the Theory of Induced Alternate Polarities in a Chain of Atoms." *Trans.Far.Soc.* 19 (1923): 503–505.

Latour, Bruno. *The Pasteurization of France*. Trans. Alan Sheridan and John Law. Cambridge: Harvard University Press, 1988.

Latour, Bruno, and Steve Woolgar. *Laboratory Life: The Construction of Scientific Facts*. 2d ed. Princeton: Princeton University Press, 1986.

Laugier, Henri, et al., eds. *Hommage national à Paul Langevin et Jean Perrin*. Paris: Orléans, 1948.

Laurent, Auguste. *Méthode de chimie*. Paris: Mallet-Bachelier, 1854.

Lavoisier, Antoine. *Traité élémentaire de chimie*. 2 vols. Paris: Cuchet, 1789.

———. "Considérations générales sur la dissolution des métaux dans les acides." In *Oeuvres de Lavoisier*. Paris: Imprimerie Impériale, 1862. II: 509–527.

———. *Opuscules physiques et chimiques*. I. *Oeuvres*. Paris: Imprimerie Impériale, 1864.

———. *Elements of Chemistry*. Reprint ed. New York: Dover, 1965.
Leatherdale, W. H. *The Role of Analogy, Model and Metaphor in Science*. Amsterdam: North Holland, 1974.
Lemaine, Gerard, et al., eds. *Perspectives on the Emergence of Scientific Disciplines*. The Hague: Mouton, 1976.
Lennard-Jones, J. E. "The Electronic Structure of Some Diatomic Molecules." *Trans.Far.Soc.* 25 (1929): 668–682.
Lennard-Jones, J. E., and C. A. Coulson. "The Structure and Energies of Some Hydrocarbon Molecules." *Trans.Far.Soc.* 35 (1939): 811–823.
Lespieau, Robert. "Poids moléculaires et formules dévelopées." *JP* (1901) 8 pp.
———. *Notice sur les travaux scientifiques*. Paris: Gauthier-Villars, 1910.
———. "Sur les notations chimiques." *Revue du Mois* 16 (1913): 257–278.
———. *La molécule chimique*. Paris: Alcan, 1920.
———. *Notice sur les travaux scientifiques*. Paris: Gauthier-Villars, 1925.
———. "Notice sur les travaux de Maurice Bourguel." *Bull.SCF Memoires* 53 (1933): 1145–1153.
———. "Sur l'ebullioscope de M. Raoult." *Bull.SCF* 3 (1890): 855–858.
Levere, Trevor H. *Affinity and Matter: Elements of Chemical Philosophy, 1800–1865*. Oxford: Oxford University Press, 1971.
Lewis, Gilbert N. "Valence and Tautomerism." *JACS* 35 (1913): 1448–1455.
———. "The Atom and the Molecule." *JACS* 38 (1916): 762–785.
———. "The Static Atom." *Science* 46 (1917): 297–302.
———. *Valence and the Structure of Atoms and Molecules*. Washington, D.C.: Chemical Catalogue Company, 1923.
———. *The Anatomy of Science*. Washington, D.C.: American Chemical Society, 1926.
Lewis, Gilbert N., and Joseph E. Mayer. "A Disproof of the Radiation Theory of Chemical Activation." *Proc.NAS* 13 (1927): 623–625.
Lewis, Gilbert N., and Merle Randall. *Thermodynamics and the Free Energy of Chemical Substances*. New York: McGraw-Hill, 1923.
Lewis, William C. McCullagh. "Studies in Catalysis. Pt. VII. Heat of Reaction, Equilibrium Constant, and Allied Quantities, from the Point of View of the Radiation Hypothesis." *JCS* 111 (1917): 457–469.
Libes, Antoine. *Traité élémentaire de physique*. Paris: Crapelet, 1801.
Liebig, Justus von. *Introduction à l'étude de la chimie*. Paris: Mathias, 1837.
Lindauer, Maurice. "The Evolution of the Concept of Chemical Equilibrium from 1775 to 1923." *JChem.Ed.* 39 (1962): 384–390.
Lindroth, Sten. "Urban Hiärne and the 'Laboratorium Chymicum.'" *Lynchos* (1946–47): 51–112.
Lowry, Thomas Martin. "Dynamic Isomerism." *BAAS Rep.* Winnipeg, 1909 (1910): 135–143.
———. "On Chemical Change." *The Central* 13 (1916): 25–41.
———. "Is a True Monomolecular Action Possible?" *Trans.Far.Soc.* 17 (1921–1922): 596–597.
———. "Introductory Address to Part II: Applications in Organic Chemistry of the Electronic Theory of Valency." *Trans.Far.Soc.* 19 (1923): 485–487.

———. "Intramolecular Ionisation in Organic Compounds." *Trans.Far.Soc.* 19 (1923): 487–496.

———. "The Transmission of Chemical Affinity by Single Bonds." *Trans.Far.Soc.* 19 (1923): 497–502.

———. "Studies of Electrovalency. Pt. I. The Polarity of Double Bonds." *JCS* 123 (1923): 822–831.

———. "Nouveaux aspects de la théories de la valence." *Bull.SCF* 35 (1924): 815–837.

———. "Le dispersion rotatoire optique: Hommage à la memoire de Biot (1774–1862)." *JCP* 23 (1926): 565–585.

———. "Preuves expérimentales de l'existence des doubles liaisons semi-polaires." *Bull.SCF* 39 (1926): 203–206.

———. *Optical Rotatory Power.* London: Longmans, 1935.

Lucas, Howard J., and Archibald Y. Jameson. "Electron Displacement versus Alternate Polarity in Aliphatic Compounds." *JACS* 46 (1924): 2475–2482.

Lucas, Howard J., and H. W. Moyse. "Electron Displacement in Carbon Compounds. II. HBr and 2-Pentene." *JACS* 47 (1925): 1459–1461.

Lucas, Howard J., T. P. Simpson, and J. M. Carter. "Electron Displacement in Carbon Compounds. III. Polarity Differences in C-H Unions." *JACS* 47 (1925): 1462–1469.

Lundgren, Anders. "The Changing Role of Numbers in Eighteenth-Century Chemistry." In *The Quantifying Spirit in the Eighteenth Century.* Ed. J. L. Heilbron, Robin Rider, Tore Frängsmyr. Berkeley, Los Angeles, and Oxford: University of California Press, 1990. Pp. 245–266.

McKie, Douglas. " 'Observations' of the Abbé François Rozier (1734–93)." *Annals of Science* 13 (1958): 73–89.

MacLean, Hugh, and Ida Smedley MacLean. *Lecithin and Allied Substances.* New York: Longmans, Green and Co., 1918.

McLeod, Herbert. "Edward Frankland." *JCS* 87 (1905): 574–590.

Macquer, P. J. *Elemens de chymie pratique.* Paris: Hérissant, 1749.

———. *Elemens de chymie théorique.* Paris: Hérissant, 1751.

Marcelin, René. *Contribution à la cinétique physico-chimique.* Paris: Gauthier-Villars, 1914.

Marcet, Jane. *Conversations on Chemistry.* 6th ed. 2 vols. London: Longman, 1819.

Margenau, Henry. *The Nature of Physical Reality: A Philosophy of Modern Physics.* New York: McGraw-Hill, 1950.

Maxwell, James Clerk. "Physical Science." *Encyclopedia Britannica.* 9th ed. 1875. 19: 1–3.

Meadows, A. J., ed. *The Development of Science Publishing in Europe.* Amsterdam: Elsevier, 1980.

Mehra, Jagdish. *The Solvay Conferences in Physics: Aspects of the Development of Physics since 1911.* Dordrecht: Reidel, 1975.

Meinel, Christoph. "Theory or Practice? The Eighteenth-Century Debate on the Scientific Status of Chemistry." *Ambix* 30 (1983): 121–132.

Melhado, Evan M. "Chemistry, Physics, and the Chemical Revolution." *Isis* 76 (1985): 195–211.

———. "Metzger, Kuhn and Eighteenth-Century Disciplinary History." In *Etudes sur Hélène Metzger.* Ed. Gad Freudental. Leiden: E. J. Brill, 1990. Pp. 111–134.

Merritt, Ernest. ''Early Days of the Physical Society.'' *The Review of Scientific Instruments* 5 (1934): 143–148.

Metzger, Hélène. ''Introduction à l'étude du rôle de Lavoisier dans l'histoire de chimie.'' *Archeion* 14 (1932): 21–50.

———. *La philosophie de la matière chez Lavoisier*. Paris: Hermann, 1935.

Meyer, Lothar. *Die Modernen Theorien der Chemie und ihre Bedeutung für die Chemische Statik*. 3d ed. Breslau: Maruschke, 1877.

———. *Modern Theories of Chemistry*. 5th ed. London: Longmans, Green and Co., 1888.

———. *Les théories modernes de la chimie et leur application à la mécanique chimique*. 5th German ed. 2 vols. Paris: Carré, 1889.

Miller, William Allen. *Elements of Chemistry: Theoretical and Practical*. London: John Parker, 1855.

Mislow, Kurt, and Paul Bickart. ''An Epistemological Note on Chirality.'' *Israel Journal of Chemistry*, 15 (1976–1977): 1–6.

Moore, Walter. *Schrödinger: Life and Thought*. Cambridge: Cambridge University Press, 1991.

Morrell, J. B. ''The Chemist Breeders: The Research Schools of Liebig and Thomas Thomson.'' *Ambix* 19 (1972): 1–46.

Mott, N. F. ''John Edward Lennard-Jones (1894–1954).'' *Biographical Memoirs of the Royal Society* 1 (1955): 175–184.

Muir, M. M. Pattison. *A History of Chemical Theories and Laws*. New York: Wiley, 1907.

Mulkay, Michael. ''Three Models of Scientific Development.'' *Sociological Review* 23 (1975): 509–526, 535–537.

Mulliken, Robert S. ''The Assignment of Quantum Numbers for Electrons in Molecules. II. Correlation of Molecular and Atomic Electron States.'' *Physical Review* 32 (1928): 761–772.

———. ''Electronic Structures of Polyatomic Molecules and Valence.'' *Physical Review* 41 (1932): 49–71.

———. ''Quelques aspects de la théorie des orbitales moléculaires.'' *JCP* 46 (1949): 497–542, 675–713.

———. ''Spectroscopy, Quantum Chemistry and Molecular Physics.'' *Physics Today* 21 (1968): 52–57.

———. ''Spectroscopy, Molecular Orbitals, and Chemical Bonding.'' *Nobel Lectures: Chemistry, 1963–1970*. Amsterdam: Elsevier, 1972. Pp. 131–160.

———. *Life of a Scientist. An Autobiographical Account of the Development of Molecular Orbital Theory*. Ed. Bernard J. Ransil. Berlin: Springer-Verlag, 1989.

Muspratt, James S., and A. W. Hofmann. ''On Certain Processes in Which Aniline is Formed.'' *Mem.London CS* 2 (1845): 249–254.

Mysels, Karol J. ''René Marcelin: Experimental and Surface Scientist,'' *JChem. Ed.* 63 (1986): 740.

Nernst, Walther. *Die Ziele der physikalischen Chemie*. Göttingen: Vandenhoek und Ruprecht, 1896.

———. *Theoretical Chemistry from the Standpoint of Avogadro's Rule and Thermodynamics*. Rev. from 4th German ed. New York: Macmillan, 1904.

———. *Theoretische Chemie.* 7th ed. Stuttgart: Ferdinand Enke, 1913.
Nicolson, William. *Introduction to Natural Philosophy.* 2d ed. 2 vols. London: J. Johnson, 1787.
Nier, Keith. "The Emergence of Physics in Nineteenth-Century Britain as a Socially Organized Category of Knowledge: Preliminary Studies." Ph.D. dissertation, Harvard University, 1975.
Nye, Mary Jo. *Molecular Reality: A Perspective on the Scientific Work of Jean Perrin.* London and Amsterdam: Macdonald and Elsevier, 1972.
———. "The Nineteenth-Century Atomic Debates and the Dilemma of an 'Indifferent Hypothesis.' " *SHPS* (1976): 245–268.
———. "Berthelot's Anti-Atomism: A 'Matter of Taste'?" *Annals of Science* 38 (1981): 585–590.
———, ed. *The Question of the Atom: From the Karlsruhe Congress to the First Solvay Conference, 1860–1911.* Los Angeles: Tomash, 1984.
———. *Science in the Provinces. Scientific Communities and Provincial Leadership in France, 1870–1930.* Berkeley, Los Angeles, and London: University of California Press, 1986.
———. "Chemical Explanation and Physical Dynamics: Two Research Schools at the First Solvay Chemistry Conferences, 1922–1928." *Annals of Science* 46 (1989): 461–480.
Odling, William. "Presidential Address." *BAAS Rep.* (1864): 21–24.
Oesper, Ralph. "Walter Hückel." *JChem.Ed.* 27 (1950): 625.
Olesko, Kathryn. "Tacit Knowledge and School Formation: Exact Experimental Physics at Göttingen." Ed. Gerald Geison and F. L. Holmes. *Osiris*, 2d ser., 8 (in press).
Ostwald, Wilhelm. *Lehrbuch der allgemeine Chemie.* 2 vols. Leipzig: Engelmann, 1885–1887.
———. *Elektrochemie: Ihre Geschichte und Lehre.* Leipzig: Verlag von Veit, 1896.
———. "Elements and Compounds." *Faraday Lecture, 1869–1928.* Ed. C. S. Gibson and A. J. Greenaway. London: The Chemical Society, 1928. Pp. 185–201.
Paley, William. *Natural Theology.* Rev. Amer. ed. New York: Sheldon, 1854.
Paneth, F. A. "The Epistemological Status of the Chemical Concept of Element." *British Journal for the Philosophy of Science* 13 (1962): 1–14, 144–160.
Pariselle, Henri. "Georges Dupont." *Ann.AEENS* (1958): 29–30.
Parr, Robert G., and Bryce L. Crawford, Jr. "National Academy of Sciences Conference on Quantum-Mechanical Methods in Valence Theory." *Proc.NAS* 38 (1952): 547–553.
Partington, J. R. *A History of Chemistry.* IV. London: Macmillan, 1964.
Paul, Harry W. *From Knowledge to Power: The Rise of the Science Empire in France, 1860–1939.* Cambridge: Cambridge University Press, 1985.
Pauli, Wolfgang. "Ueber den Einfluss der Geschwindigkeitsabhängigkeit der Elektronenmasse auf den Zeemaneffekt," *ZP* 31 (1925): 373–386.
Pauling, Linus. "The Nature of the Theory of Resonance." In *Perspectives in Organic Chemistry.* Ed. [Sir] Alexander Todd. New York: Interscience Publishers, 1926. Pp. 1–8.
———. "The Application of the Quantum Mechanics to the Structure of the Hydrogen Molecule and Hydrogen Molecule-Ion and Related Problems." *Chemical Reviews* 5 (1928): 173–213.

———. ''The Nature of the Chemical Bond: Applications of Results Obtained from Quantum Mechanics and from a Theory of Paramagnetic Susceptibility to the Structure of Molecules.'' *JACS* 53 (1931): 1367–1400.

———. ''The Nature of the Chemical Bond. II. The One-Electron Bond and Three-Electron Bond.'' *JACS* 53 (1931): 3225–3237.

———. ''The Nature of the Chemical Bond. III. The Transition from One Extreme Bond Type to Another.'' *JACS* 54 (1932): 98–1003.

———. ''Interatomic Distances in Covalent Molecules and Resonance between Two or More Lewis Electronic Structures.'' *Proc.NAS* 18 (1932): 293–297.

———. ''The Nature of the Chemical Bond. IV. The Energy of Single Bonds and the Relative Electronegativity of Atoms.'' *JACS* 54 (1932): 3570–3582.

———. ''The Calculation of Matrix Elements for the Lewis Electronic Structure of Molecules.'' *Journal of Chemical Physics* 1 (1933): 280–283.

———. *The Nature of the Chemical Bond.* Ithaca: Cornell University Press, 1939.

Pauling, Linus, and J. Albert Sherman. ''The Nature of the Chemical Bond. VI. Calculation from Thermochemical Data of the Energy of Resonance of Molecules among Several Electronic Structures.'' *J.Chem.Physics* 1 (1933): 606–617.

———. ''The Nature of the Chemical Bond. VII. The Calculation of Resonance Energy in Conjugated Systems.'' *J.Chem.Physics* 1 (1933): 679–686.

Pauling, Linus, and George W. Wheland. ''The Nature of the Chemical Bond. V. The Quantum-Mechanical Calculation of the Resonance Energy of Benzene and Naphthalene and the Hydrocarbon Free Radicals.'' *J.Chem.Physics* 1 (1933): 362–374.

Pauling, Linus, and E. Bright Wilson, Jr. *Introduction to Quantum Mechanics with Applications to Chemistry.* New York: McGraw-Hill, 1935.

Pearson, Karl. ''Ether Squirts: Being an Attempt to Specialize the Form of Ether Motion Which Forms an Atom in a Theory Propounded in Former Papers.'' *American Journal of Mathematics* 13 (1891): 309–362.

Perrin, Jean. *Traité de chimie physique: Les principes.* Paris: Gauthier-Villars, 1903.

———. *Les atomes.* Paris: Alcan, 1913.

———. ''Matière et lumière: Essai de synthèse de la mécanique chimique.'' *Annales de Physique* 11 (1919): 1–108.

———. ''Radiation and Chemistry.'' *Trans.Far.Soc.* 17 (1921–22): 546–572.

———. ''La chimie physique.'' In *L'orientation actuelle des sciences.* Ed. Jean Perrin et al. Paris: Alcan, 1930. Pp. 18–28.

Perrin, Jean, and Georges Urbain, eds. *Formes chimiques de transition.* Paris: Société d'Editions Scientifiques, 1931.

Pestre, Dominique. *Physique et physiciens en France, 1918–1940.* Paris: Editions des Archives Contemporaines, 1984.

Picard, Emile. *La vie et l'oeuvre de Pierre Duhem.* Paris: Gauthier-Villars, 1922.

Piganiol, Pierre. ''Charles Prévost.'' *Ann.AEENS* (1985): 46–48.

Poggendorff, Johann Christian. *Geschichte der Physik.* Leipzig: Barth, 1879.

Pope, W. J. ''Thomas Martin Lowry.'' *Obituary Notices of Fellows of the Royal Society* 2 (1936–1938): 287–293.

Prévost, Charles. ''La transposition allylique et les dérivés d'addition des carbures érythréniques.'' *Annales de Chimie* 10 (1928): 113–146, 147–181, 356–439.

———. *Leçons de chimie organique.* 4 vols. Paris: Société d'Enseignement Supérieure, 1949–1953.

———. ''La valence et l'enseignement.'' *L'Information Scientifique* 6 (1951): 14–18.

———. ''Notice nécrologique: Albert Kirrmann.'' *Bull.SCF* (1957): 1451–1454.

———. *Notice sur les titres et travaux scientifiques.* Paris: Société d'Edition d'Enseignement Supérieure, 1967.

Prévost, Charles, and Albert Kirrmann. ''Essai d'une théorie ionique des réactions organiques. Premier memoire.'' *Bull.SCF* 49 (1931): 194–243.

———. ''Essai d'une théorie ionique des réactions organiques: Deuxième memoire.'' *Bull.SCF* 49 (1931): 1309–1368.

———. ''La tautomérie anneau-chaine, et la notion de synionie: Troisième communication sur la théorie ionique des réactions organiques.'' *Bull.SCF* 53 (1933): 253–260.

Prigogine, Ilya, and Isabelle Stengers. *Order Out of Chaos.* London: New Science Library, 1984.

Primas, Hans. *Chemistry, Quantum Mechanics, and Reductionism.* New York: Springer Verlag, 1981.

Pullman, Bernard, and Alberte Pullman. *Les théories électroniques de la chimie organique.* Paris: Masson, 1952.

Pynchon, Thomas R. *Introduction to Chemical Physics.* 3d rev. ed. Philadelphia: Van Nostrand, 1881.

Rabkin, Yakov. ''Technological Innovation in Science: The Adoption of Infrared Spectroscopy by Chemists.'' *Isis* 78 (1987): 31–54.

Ramsay, William. ''The Electron as an Element.'' *JCS* (1908): 774–788.

Ramsay, D. A., and J. Hinze, eds. *Selected Papers of Robert S. Mulliken.* Chicago: University of Chicago Press, 1975.

Ramsey, O. Bertrand. *Stereochemistry.* London: Heyden, 1981.

Raveau, C. ''Robert Lespieau.'' *Ann.AEENS* (1949): 25–26.

Richards, Joan L. *Mathematical Visions: The Pursuit of Geometry in Victorian England.* San Diego: Academic Press, 1988.

Richards, Robert J. *Darwin and the Emergence of Evolutionary Theories of Mind and Behavior.* Chicago: University of Chicago Press, 1987.

Rider, Robin E. ''Alarm and Opportunity: Emigration of Mathematicians and Physicists to Britain and the United States.'' *HSPS* 15 (1984): 107–176.

Ringer, Fritz. ''The Intellectual Field, Intellectual History, and the Sociology of Knowledge.'' *Theory and Society* 19 (1990): 269–294.

Roberts, Gerrylynn K. ''The Establishment of the Royal College of Chemistry: An Investigation of the Social Context of Early Victorian Chemistry.'' *HSPS* 7 (1976): 437–485.

Robinson, Robert. ''The Conjugation of Partial Valencies.'' *Mem.Manchester LPS* 64.4 (1920): 1–14.

———. ''Polarisation of Nitrosobenzene.'' *Journal of the Society of Chemistry and Industry [Chemistry and Industry]* 44 (1925): 456–458.

———. *Memoirs of a Minor Prophet: Seventy Years of Organic Chemistry.* London: Elsevier, 1976.

Robinson, Robert, and Gertrude M. Robinson. ''Researches on Pseudo-Bases. Pt. II. Note on some Berberine Derivatives and Remarks on the Mechanism of the Condensation Reactions of Pseudo-Bases,'' *JCS* 111 (1917): 958–969.

Rocke, Alan J. ''Atoms and Equivalents: The Early Development of the Chemical Atomic Theory.'' *HSPS* 9 (1978): 225–263.

———. *Chemical Atomism in the Nineteenth Century: From Dalton to Cannizzaro.* Columbus: Ohio State University Press, 1984.

———. ''Hypothesis and Experiment in the Early Development of Kekulé's Benzene Theory.'' *Annals of Science* 42 (1985): 355–381.

———. ''Kolbe versus the 'Transcendental Chemists': The Emergence of Classical Organic Chemistry.'' *Ambix* 34 (1987): 156–168.

———. ''Kekulé's Benzene Theory and the Appraisal of Scientific Theories.'' In *Scrutinizing Science.* Ed. Arthur Donovan et al. Dordrecht: Kluwer, 1988. Pp. 45–61.

———. ''The 'Quiet Revolution' of the 1850s: Scientific Theory as Social Production and Empirical Practice.'' In *Chemical Sciences in the Modern World.* Ed. Seymour Mauskopf. Philadelphia: University of Pennsylvania Press, in press.

Root-Bernstein, Robert Scott. ''The Ionists: Founding Physical Chemistry, 1872–1890.'' Ph.D. dissertation, Princeton University, 1980.

Roscoe, Henry E. *The Life and Experiences of Sir Henry Enfield Roscoe.* London: Macmillan, 1906.

Roscoe, Henry E., and Carl Schorlemmer. *A Treatise on Chemistry.* London: Macmillan, 1880.

Rosenberg, Charles. ''Toward an Ecology of Knowledge: On Discipline, Context, and History.'' In *The Organization of Knowledge in Modern America, 1860–1920.* Ed. Alexandra Oleson and John Voss. Baltimore: Johns Hopkins University Press, 1979.

Rosenberger, Ferdinand. *Geschichte der Physik.* 2 vols. Braunschweig: Vieweg, 1882–1884.

Rossiter, Margaret. *Women Scientists in America: Struggles and Strategies to 1940.* Baltimore: Johns Hopkins University Press, 1982.

Rousseau, G. S., and Roy Porter, eds. *The Ferment of Knowledge: Studies in the Historiography of Eighteenth-Century Science.* Cambridge: Cambridge University Press, 1980.

Russell, A. S. ''Lord Rutherford: Manchester, 1907–1919: A Partial Portrait.'' In *Rutherford at Manchester.* Ed. J. B. Birks. London: Heywood, 1962. Pp. 87–101.

Russell, Colin A. *A History of Valency.* New York: Humanities Press, 1971.

———. ''Specialism and Its Hazards.'' In *The Structure of Chemistry.* Unit III. Ed. C. A. Russell. Milton Keynes: Open University Press, 1976.

———. *Lancastrian Chemist: The Early Years of Sir Edward Frankland.* Milton Keynes: Open University Press, 1986.

Sachtleben, R. ''Nobel Prize Winners Descended from Liebig.'' *JChem.Ed.* 35 (1958): 73–75.

Saltzmann, Martin D. ''Arthur Lapworth: The Genesis of Reaction Mechanism.'' *JChem.Ed.* 49 (1972): 750–752.

———. ''The Robinson-Ingold Controversy: Precedence in the Electronic Theory of Organic Reactions.'' *JChem.Ed.* 57 (1980): 484–488.

———. ''Sir Robert Robinson—A Centennial Tribute.'' *Chemistry in Britain* (1986): 543–548.

———. ''The Development of Physical Organic Chemistry in the United States and the United Kingdom: 1919–1939. Parallels and Contrasts.'' *JChem.Ed.* 63 (1986): 588–593.

Schaffer, Simon. "Natural Philosophy." In *The Ferment of Knowledge: Studies in the Historiography of Eighteenth-Century Science*. Ed. G. S. Rousseau and Roy Porter. Cambridge: Cambridge University Press, 1980. Pp. 55–92.

———. "History of Physical Sciences." In *Information Sources in the History of Science and Medicine*. Ed. Pietro Corsi and Paul Weindling. London: Butterworth, 1983. Pp. 285–314.

Schorlemmer, Carl. *The Rise and Development of Organic Chemistry*. London: Macmillan, 1879.

———. *The Rise and Development of Organic Chemistry*. Ed. Arthur Smithells. Rev. ed. London: Macmillan, 1894.

Schütt, Hans-Werner. "Chemiegeschichtsschreibung—'Zu welchem Ende'?" *Chemie in Unserer Zeit* 22 (1988): 139–145.

Schweber, S. S. "Shelter Island, Pocono, and Oldstone: The Emergence of American Quantum Electrodynamics after World War II." *Osiris* , 2d ser., 2 (1986): 265–302.

———. "The Young Clarke Slater and the Development of Quantum Chemistry." *HSPS* 20 (1990): 339–406.

Schofield, Robert E. *Mechanism and Materialism: British Natural Philosophy in an Age of Reason*. Princeton: Princeton University Press, 1970.

Seddon, Jennifer. "The Development of Electronic Theory in Organic Chemistry." Honours thesis, St. Hugh's College, Oxford University, 1972.

Segrè, Emilio. *From X-rays to Quarks: Modern Physicists and Their Discoveries*. San Francisco: Freeman, 1980.

Serafini, Anthony. *Linus Pauling: A Man and His Science*. New York: Paragon House, 1989.

Servos, John W. *Physical Chemistry from Ostwald to Pauling: The Making of a Science in America*. Princeton: Princeton University Press, 1990.

Shapin, Steven. "Property, Patronage and the Politics of Science: The Founding of the Royal Society of Edinburgh." *BJHS* 7 (1974): 1–41.

Shapin, Steven, and Simon Schaffer. *Leviathan and the Air-Pump: Hobbes, Boyle and the Experimental Life*. Princeton: Princeton University Press, 1985.

Sherman, J. Albert, and J. H. Van Vleck. "The Quantum Theory of Valence." *Reviews of Modern Physics* 7 (1935): 167–228.

Shimmin, A. N. *The University of Leeds: The First Half-Century*. Cambridge: Cambridge University Press, 1954.

Shoppee, Charles W. "Christopher Kelk Ingold (1893–1970)." *Biographical Memoirs of Fellows of the Royal Society* 18 (1972): 349–411.

Shorter, J. "Electronic Theories of Organic Chemistry: Robinson and Ingold." *Natural Products Reports. Royal Society of Chemistry* 4 (1987): 61–66.

Silliman, Benjamin, Jr. *First Principles of Chemistry*. Philadelphia: Loomis and Peck, 1847.

Silliman, Robert H. "Fresnel and the Emergence of Physics as a Discipline." *HSPS* 4 (1972): 137–162.

Slater, John Clarke. "Directed Valence in Polyatomic Molecules." *Physical Review* 37 (1931): 481–489.

———. *Introduction to Chemical Physics*. New York: McGraw-Hill, 1939.

———. *Solid State and Molecular Theory: A Scientific Biography*. New York: Wiley Interscience, 1975.

Smith, Crosbie. " 'Mechanical Philosophy' and the Emergence of Physics in Britain: 1800–1850." *Annals of Science* 33 (1976): 3–29.

———. "A New Chart for British Natural Philosophy: The Development of Energy Physics in the Nineteenth Century." *History of Science* 16 (1978): 231–279.

Snelders, H. A. M. "J. H. van't Hoff's Research School in Amsterdam (1877–1895)." *Janus* 71 (1984): 1–30.

Solvay, Institut International de Chimie. *Rapports et discussions sur cinq questions d'actualité: Premier conseil de chimie, 21 au 17 avril 1922*. Paris: Gauthier-Villars, 1925.

———. *Structure et activité chimique. Rapports et discussions. Deuxième conseil de chimie. 16 au avril 1925*. Paris: Gauthier-Villars, 1926.

———. *Rapports et discussions sur des questions d'actualité. Troisième conseil de chimie. 12 au 18 avril 1928*. Paris: Gauthier-Villars, 1928.

———. *Rapports et discussions relatifs à la constitution et à la configuration des molécules organiques. Quatrième Conseil de Chimie tenu à Bruxelles du 9 au 14 avril 1931*. Paris: Gauthier-Villars, 1931.

Sommerfeld, Arnold. *Atombau und Spektrallinien*. 3d ed. Braunschweig: Teubner, 1922.

———. *Atomic Structures and Spectral Lines*. Trans. Henry Brose. 3d German ed. London: Methuen, 1923.

Sprat, Thomas. *History of the Royal Society*. Ed. Jackson I. Cope and Harold W. Jones. London: Routledge, Kegan and Paul, 1959.

Stark, Johannes. "Die Valenzlehre auf atomistisch elektrischer Basis." *Jahrbuch der Radioaktivität und Elektronik* 5 (1908): 124–153.

———. *Die Elektrizität in chemischen Atom*. Leipzig: Hirzel, 1915.

Stewart, A. W. *Recent Advances in Organic Chemistry*. London: Longman, Green, 1908.

Stewart, A. W. *Recent Advances in Organic Chemistry*. 6th ed. 2 vols. London: Longmans, Green, 1931.

Stout, J. W. "The 'Journal of Chemical Physics': The First 50 Years." *Ann.Rev.P.Chem.* 37 (1986): 1–23.

Stranges, Anthony N. *Electrons and Valence: Development of the Theory, 1900–1925*. College Station: Texas A&M University Press, 1982.

Suchet, J. P. "De quels modèles les chimistes ont-ils besoin?" *Bull.SCF* (1975): 3–5.

Suckling, Colin J., Keith E. Suckling, and Charles W. Suckling. *Chemistry through Models: Concepts and Applications of Modelling in Chemical Science, Technology and Industry*. Cambridge: Cambridge University Press, 1978.

Sullivan, Walter. "Dr. Alfred Kastler, 81, Nobel Prize Winner, Dies." *New York Times* (6 January 1984).

Sviedrys, Romualdas. "The Rise of Physics Laboratories in Britain." *HSPS* 7 (1976): 405–436.

Tarbell, D. Stanley. "Organic Chemistry: The Past 100 Years." *CENews* 54 (1976): 110–123.

Taylor, Hugh. "Fifty Years of Chemical Kineticists." *Ann.Rev.P.Chem.* 13 (1962): 1–18.

Thackray, Arnold. *Atoms and Powers: An Essay on Newtonian Matter-Theory and the Development of Chemistry.* Cambridge: Harvard University Press, 1970.
Theobald, D. W. ''Some Considerations on the Philosophy of Chemistry.'' *Chemical Society Reviews* 5 (1976): 203–213.
Thiele, F. K. Johannes. ''Zur Kenntnis der unsättigten Verbindungen.'' *Liebig's Ann.* 306 (1899): 87–142.
Thomson, Joseph John. *A Treatise on the Motion of Vortex Rings.* London: Macmillan, 1883.
———. *Applications of Dynamics to Physics and Chemistry.* London: Macmillan, 1888.
———. *Electricity and Matter.* Westminster: Constable, 1904.
———. ''The Forces between Atoms and Chemical Affinity.'' *Philosophical Magazine* 27 (1914): 758–789.
———. ''On the Origin of Spectra and Planck's Law.'' *Philosophical Magazine* 37 (1919): 419–446.
———. *The Electron in Chemistry.* London: Chapham and Hall, 1923.
Thomson, Thomas. *The History of Chemistry.* 2 vols. London: H. Colburn and R. Bentley, 1830–31.
Thomson, William [Lord Kelvin]. ''On Vortex Atoms.'' *Philosophical Magazine* 34 (1867): 15–24.
Thorpe, Jocelyn, and C. K. Ingold. ''Quelques nouveaux aspects de la tautomérie par MM. Jocelyn Field Thorpe et Christopher Kelk Ingold.'' *Bull.SCF* 33 (1923): 1342–1391.
Todd, [Lord] Alexander, and [Sir] John Cornworth. ''Robert Robinson.'' *Biographical Memoirs of Fellows of the Royal Society* 22 (1976): 465–478.
———. ''Summing Up.'' In *Further Perspectives in Organic Chemistry. Ciba Foundation Symposium 53 (new series) to Commemorate Sir Robert Robinson and His Research.* Amsterdam: Elsevier, 1978. Pp. 203–204.
———. *A Time to Remember: The Autobiography of a Chemist.* Cambridge: Cambridge University Press, 1983.
Toulmin, Stephen E. ''The Evolutionary Development of Natural Science.'' *American Scientist* 55 (1967): 456–471.
Turner, Edward. *Elements of Chemistry.* 4th ed. London: Printed for J. Taylor, Bookseller, 1833.
Turner, R. Steven. ''Justus Liebig versus Prussian Chemistry: Reflections on Early Institute-Building in Germany.'' *HSPS* 13 (1982): 129–162.
———. ''Scientific Schools and Scientific Controversy: The Case of Vision Studies in Germany.'' Ed. Gerald Geison and F. L. Holmes. *Osiris* 2nd series, 8 (to appear in 1993).
Uhlenbeck, G. E., and Samuel Goudsmit. ''Zuschriften und vorläufige Mitteilungen.'' *Naturwissenschaften* 13 (1925): 953–954.
———. ''Spinning Electrons and the Structure of Spectra.'' *Nature* 117 (1926): 264–265.
Urbain, Georges. *La coordination des atomes dans la molécule: la symbolique chimique.* Paris: Hermann, 1933.
van't Hoff, Jacobus Henricus. *Ansichten über die organische Chemie.* 2 vols. Braunschweig: F. Vieweg, 1878–1881.

———. "Sur les formules de structure dans l'espace." *Archives néerlandaises des sciences exactes et naturelles* 9 (1874): 445–454.

———. *Etudes de dynamique chimique*. Amsterdam: Frederik Muller, 1884.

———. *Physical Chemistry in the Service of the Sciences*. Chicago: University of Chicago Press, 1903.

Venel, François. "Chymie." *Encyclopédie ou Dictionnaire Raisonné des Arts et Métiers*. Facsimile ed. Stuttgart: Frederich Frommann, 1966. Pp. 408–437.

Volhard, Jakob. "Die Begrundung der Chemie durch Lavoisier." *Journal für praktische Chemie* 110 (1870): 1–47.

Walker, Adam. *Analysis of a Course of Lectures on Natural and Experimental Philosophy*. London: Printed for the Author, 1766.

———. *A System of Familiar Philosophy in Twelve Lectures*. London: Printed for the Author, 1799.

Waters, William A. *Physical Aspects of Organic Chemistry*. London: Routledge, 1935.

Watts, Henry, ed. *A History of Chemistry and the Allied Branches of Other Sciences*. Vols. 1–2. London: Longman, Green, 1870, 1872.

Weart, Spencer. "The Physics Business in America, 1919–1940: A Statistical Reconnaissance." In *The Sciences in the American Context: New Perspectives*. Ed. Nathan Reingold. Washington, D.C.: Smithsonian Institution Press, 1979. Pp. 295–358.

Weininger, Stephen J. "The Molecular Structure Conundrum: Can Classical Chemistry be Reduced to Quantum Chemistry?" *JChem.Ed.* 61 (1984): 939–944.

Werner, Alfred. *Neuere Anschauungen auf den Gebiete der anorganischen Chemie*. Braunschweig: F. Vieweg,1905.

Whewell, William. *Selected Writings on the History of Science*, ed. Yehuda Elkana. Chicago: University of Chicago Press, 1984.

Williams, Trevor I. *Robert Robinson: Chemist Extraordinary*. Oxford: Oxford University Press, 1990.

Williamson, A. W. "Suggestions for a Dynamics of Chemistry." *Notices and Proceedings of the Royal Institution* 1 (1851–1854).

Willstätter, Richard. *From My Life*. Trans. Lili S. Hornig. New York: W. A. Benjamin, 1965.

Wood, Alexander. *The Cavendish Laboratory*. Cambridge: Cambridge University Press, 1946.

Woolley, R. G. "Must a Molecule Have a Shape?" *JCS* 100 (1978): 1073–1078.

———. "Further Remarks on Molecular Structure in Quantum Theory." *Chemical Physics Letters* 55 (1978): 443–446.

———. ed. *Quantum Dynamics of Molecules: The New Experimental Challenge to Theorists*. New York: Plenum, 1980.

Wotiz, John H., and Susanna Rudofsky. "The Unknown Kekulé." In *Essays on the History of Organic Chemistry*. Ed. James G. Traynham. Baton Rouge: Louisiana State University Press, 1987. Pp. 21–34.

Wurtz, Adolphe. *Leçons élémentaires de chimie moderne*. Paris: Masson, 1867.

———. *A History of Chemical Theory*. Trans. Henry Watts. London: Macmillan, 1869.

———. *Dictionnaire de chimie pure et appliquée*. Paris: Hachette, 1869–1874.

———. *Introduction à l'étude de la chimie*. Paris: Masson, 1885.

Zuckermann, Jerald J. "The Chemist as Teacher of History." *JChem.Ed.* (1987): 828–835.

Index of Names

Index of Subjects

Designer: U.C. Press Staff
Compositor: Impressions, A Division of
Edwards Brothers, Inc.
Text: 10/12 Times Roman
Display: Helvetica Bold
Printer: Edwards Brothers, Inc.
Binder: Edwards Brothers, Inc.